W9-CCL-810

DISCARD

Ecological Studies

Analysis and Synthesis

Edited by

W.D. Billings, Durham (USA) F. Golley, Athens (USA)
O.L. Lange, Würzburg (FRG) J.S. Olson, Oak Ridge (USA)
H. Remmert, Marburg (FRG)

Volume 38

The salt marshes of Sapelo Island, with the Duplin River in the foreground.

The Ecology of a Salt Marsh

Edited by
L.R. Pomeroy and R.G. Wiegert

With 57 Figures

Springer-Verlag
New York Heidelberg Berlin

577.69
P771e

Lawrence R. Pomeroy
Richard G. Wiegert
University of Georgia
Department of Zoology
Institute of Ecology
Athens, Georgia 30602
U.S.A.

Library of Congress Cataloging in Publication Data
The Ecology of a Salt Marsh

(Ecological studies; v. 38)
Bibliography: p.
Includes index.
1. Tidemarsh ecology. I. Pomeroy, Lawrence R., 1925-
II. Wiegert, Richard G. III. Series.
QH541.5.S24E26 574.5'2636 80-29676 AACR1

© 1981 by Springer-Verlag New York Inc.
All rights reserved.
No part of this book may be translated or reproduced in any form without written permission from Springer-Verlag, 175 Fifth Avenue, New York, New York, 10010, USA. The use of general descriptive names, trade names, trademarks, etc. in this publication, even if the former are not especially identified, is not to be taken as a sign that such names, as understood by the Trade Marks and Merchandise Marks Act, may accordingly be used freely by anyone.

Printed in the United States of America
9 8 7 6 5 4 3 2 1
ISBN 0-387-90555-3 Springer-Verlag New York Heidelberg Berlin
ISBN 3-540-90555-3 Springer-Verlag Berlin Heidelberg New York

Foreword

Ecologists have two long-standing ways to study large ecosystems such as lakes, forests, and salt-marsh estuaries. In the first, which G. E. Hutchinson has called the *holological* approach, the whole ecosystem is first studied as a "black box," and its components are investigated as needed. In the second, which Hutchinson has called the *merological* approach, the parts of the system are studied first, and an attempt is then made to build up the whole from them. For long-term studies, the holological approach has special advantages, since the general patterns and tentative hypotheses that are first worked out help direct attention to the components of the system which need to be studied in greater detail. In this approach, teams of investigators focus on major functions and hypotheses and thereby coordinate their independent study efforts. Thus, although there have been waves, as it were, of investigators and graduate students working on different aspects of the Georgia salt-marsh estuaries (personnel at the Marine Institute on Sapelo Island changes every few years), the emphasis on the holological approach has resulted in a highly differentiated and well-coordinated long-term study.

Very briefly, the history of the salt-marsh studies can be outlined as follows. First, the general patterns of food chains and other energy flows in the marshes and creeks were worked out, and the nature of imports and exports to and from the system and its subsystems were delimited. Next, a number of general hypotheses were formulated and subsequently tested by detailed studies of key components, studies designed to prove or disprove the general hypotheses. This approach lent itself naturally to experiments and modeling, and efforts along these lines dominated studies for the next decade or so. Finally, the group of investigators who were most active in the study in recent years got together to prepare a composite analysis of the salt-marsh estuary as an ecosystem and of the major populations that are most vital to its function. This volume is the result of that synthesis.

When Sidney Lanier sat under the oak tree on the edge of the Georgia marsh at Brunswick, in Glynn County, and wrote his well-known poem, "The Marshes of

872615

Glynn," he was obviously very much impressed with the marsh grasses;

In a league and a league of marsh grass, waist high, broad in the blade,
Green, and all of a height, and unflecked with a light or a shade,
Stretch leisurely off, in a pleasant plain,
To the terminal blue of the main.
Oh, what is abroad in the marsh and the terminal sea?
Somehow my soul seems suddenly free
From the weighing of fate and sad discussion of sin,
By the length and the breadth and the sweep of the marshes of Glynn.

If Lanier had walked out into the marsh, as he may very well have done, he would have been impressed with two other major aspects of the marshes of Glynn–the powerful ebb and flow of brown tidal waters and the vast banks and flats of mud, which at low tide glisten golden in the sun due to the myriad diatoms and other living organisms which populate these sediments. These three major components, the marsh grass, the ever-moving water masses, and the soils and sediments, are bound together as a functional input-output ecosystem by the energy of the sun and the tides. It has taken the scientists at Sapelo Island some 30 years to partially answer Lanier's question, "What is abroad in the marsh?" and, we might add, the effect of the marsh on the "terminal sea" is even today not fully understood.

When the late R. J. Reynolds, Jr. invited the University of Georgia to begin studies at Sapelo Island in 1954, he set up a foundation to support these studies. There were sufficient funds at that time to hire three young PhD.s to form the core of a resident staff. It was only natural that we looked for persons with interest and expertise in each of the three major components of the ecosystem. Thus, we hired a zoologist-ecologist to study the food chains and related aspects of the marshes, a hydrologist-limnologist to investigate the water masses, and a microbiologist to study the golden mud. As it turned out, the early microbiological efforts were premature, since effective methods of studying microbial activities in such a complex and difficult situation as the marshes had not yet been developed. The conventional techniques that involve isolating organisms and culturing them in the laboratory are not by themselves sufficient for the study of natural ecosystems. Not until a dozen or so years later were techniques developed to study microorganisms *in situ* by measuring and assessing their products and their activities, rather than by counting or isolating the organisms themselves. It became evident in the early stages of the work that the microbial transformations of the primary production of grasses and algae are the key to the function of the Sapelo Island salt-marsh ecosystem, a system that is, in large measure, a detritus-based system, rather than the better understood grazing-based. If I were asked to single out the major contributions which our long-term studies at Sapelo Island have made to the general field of ecology, I would certainly rate the work in microbial ecology as being of a breakthrough variety.

It is with great personal pleasure and satisfaction that I introduce to you this synthesis of some 25 years of study on the marshes of Sapelo Island.

Eugene P. Odum

Preface

The organization of this volume and the research it synthesizes is testimony to our long-standing belief that the study of either macroorganisms or microorganisms in isolation from each other cannot yield a true synthesis of the structure and function of any ecosystem. Our goal is not only to conduct a balanced analysis of the roles of both the macrocommunity and the microbial community, but to understand how these communities interact. Because of the prominence of the detritus food web in salt marshes, that aspect of the interaction is quite evident. However, we find other, perhaps equally pivotal, interactions in the relationship of algae to their grazers and in the effects of both higher plants and higher animals on microbial communities of marsh soils. Therefore, we intend not to stress any single aspect of the salt marsh ecosystem for its own sake, or to only redress former oversights, but to present, in so far as possible, a balanced view of the system and how it works.

Through the use of a model of the flux of carbon through the salt marsh we achieved an efficient research plan and brought together the investigators who carried out the necessary interdisciplinary work. The choice of carbon was arbitrary. We could have worked with another element or in energy units. The choice of the kind of model, however, was important. Building the ecosystem model together, we became participants in the quest for missing pieces of the system as we perceived it. Throughout the process, the model continuously told us unexpected things about how the system might operate, and served both as a unifying force in our work and as an arbiter of balance in our approach. We studied not whatever struck the fancy of any one of us, but rather what modeling showed us to be potentially significant. The word *potentially* must be emphasized because the modeling process did not tell us how the system worked. We could only discover that by studying the real-world salt marsh, thereby validating, or invalidating the results of our modeling exercise. We proceeded in this way over six years, years which began and ended with modeling, but which were spanned by continuing interaction between simulated experiments and experiments in the real salt marsh.

The importance of good, basic, descriptive science to the success of an interdisciplinary study such as this can hardly be overstated. We began with the secure feeling that we were building upon an established description of the structure and function of the marshes and estuaries of Sapelo Island. Yet, the early results of our modeling almost immediately began to challenge our current beliefs. As a result, we repeated some measurements of such basic processes as photosynthesis, with results which did indeed alter our view of how the salt marsh operates.

Our study of the salt marsh developed in directions quite different from those taken by most other studies of highly stressed ecotones, such as rocky intertidal zones. Like the rocky intertidal, the salt marsh is hospitable to a limited set of species populations that clearly interact and coexist. Rather than focus solely on these interactions, however, we also looked at the salt marsh through the other end of the telescope, viewing the geological and physical setting and observing the chemical, biochemical, and biological processes within that setting. Viewed on this level, what we have seen is a system controlled to a great degree by its physical and chemical properties. Indeed, some of the most interesting and intricate population interactions proved to be the biochemical ones of the anaerobic microorganisms in the salt-marsh soil. Primary and secondary production are the biological processes which drive the system. We believe that a study of the salt marsh from this perspective produced a more complete and balanced analysis than would have been possible with a less extensive approach. We hope that this account may be of interest not only to salt-marsh ecologists but to others whose favorite ecosystem contains, albeit in different structural configurations, comparable controls and interactions.

Acknowledgments. The 20 years of research at Sapelo Island which laid the foundation for this work would not have been possible without the support of the late R. J. Reynolds, Jr., who not only provided financial aid but also accepted us as permanent guests on his plantation. The present work has enjoyed the continuing support of the Sapelo Island Research Foundation. The National Science Foundation supported the major portion of the work synthesized in this volume through the following grants:

DES72-01605 to L. R. Pomeroy and Dirk Frankenberg,
OCE75-20842 to L. R. Pomeroy and R. G. Wiegert,
GA-41189, OCE74-00148, and OCE77-26920 to W. M. Darley and D. M. Whitney.

Drawings and drafting were prepared by Rainer Krell.

Martha Hoak helped prepare the Index.

Finally, we wish to dedicate this volume to the permanent residents of Sapelo Island, who have been loyal friends and who have contributed in many ways to the success of this undertaking.

August 1980
Sapelo Island, Georgia

L. R. Pomeroy
R. G. Wiegert

Contents

Contributors

BUNKER, S. M.	Chesapeake Bay Laboratory, Solomons, Maryland, U.S.A.
CHALMERS, ALICE G.	University of Georgia Marine Institute, Sapelo Island, Georgia, U.S.A.
CHRISTIAN, ROBERT R.	Drexel University, Philadelphia, Pennsylvania, U.S.A.
DARLEY, W. MARSHALL	Department of Botany, University of Georgia, Athens, Georgia, U.S.A.
DUNN, E. LLOYD	Division of Biology, Georgia Institute of Technology, Atlanta, Georgia, U.S.A.
GALLAGHER, JOHN L.	University of Delaware College of Marine Studies, Lewes, Delaware, U.S.A.
HAINES, EVELYN B.	University of Georgia Marine Institute, Sapelo Island, Georgia, U.S.A.
HALL, JOHN R.	U.S. Department of Commerce, NOAA, NMFS, Washington, D.C., U.S.A.
HANSEN, JUDITH A.	CSIRO Division of Fisheries and Oceanography, North Beach, Western Australia
HANSON, ROGER B.	Skidaway Institute of Oceanography, Savannah, Georgia, U.S.A.

IMBERGER, JÖRG	Department of Civil Engineering, University of Western Australia, Nedlands, Western Australia
KING, GARY M.	Kellogg Biological Station, Michigan State University, Hickory Corners, Michigan, U.S.A.
MONTAGUE, CLAY L.	Department of Environmental Engineering Sciences, University of Florida, Gainesville, Florida, U.S.A.
PACE, MICHAEL L.	Department of Zoology, University of Georgia, Athens, Georgia, U.S.A.
PFEIFFER, WILLIAM J.	Department of Zoology, University of Georgia, Athens, Georgia, U.S.A.
SHERR, BARRY D.	Kinneret Limnological Laboratory, Tiberias, Israel
SKYRING, GRAHAM	Baas-Becking Geobiological Laboratory, Canberra, Australia
WETZEL, RICHARD L.	Virginia Institute of Marine Sciences, Gloucester Point, Virginia, U.S.A.
WHITNEY, DAVID M.	University of Georgia Marine Institute, Sapelo Island, Georgia, U.S.A.
WIEBE, WILLIAM J.	Department of Microbiology, University of Georgia, Athens, Georgia, U.S.A.

Ecosystem Structure and Function

1. Ecology of Salt Marshes: An Introduction

R.G. WIEGERT, L.R. POMEROY, and W. J. WIEBE

Salt marshes are fascinating yet frustrating places in which to study ecology–fascinating because of the wealth of physical and biological interactions present in this blend of terrestrial, aquatic, and marine communities, yet frustrating because of the many methodological problems peculiar to tidal communities. Quantitative measurement of flows of energy and matter, simple and straightforward in many other ecosystems, becomes more difficult in a complex tidal estuary. Perhaps these common problems fostered in us a greater spirit of cooperation than is common among other ecologists. At any rate, our group at Sapelo Island developed a deep appreciation of the value of each other's approach to scientific problems. In particular, we quickly came to realize the artificiality of separating ecosystem ecology from population ecology. Instead of perpetuating a sterile argument over the wisdom of measuring the whole or its parts, we adopted the more logical and scientifically defensible maxim that any explanation, as opposed to an observation, of the behavior of an entire system is impossible without some knowledge of its parts, and, in turn, any observed behavior of a single part can only be explained within the context of the system in which that part is functioning. This has been our philosophical approach to understanding ecological processes in the salt marshes.

1.1. Salt Marsh Ecology on Sapelo Island

The salt marshes of Sapelo Island have been the site of both ecosystem and population studies since 1954 (Figures 1.1 and 1.2). Early work at the ecosystem level emphasized flows of energy and materials through trophic levels or other aggregated components. Although this approach permitted the early investigators to deal with a complex system in a holistic manner, it was a "black-box" approach in which they measured inputs and outputs with little knowledge of the processes mediating the fluxes. As a result, concurrent studies of individual populations were initiated in an attempt

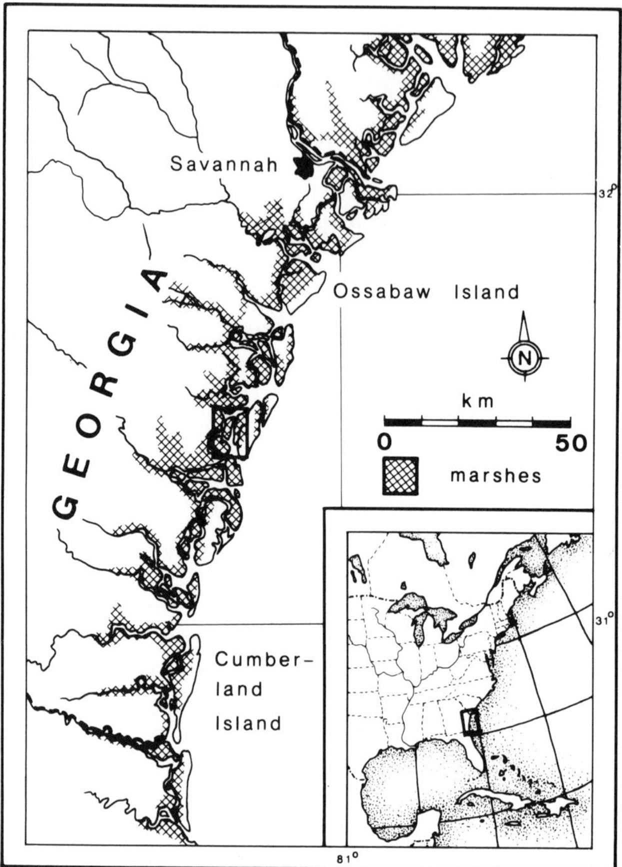

Figure 1.1. Location of the study area, the Duplin River watershed, which is immediately west of Sapelo Island (indicated by the square midway between Ossabaw and Cumberland Islands).

to unpack, one at a time, the black-box compartments of the salt marsh and to explain their internal function in relation to the system as a whole. As a practical matter this could not be done for all populations. Although population studies are often done in isolation from ecosystem analyses, the populations studied in the early years of work at Sapelo Island were determined to a large extent by the significance of their part in what we then perceived to be the ecosystem: an intertidal region dominated by a single species of grass, containing a food chain composed of grazing and sucking insects and their predators.

The emphasis on the ecosystem, with its quantitative description of processes, revealed over the first decade of research that only a small fraction of the energy captured by the grass was being grazed; most was going down some other pathway. Teal's (1962) early synthesis of energy flow through the Sapelo salt marsh suggested the presence of a large microbial component in the system which used a large fraction of the total energy available. A short time later, the pioneering work of Odum and de la

Cruz (1967) suggested how microorganisms gain access to so much of the organic matter in the system and how energy from that matter flows back in part to organisms higher in the food chain. From that study came the new perception of the salt-marsh ecosystem as one in which microbial transformations of energy and materials are quantitatively dominant and are key links in the food web.

At the end of the first 15 years of study at Sapelo Island the detritus food web, especially that part of it in the salt marsh, as opposed to the subtidal portion in the estuary, remained a series of black boxes in our conceptual model of the ecosystem. In 1972 we undertook an interdisciplinary study of the salt marshes at Sapelo Island with transformation of organic matter, particularly by microorganisms, as a central focus. In the next six years we reexamined ecosystem structure and function, developed budgets and models of the flux of carbon, and, at the same time, completed additional population studies that focused on interactions between the microbial and macroorganismal components of the detritus food web. Our goal was to understand and to explain ecosystem function through a combination of ecosystem modeling and experimental studies of populations of both macroorganisms and microorganisms. While some of the detailed findings are peculiar to intertidal marshes, or even to the marshes of Georgia, some of the principles appear to have wider application in ecological theory.

Like an iceberg, much of the salt marsh is hidden beneath the surface. Under the patina of green grass, grazers, and predators lies a dominant but unseen part of the salt marsh ecosystem, the microbial populations of the water, sediments, and soils. These populations transform dissolved and particulate organic matter and account for much of the flow of energy and materials. In salt marshes the effects of microorganisms stand out, perhaps because the ecologist is less distracted by the macroflora and macrofauna, luxuriant in numbers and biomass, but depauperate in species when compared to most other communities. Preliminary analysis laid the groundwork for our concept of an ecosystem-level intermediary metabolism at work in the salt marsh, a metabolism consisting of the transformation and degradation of nonliving particulate organic carbon (POC) and dissolved organic carbon (DOC). While most of the metabolism is that of the microorganisms, the processes are facilitated and controlled in vital ways by higher organisms. In turn, many of these dominant higher organisms of the salt marsh and estuary depend for sustenance upon microorganisms.

To understand the intermediary processes in the transformation of energy and materials in the salt marsh, we devoted much effort to the quantitative study of the rates of transformation of organic substrates under natural conditions in the marsh. Realizing that about 10% of primary production was going to direct grazers and 90% was going into the detritus food web, we attempted a balanced study, placing emphasis on particular groups in proportion to their impact, either as producers and consumers or as controlling species. To this end we brought together field and laboratory studies of both macroorganisms and microorganisms, giving some attention to measuring rates of microbial processes in the marsh itself.

Biological transformations of organic matter release a large number of organic chemical species, species which are products of microbial metabolism, of macroorganismal excretion, of chemical interactions of those compounds (e.g., humates and fulvates), and of the interactions of all of these materials with clay minerals. Faced with this array, we were tempted to launch a major effort to identify the dissolved organic compounds in the system, but decided against this for two reasons. First, no fixed relationship exists between the abundance of a compound and its quantitative or qualitative importance to a system, since measurement accounts only for instantaneous abundance of a compound, and not for its generation and utilization. Thus, even if we could successfully identify the compounds, their identification would provide only minimal insight into their function in the ecosystem. Second, because adequate analytical techniques for many compounds in salt water are either nonexistent or tedious to perform, examining enough of them to fully understand the system is impossible.

In this phase of the study our primary objective was to examine those biological processes related to the organic chemistry of the system. Since our goal was to understand the real ecosystem, we did not attempt to isolate microorganisms. Instead, we used the following strategy to examine microbial activities. First, we identified *in situ* a process and the magnitude of the fluxes involved; second, we performed field and laboratory manipulations to discover what controlled the process, and, finally, we isolated organisms that appeared to manifest unique capabilities. In this way, we worked "down" to isolation of organisms.

We also considered how all of these studies were to be linked, so that relationships between phenomena could be examined. We decided to use modeling to accomplish this, and it became a central activity, integrating the work from its inception. At the beginning of our six-year effort, the group constructed a simplified model of carbon flux in the estuary, a model based on existing knowledge. The formal exercise of modeling inevitably revealed the areas about which we had insufficient information and also provided us with some clues as to the relative importance of the myriad subjects we might study. New studies were initiated, and some areas which were already being studied, such as primary production, were reexamined or refocused in order to allow us to consider new questions.

Throughout our study, we used modeling as a heueristic device. Although predictive modeling was not a goal in itself, we used this approach whenever we could, because prediction offered at least one kind of independent validation of the model. Simulation modeling, a potentially powerful ally, is indeed at present the only method by which one can order and integrate the large number of variables that emerge from an ecosystem study in a manner that permits the ecologist to develop insightful, testable hypotheses about the behavior of the system as a whole.

1.2. Development of Salt Marshes

Salt marshes are intertidal ecosystems, backed up against the land on one side while opening to the estuary and the sea on the other. Not surprisingly, these ecosystems contain elements and attributes of both terrestrial and marine communities. The fundamental unit in space is the drainage basin, and many of the obvious organisms, especially the plants, are terrestrial species, siblings of those found in salt deserts of the continental interior. In the upper intertidal zone the terrestrial plants and the processes associated with them lead to the development of soils with horizons like those of terrestrial soils. The intertidal zone is in direct contact with the atmosphere and is influenced directly by rain; yet, in some parts of this zone, evaporation and the subsequent high salinities make water a limiting factor.

Complementing these terrestrial attributes are equally pivotal aquatic ones. In the salt marsh, water is the active medium for circulation of organic and inorganic nutrients and the medium in which most organisms live. Because salt marshes are connected with the sea and with rivers, their biogeochemical processes are more like those of water bodies than like those of purely terrestrial watersheds. While flow in terrestrial watersheds is a one-way process driven by Earth's gravity, flow in estuaries is a two-way process driven by tides. Thus, the ocean contributes significantly to the inward flux of materials, a fact demonstrated by the salinity of estuarine water, salinity usually greater than half that of seawater. The sediments of the subtidal, permanently submerged parts of the estuary are also more like those of water bodies than those of the marsh. Moreover, the system abounds with aquatic organisms, from plankton to fishes and porpoises. Many are daily tidal migrants that even forage among the grasses at high tide.

Salt marshes are plastic coastal features, shaped by the interaction of water, sediments, and vegetation. For stability they require protection from high energy waves and therefore usually develop in the lagoon behind a barrier island or in the protection of an estuary. In such protected situations deposition of sediment creates extensive, gently sloping formations. Perhaps salt-marsh vegetation occurs on protected coastlines not so much because the mature marsh is susceptible to erosion, but because waves prevent the establishment of seedlings in stable sediments. Indeed, once a marsh is mature, the vegetation can substantially modify the physical environment. Frey and Basan (1978) report an instance of *Spartina alterniflora* marsh that was able to withstand the full force of waves along the exposed shore of Cape Cod, once the grass was firmly rooted in peat. Thus, if the salt-marsh vegetation can establish itself and form the proper sediment base in a protected locale, subsequent shifting of protective barriers and exposure to the open sea need not result in its immediate destruction. Such a marsh can persist perhaps for centuries. These communities are analogous with relict communities, so commonly described in terrestrial situations.

Occasionally along a low-energy coast, salt marshes front the open sea. The marshes of the north coast of the Gulf of Mexico in western Florida (Tanner, 1960) and parts of Louisiana, the north Norfolk coast in Britain, and the coast of the Netherlands all are of this type. In the latter two sites the large, gently sloping marshes end abruptly landward at dikes built to reclaim land from the sea. These reclamations have gone on sporadically since the time of the Romans; thus, the physiography of the coastline, the hydrology, the deposition, and the resulting marsh are all, to a significant degree, the creations of humans. When these modifications were made, the natural buffering capacity of the marshes to protect the land from storms was sacrificed. Even in unmodified marshes, the transition zone between the marsh proper and the adjacent upland may be abrupt. In this case the ecotone is often dominated by shrubby species. If the transition zone is extensive and changes gradually, as, for example, when a gradient of salinity exists, a dominant higher plant less tolerant of salinity may take over, leaving two abrupt boundaries separating the marsh-dominant zone, the ecotone, and the upland. Vegetational zonation within the salt marsh itself is common, but there is disagreement about whether this zonation should be regarded as successional or simply as a reflection of underlying differences in elevation, salinity, tidal inundation, or sediment type (Chapman, 1977; Randerson, in press).

Most salt marshes are dominated by a single species. In marshes on the northwest coast of North America, from Oregon to Alaska, marshes dominated by the sedge, *Carex Lyngbyei*, predominate. In salt marshes worldwide, the most frequent dominant is the genus *Spartina*. In North America, large expanses of intertidal marsh, particularly along the southeastern coast, are dominated by the smooth cordgrass, *S. alterniflora* Loisel. The Mississippi River delta and the coastal plain to the west of it contain some 500,000 hectares of salt marsh. A marsh of approximately equal area lies along the east coast of North America from Virginia to Georgia (Chapman, 1977). The *Spartina* marshes represent an extensive ecosystem dominated by a single genus of higher plant. Not all salt marshes follow this pattern, however. Some marshes are dominated by species of Compositae, for example, those of Northeastern Britain, the Netherlands, and Southern California. Species diversity in such marshes is relatively high (Randerson, in press; Zedler, 1977).

The marshes of southeastern North America, which include those of Sapelo Island, are a mixture of deltaic and lagoonal development. Several theories have been proposed to explain the formation of the barrier islands like those along the coast of Georgia and South Carolina (Hoyt, 1967). During the Pleistocene and up to the present time, substantial variations in sea level occurred; at times large dunes formed along the shore through the combined action of tides, waves, and prevailing winds. These dunes were partially submerged and cut off from shoreward contact during periods of rising sea level. The resultant formation of a lagoon separated the dunes from the mainland and made them, in effect, barrier islands. As the lagoon accumulated fine sediments, salt-tolerant grasses invaded. We are now in such an interglacial period, with sea level still rising. A number of distinct belts of old barrier islands, now reduced to ridges, can be traced along the coast of Georgia (Hails and Hoyt, 1969). Excluding a few small seaward islands or extensions of islands of more recent (Holocene) origin (e.g., Sea Island and Blackbeard Island), all of these, including the present barrier islands, are of Pleistocene age.

The Georgia coastal islands are located about seven kilometers offshore, with their long axis parallel to the coast (Figure 1.2). Behind the islands, the coastal lagoons are three-fourths filled by intertidal salt marshes, with estuaries and river deltas opening between the barrier islands. A diagrammatic cross section of a typical barrier island (Figure 1.3) shows the relative position of ocean beach, protective dunes, stabilized dunes, forest, and tidal marsh areas. Sediment deposited in this area, 10 meters in thickness in the center of the lagoon and mostly of Pleistocene age or older, has formed extensive, flat or gently sloping topography essential for the development of a salt marsh (Johnson et al., 1974; Chapman, 1977; Howard et al., 1975; Frey and Basan, 1978). The lagoonal marshes, which are the subject of this study, are brackish to saline, with the interstitial salinity around the roots of the plants varying from 15 to 70‰. In contrast, the deltaic marshes of the nearby Altamaha River usually have interstitial salinities of less than 15‰. Most of those lower-salinity marshes are dominated by the rush, *Juncus roemerianus* L., with stands of *Spartina cynosuroides* along the margins.

Although the Georgia coast has only moderate surf energy, the tidal amplitude is 2 to 3 meters, an amplitude larger than that in any other coastal area of southeastern North America. The relatively large daily fluctuation in the height of the tide affects

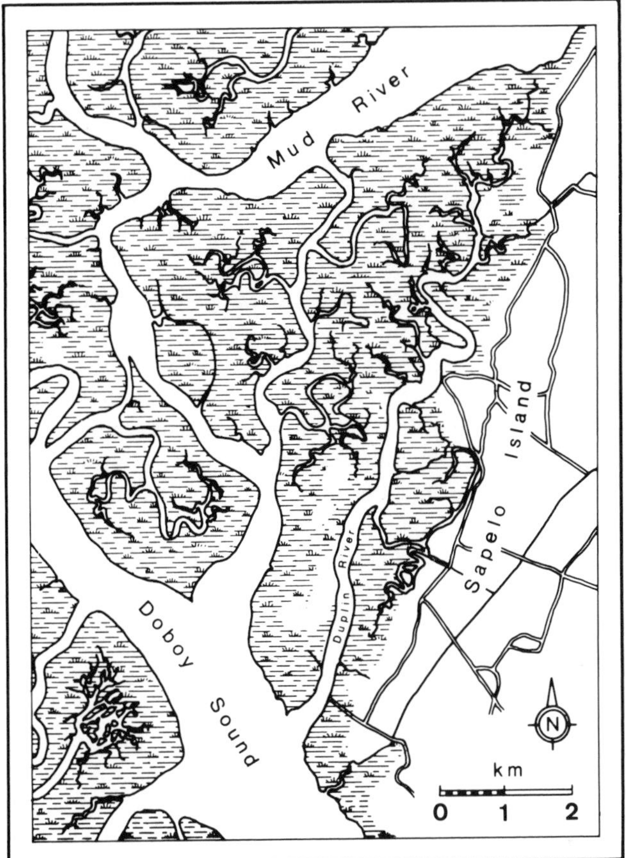

Figure 1.2. The marshes of Sapelo Island (expansion of square in Figure 1.1.). Most of the work discussed was done in the tidal watershed of the Duplin River, a brackish tidal stream which is tributary to Doboy Sound.

Figure 1.3. Schematic cross section through one of the sea islands of Georgia, such as Sapelo Island, showing the types of land development and vegetation. Level reference line is approximately at spring high tide. Adapted from Clement, C., *The Georgia Coast: Issues and Options for Recreation* (Athens, Ga.: Division of Research, College of Business Administration, University of Georgia, 1971).

marsh development, notably in the formation of natural levees much more pronounced than those of the marshes of South Carolina. There are extensive areas reached only by the fortnightly spring tides, in which increased salinity due to evaporation from the marsh surface profoundly influences the size and distribution of dominant plants.

The development of prominent natural levees influences, in turn, the tidal movement of water over the marsh. In a fully developed system of natural levees, water does not flow laterally from the tidal streams into and out of the marsh, but is instead guided by the levees up the tidal creeks to central distributary points in the interior of the marsh. From these points onward, water speed decreases enough to permit laminar flow, and sediments drop out. Although the marsh is riddled with crab burrows, a perched water table behind the levees holds interstitial water that is more saline than tidal water and largely anaerobic. Some of this water percolates through the levees during low tide, but enough remains to create a permanent supply of anaerobic ground water in the central part of any large expanse of marsh (Nestler, 1977,1977a). The course of microbial processes and the success of the dominant plants is influenced greatly by this anaerobic pool.

1.3. Ecological Processes in the Marsh

An ecological process, as the term is used here, encompasses any transformation of matter or energy initiated by living organisms and modified or controlled in some manner by physical conditions of the environment or by interactions with other organisms or with abiotic materials. Thus, primary and secondary production, respiratory catabolism, predation, and seasonal mortality are all examples of ecological processes important in the salt-marsh ecosystem. Organisms and their ecological processes are responsible for generating flows of matter and energy. For example, photosynthesis and microbial fixation of carbon in the marsh remove inorganic carbon from the air or from solution in the water and transform it into organic carbon in living organisms. These living organic compounds are later degraded to nonliving dissolved and particulate carbon compounds in water and sediment. Compounds in the sediment exchange with those in the water by erosion, deposition, and diffusion. Ecological interactions, resource availability (essential elements and energy), and physical forces and constraints (light, temperature, tidal regime) all influence the magnitude of the flows of organic materials.

In our modeling studies we divided the salt marsh into three physically separate regions: the region of emergent shoots of grass in air, the region of intertidal and subtidal water, and the region of soils and sediments. Although these divisions lead to some arbitrary separations, they form the basis of the discussion which follows. In the region of the emergent grass, terrestrial-type processes and organisms predominate. In the tidal creeks and in the water flowing over the marsh at high tide, aquatic organisms and predominantly aerobic processes are found. The soils and sediments are the site of both aerobic and anaerobic microbial and meiofaunal processes, and they have attributes of both terrestrial soils and aquatic sediments. These simplified divisions serve to emphasize an important distinction between the intertidal system and most other systems: the air-water and sediment-water interfaces, which normally form the boundaries between distinct systems, are here compressed into such a dynamically interacting mosaic that they cannot be easily separated, even for purposes of description and study. Diagrammatic representations of the spatial (Figure 1.4) and temporal (Figure 1.5) heterogeneity of the marsh show both the major organismal groups and the dominant ecological processes that will be discussed in subsequent chapters.

The use of some conserved unit, in this study, carbon, facilitates discussion of flows of materials, whether the rates are expressed in instantaneous form (materials x area^{-1} x time^{-1}) or specific form (time^{-1}). Flows of other conserved elements and of energy will be introduced only where those units influence the rate of specific flow of carbon, or where they provide more insight into an ecological process. The selection of the organisms and flows for study was conditioned to a large extent by the significance of their relation to the system diagrammed in Figure 1.5.

Consider the fate of the annual crop of *Spartina*, a marsh grass. Little of the production of *Spartina* is grazed. Most of it dies and is transformed into detritus and then into more readily assimilated compounds. Microorganisms play a major role in these transformations, and the result is an array of dissolved and particulate organic materials in the water and sediments. The residence times of these compounds range from minutes to millenia (Hanson and Wiebe, 1977; Sottile, 1973; Williams et al., 1969). The quantitative importance of any ecological process associated with these transformations is defined either by the magnitude of the flow or by its direct influence in regulating another large carbon flow. For example, the microbial transformation of cellulose causes a quantitatively large flow of carbon in the marsh, and cellulolytic bacteria are therefore a significant population in terms of carbon flow. By contrast, although ammonification involves modest transformations of carbon, the availability of nitrogen, as ammonia, can limit or regulate the activities of cellulolytic bacteria, and the absence of a sufficient supply of nitrogen can retard degradation of detritus (Tenore et al., 1977). Therefore, the ammonifying bacteria are a keystone group, controlling carbon flow and physical degradation of detritus by affecting the activities

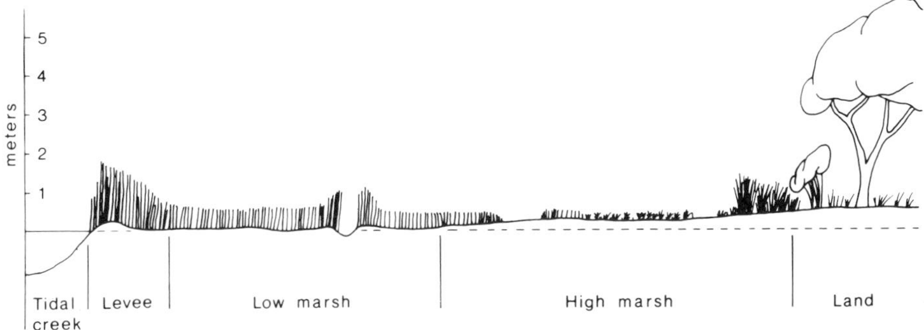

Figure 1.4. Schematic cross section through the Georgia salt marsh, showing levee and low marsh with tall *Spartina*, and high marsh with short *Spartina*. The level reference line is approximately at mean high water. Drawing by Krell from sketch by Wiegert.

of cellulolytic bacterial populations. This, in turn, affects the higher organisms that utilize those populations as sources of energy and materials.

1.3.1 Processes in the Emergent Grass

The major process by which organic carbon reaches the aboveground region of the marsh is the annual growth of shoots of *S. alterniflora*. In a Georgia lagoonal salt marsh, this species of grass so dominates the view that the marsh seems to be a sea of waving grass. On closer inspection, as indicated in Figure 1.6, this apparent homogeneity gives way to a spatially heterogeneous pattern of tidal creeks, mudbanks, oyster reefs, dead *Spartina* stems, surface detritus, animal burrows, and clumps of mussels. This heterogeneity extends even to the production of *Spartina*, which varies by almost twofold between stream bank and central marsh.

The primary cause of spatial heterogeneity in the production of *Spartina* is the change in the height of the sediments above mean low tide and the related process of levee formation. Wherever the natural levee is well developed, there is an abrupt rise of the sediment surface from the tidal creek, then a gradual drop to a low point in the interior of the marsh, from which point the surface rises landward very gradually until the limit of the highest spring tide is reached (Figure 1.4). This limit represents the boundary of the marsh.

In general, the mean height of *Spartina* shoots decreases with distance from a tidal creek, but the change is greatest and most abrupt near the levee. An area of tall *Spartina* (≅2m), occupying 8% of the total marsh surface (Reimold et al., 1975) includes the creek banks, the levees, and the marsh surface immediately behind the levees. The remainder of the area dominated by *Spartina* is given to grass which is graded from heights of 60 cm behind the levees to heights of 30 cm in the most isolated interior or landward portions of the marsh. Where the marsh abuts the high ground of the land and wherever the salinity of interstitial water is lowered, *Spartina* gives way to the black needle rush, *Juncus roemerianus*. A decrease in salinity may be caused by surface drainage, by intrusion of ground water near the landward edge of the marsh, or by local elevation, as where *Juncus* grows on beds of mussels in the marsh. Extensive stands of *J. roemerianus* also occupy the nearby river delta. Where ground water does not intrude from the land, or where there is a large, isolated expanse of marsh, evaporation leads to salinities high enough to exclude *Spartina*, leaving relatively bare expanses with a thin population of *Salicornia spp*. Tall *Spartina*, which supplies approximately 2500 g C m^{-2} yr^{-1}, is more productive than short *Spartina*, which supplies approximately 1500 g C m^{-2} yr^{-1} (Gallagher et al., 1973; Reimold et al., 1975; Giurgevich and Dunn, in press). The inequality of production in the short and tall forms seems definitely to be a positional and not an ecotypic effect, related primarily to interstitial flow and thus to height of the soil surface or to isolation from creeks. Transplants and other manipulations have transformed each type into the other, although the time needed to affect a complete change may be several years (Nestler, 1977; Chalmers, 1977; Valiela and Teal, 1979).

For the Sapelo marshes, allowing for open water and mud bank areas, the net input of fixed organic carbon derived from *S. alterniflora* is 1539 g C m^{-2} yr^{-1} (computed from Gallagher et al., 1973, and Giurgevich and Dunn, in press). Of this total, 50% is

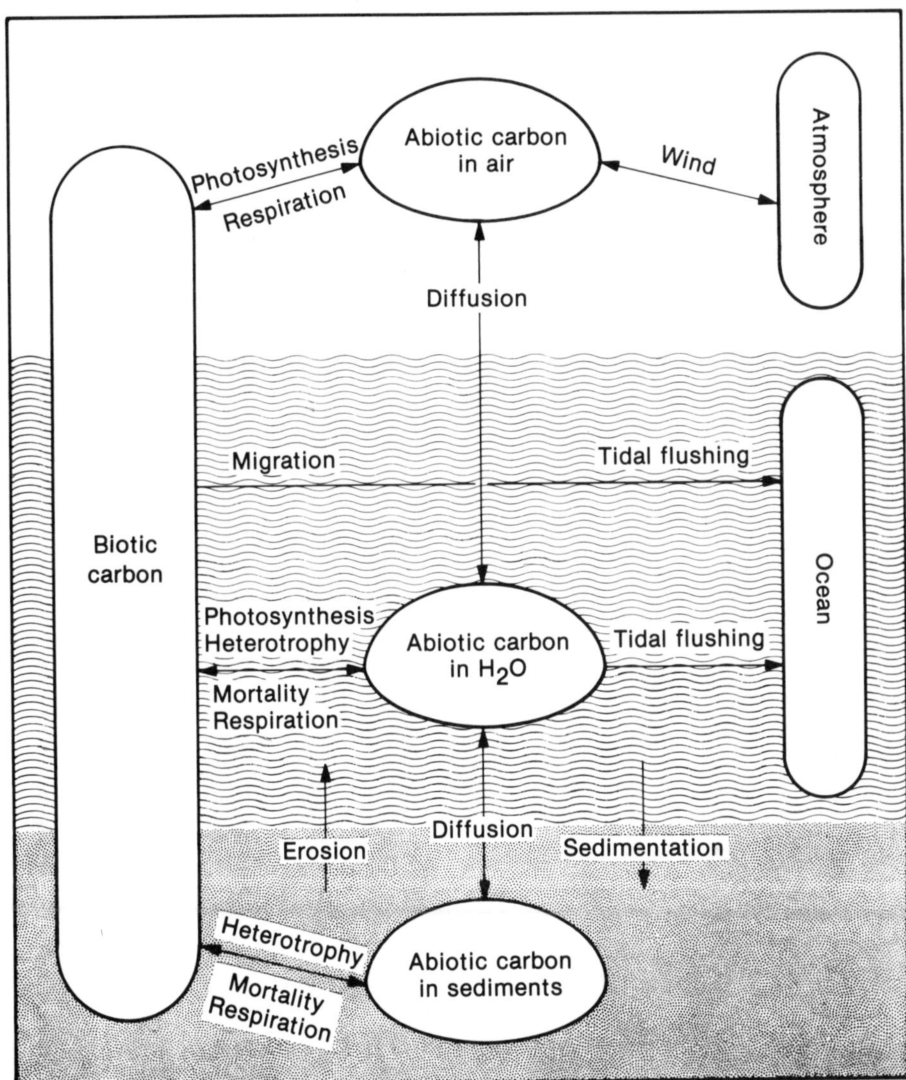

Figure 1.5. Compartmental model of carbon fluxes associated with air, water, and sediments in the salt marsh at Sapelo Island. Adapted from Wiegert and Wetzel (1978) in *Marsh-Estuarine Systems Simulation*, edited by Richard F. Dame, Number 8 in the Belle W. Baruch Library in Marine Science, by permission of the University of South Carolina Press. Copyright © University of South Carolina 1979.

Figure 1.6. View of the Sapelo Island salt marsh, showing the diversity of habitats.

estimated to remain as shoot production and 50% as production of roots and rhizomes in the sediments. The rate of input is largely controlled by nutrient limitation, or by the effect of high interstitial salinity in the short *Spartina* and by limitations of available light in the tall *Spartina*. There is abundant available phosphate in the marsh, in both water and sediments (Pomeroy et al., 1972). Available nitrogen is present but the amount is limited relative to demand, for nitrogen enrichment will stimulate production (Chalmers et al., 1976; Chalmers, 1977; see Chapter 8).

The emergent shoots of *Spartina*, and to a lesser extent those of *Juncus*, support a food web of grazers and predators having terrestrial origins. Among the most common insects that obtain energy and nutrients from *Spartina* shoots are planthoppers of the genus *Prokelesia* and grasshoppers of the genus *Orchelimum* (Smalley, 1960; Teal, 1962). A number of arthropods and vertebrates in turn feed upon the insects. Spiders constitute the major group of arthropod predators; two of the most abundant are of the genera *Clubiona* and *Grammamota*. Both live in the leaf axils and hollow stems of *Spartina*. Birds take a heavy toll of both herbivorous and predatory arthropods. Long-billed marsh wrens are abundant in the marsh and account for much of the arthropod mortality, particularly during the nesting season (Kale, 1965). During high spring tides, when large numbers of the salt-marsh grasshopper, *Orchelimum*, are forced to climb to the extreme tips of *Spartina* shoots, seagulls are able to capture them.

Although the grazing food chain comprises some abundant and ecologically interesting species (Marples, 1966), the flow of carbon and energy through the web is small. For example, Teal (1962) summarized the existing data on energy flow through the grazers of *Spartina* shoots, and his values, converted to carbon equivalents, give 31 g C

m^{-2} yr^{-1} as the loss to herbivores. Current investigations of the populations of planthoppers suggest this to be, if anything, an overestimate (Chapter 4). Therefore, the direct removal of organic carbon compounds from *Spartina* shoots by terrestrial grazers cannot be a significant factor in the overall carbon flux of the salt marsh. The effects of the grazers, if they do have an effect, must lie in some as yet undefined control functions.

1.3.2 Processes in the Water

The aquatic component of the system (Figure 1.5) comprises the water that covers the marsh at high tide and the water that remains in the tidal creeks. At low tide this water resides in the larger tidal creeks which do not drain completely, and in the tidal arms of the estuary, such as the Duplin River (Figure 1.2). Many of the important ecological processes characteristic of this region of the marsh are associated with populations that reside permanently in the water, such as plankton, fishes, and crabs. Although these populations may move with the tide onto the surface of the marsh, they do not remain there when the tide goes out. Other groups, such as mussels, oysters, fiddler crabs, and xanthid crabs, do reside permanently in the marsh and are periodically inundated by the water. Some of these organisms have little direct interaction with the air. Oysters and mussels, for example, close their shells during exposure, respire anaerobically and make no use of gas exchange with the air. For them the intertidal zone is a refuge from aquatic predators. Others, such as fiddler crabs, are active at low tide, feeding on the sediment surface and exchanging carbon dioxide with the air.

In addition to the living components, carbon is also carried in the tidal water as bicarbonate, carbon dioxide, methane, dissolved organic compounds (DOC) and abiotic particulate organic materials (POC). Organic carbon may enter the water in a number of ways:

(1) *Spartina*, benthic micro-algae, and phytoplankton fix inorganic carbon, which is replenished by respiration and by some diffusion of CO_2 from the air to the water;

(2) benthic algae and phytoplankton release some fraction of their fixed carbon as DOC;

(3) living shoots of *Spartina* secrete some DOC into the water during high tide (Gallagher et al., 1976; Turner, 1978);

(4) standing dead and rafted *Spartina* are degraded by bacteria and fungi, with loss of both DOC and POC, the loss of DOC being about equal to that released by living plants; and

(5) dissolved organic materials may diffuse or be lifted into the water from the sediments.

Phytoplankton and benthic algae are major primary producers in the water, macroscopic algae being sparse. Some early experiments in the highly turbid tidal creeks led to the conclusion that phytoplankton production in the estuaries of Georgia is negligible (Ragotzkie, 1959). However, Chapter 3 describes recent evaluations of production of both phytoplankton and benthic algae in the estuary as a whole and in the salt marsh proper. The production of *S. alterniflora*, which occupies approximately the same parts of the marsh as does the benthic microflora, is much greater than the combined production of phytoplankton and benthic microflora. Total net primary production integrated over the wateshed is on the order of 4.6 g C m^{-2} day^{-1}; 80% of the production is rooted plants, 10% is phytoplankton, and 10% is microalgae in the sediments. These values are not only indicative of the high productivity of the ecosystem but also of the kind of food web which is possible. Most of the primary production is in the form of *Spartina* shoots and roots which are not heavily grazed, but over half of the production of *S. alterniflora* is in roots and rhizomes which do not enter directly into the aboveground food web. If only aboveground primary production is considered, 64% is *Spartina* stems and leaves, and 36% is algae. Production by algae, thus its availability to consumers, varies from time to time, but the gradual transformation of *Spartina* to detritus provides a relatively constant source of particulate organic carbon. These processes are occurring throughout the year, with seasonal peaks of shoot mortality in fall and of shoot disintegration in spring and summer. Most organic carbon from these sources enters the water as POC, but small amounts of DOC are secreted by living shoots and by the microorganisms living on the standing dead grass (Gallagher and Pfeiffer, 1977; Gallagher et al., 1976).

The processes of death and disintegration of *Spartina* are controlled by the internal physiological and phenological mechanisms of the plant and by the interactions of standing dead *Spartina* with decomposers. The amount of POC in the water has little or no effect on the rate of supply of POC; that is, this process shows little evidence of any direct negative feedback control. Some determination in the stems of *Spartina* can be attributed to semidiurnal tidal action and periodic storms, but microorganisms and macroorganisms undoubtedly play the major role in particle formation. Because there is no ice in winter in Georgia marshes, peat formation is seldom initiated through the shearing and packing of *Spartina*, as it is in more northerly marshes (Frey and Basan, 1978). The major accumulations of peat in the southeastern marshes occur when *S. alterniflora* is covered by sand moved from the eroding northern tips of barrier islands.

Transformations of carbon in the water are numerous and have by no means been fully described. All of the carbon fixed through primary production eventually becomes part of the abiotic carbon pools (Figure 1.5) through respiration (CO_2), mortality or defecation (POC), and degradation or secretion (DOC). Mortality of the phytoplankton and small heterotrophs contributes carbon directly to the water but this represents a minor flux of the element. Mortality of *Spartina* shoots, however, gives rise to standing dead and to rafted stems. From this material POC is produced through the dual action of consumer organisms and through physical disintegration. In the initial transfer of carbon from the living to the nonliving state, production of POC far outweighs the secretion of DOC. However, subsequent heterotrophic utilization of the organic compounds probably requires repeated passage of the POC through consumer organisms with accompanying production of CO_2 (Wetzel, 1976; Bunker, 1979). Lacking interaction with detritivores, microbial transformations of organic matter proceed slowly, and dead plant and animal matter tends to remain intact for months or even years. As these processes, somewhat akin to cycling, (Rigler, 1975) continue, large proportions of DOC may be produced by heterotrophs, proportions large in comparison to the POC derived secondarily from these organisms.

In contrast to POC, most of the DOC may be metabolized readily. Turnover of labile DOC is rapid, although standing stocks of labile DOC are low. Therefore, we cannot relegate DOC to a position of minor importance in the estuarine-marsh water, even on a quantitative basis. In final analysis we must consider the extent to which DOC is reassimilated into microorganisms and thence into the filter feeding and detritivore food webs. That process is revealed by the kinds and rates of transformation and assimilation of specific DOC compounds in the water.

Filter feeding, particle feeding, and microbial assimilation cause new organic compounds and CO_2 to be released into the water. The filter-feeding organisms in the marsh, including zooplankton, oysters, mussels, and worms, are responsible for incorporating some POC and DOC into the biotic pool. But the principal effect of this diverse heterotrophic group is seen mainly in their ingestion and assimilation of heterotrophic microorganisms. These microorganisms, in turn, greatly influence the subsequent transformation, if not the initial fixation, of all carbon. Particle feeders, including carbs, shrimps, and fishes, consume relatively large organisms, both living and dead. However, a number of them, including fiddler crabs, shrimps, and mullet, probably also assimilate significant quantities of algae and the microbial biomass associated with detrital POC. This has been verified in some instances (Wetzel, 1976; Odum, 1968; Odum and Heald, 1972).

Identifying trophic levels, even using the classical trophic-level concept to analyze the food web of the macroconsumers is a difficult enough task. The difficulty only becomes worse as we examine the microorganisms, some of which by definition occupy several trophic levels. The aerobic microorganisms in the water are quantitatively more significant than filter feeders and particle feeders, since the latter two groups subsist in large part on microbial biomass. Microbial oxidation of both labile and refractory organic compounds may contribute as much as three to four times the CO_2 that is produced by respiration of the larger heterotrophs. The total transformation of carbon to CO_2 in the water is unlikely to exceed 200-250 g c m^{-2} yr^{-1}.

Microbial transformation of organic materials is limited by scarcity of essential nutrients, by direct interference of competing organisms, and by death from consumption by bacteriovores. We are only now coming to understand something of these control mechanisms, but this brief evaluation shows the relatively small portion of net primary production, or even production of *Spartina* shoots alone, that is degraded to CO_2 within the water (see Chapter 4).

Carbon in the water is exchanged with the marsh sediments through settling and erosion of POC and through diffusion of DOC and gaseous CO_2 and CH_4. Only in the soils of the natural levees does water flow-through account for appreciable transfer of organic carbon in or out of the sediments. Exchange between the marsh water and the downstream estuarine water bodies may be caused by tidal mixing, by flushing of rain storms, or by the migration of motile organisms which move up and down the tidal creeks and the estuary independently of the net current flux. The role of these exchanges in moving carbon into or out of the marsh is discussed in Chapters 2 and 5.

1.3.3 Processes in Soils and Sediments

Marsh soils and sediments are the site of very substantial biological activity. Some 60% of the net carbon fixed by *Spartina* is deposited in the soil through growth of roots and rhizomes. The annual organic carbon input to the microbial community from the soil is relatively constant, although the death rate of roots and rhizomes changes with season and stage of growth. As the roots and rhizomes die, their organic compounds provide energy for oxidative and fermentative transformations involving nitrogen, sulfur, and phosphorus, as well as carbon. Microbial activities which regulate transformations of N, S, and P can have a significant regulatory effect on the primary fixation of carbon (Chapters 6 and 7). The nature and rate of these processes is influenced by the oxidative state of the soil environment. Aerobic soils are limited to the uppermost few millimeters and to a microzone around the roots of *Spartina*. The microzones around roots can be seen in sections as thin zones of red ferric oxide, the result of diffusion of oxygen down the hollow stems and rhizomes of *Spartina* and out through the surface of the roots (Teal and Kanwisher, 1966). The soil environment is, then, a complex, three-dimensional pattern of oxidized and reduced zones with gradients between them. This structure permits varied microbial transformations of both organic and inorganic compounds.

In spite of the penetration of oxygen into the soils through the roots of *Spartina*, the greater volume of the soil is anaerobic. Therefore, there is an extensive region of anaerobic, fermentative metabolicm in the salt-marsh soil. Short-term uptake of oxygen upon aeration of this anaerobic soil is 25 times the normal rate (Teal and Kanwisher, 1961). This uptake is not affected by the addition of formalin, and relatively little CO_2 is produced; both of these conditions indicate chemical rather than biological oxidation.

Not all organic carbon in marsh soils is transformed in place. At times the transformation may be a partial one. Large amounts of living or recently dead biomass are brought to the surface and deposited by the burrowing fiddler crabs (Chapter 5). Slow rates of water exchange (Nestler, 1977), mean relatively little carbon leaves the soil as DOC, but gaseous products of microbial activity–carbon dioxide (CO_2) and methane (CH_4)–do diffuse out into air and water over the interior of the marsh (King and Wiebe, 1978). An approximate estimate of the annual movement of gaseous carbon compounds from the soils into air and water can be obtained from measurements of either oxygen uptake, assuming CO_2 is produced, or CO_2 and CH_4 losses. These calculations are discussed in subsequent chapters, especially Chapters 3, 6, 9, and 10. The best estimates indicate that only about one-third of the net carbon fixed each year by plants is transformed in the soils and subsequently lost.

1.4. Carbon Mass Balance and Modeling

Adding the additional losses from degradation in the water to the gaseous carbon lost from the soils, about one-half to two-thirds of the annual net fixed carbon in the marsh can be accounted for. Defining the ultimate fate of the excess carbon is vital to an understanding of the impact of coastal marshes on both the estuarine and offshore communities. To focus our investigation, the major carbon fluxes in the marshes, those associated with the ecological processes just discussed, were cast into a food web diagram (Figure 1.5). In order to ascertain the relative importance of the controls governing the rates of transformation and transport of carbon, this simple food web was expanded into a preliminary simulation model (Wiegert et al., 1975). For the model, each carbon flux was represented by an equation that summarized our best explanation of the manner in which ecological processes controlled the rate of transfer. Nutrient cycles and nutrient availability were not explicitly included in the model, but information on the effects of nutrient scarcity and nutrient additions guided us in the construction of the simulation model and in the selection of coefficients to represent seasonal changes. Once the preliminary model was constructed, we used it to direct our research towards those ecological processes most important in regulating the carbon dynamics of the marsh and to explore various hypotheses on the interactions of marsh, estuary, and sea.

The following chapters were produced by synthesizing various approaches and disciplines. The intuitive approach to ecosystems, which has served us well, was combined with the more formal exercises of simulation modeling and testing, through simulation, the results. Most of the classical approaches to population and ecosystem analyses had already been taken in studying the marshes of Sapelo Island before this work was begun. Using those as our base, we added investigations of microbial processes, and a new, and in some respects novel, analysis of the hydrography of a marsh watershed. We also initiated studies of dominant or key populations of both macroorganisms and microorganisms. This information was incorporated into a final round of simulation modeling. We report here the overall results and final synthesis of our study of the Sapelo Island salt marsh, an ecosystem that is itself a synthesis, a blending of aquatic and terrestrial populations and processes.

2. The Physical and Chemical Environment

L.R. Pomeroy and J. Imberger

Wetlands, whether forest or heath, grassland or rush, are structurally similar to terrestrial systems. The tidal salt marshes of the Duplin River are no exception. However, here the land has been shaped by a changing sea level since the Pleistocene. Locally tidal erosion and deposition determine the pattern of marsh and creeks, but the functional unit is still the watershed. Tidal currents generated in the sea are the primary driving force in the physical environment. Fresh water input from rain may cause erosion and influence flushing of the watershed, but fresh water from the Altamaha River, which enters the mouth of the Duplin River, influences only the water chemistry.

While a salt marsh is, in the long term, a depositional environment, there is reason to doubt that deposition is the dominant feature of the present Duplin River watershed. The basin has been very nearly filled to grade (Frey and Basan, 1978). Sediments are constantly being reworked, with physical processes being the dominant force in the tidal streams (Howard et al., 1975) and biological processes being the dominant force in the marsh proper. Therefore, while the Duplin River watershed is to some degree a sink for sediments and for chemical species associated with sediments or interstitial water, the geological evidence points to a state of quasiequilibrium, perturbed on a scale of months or years by storms and on a longer time scale by continuing changes in sea level. Our concern is primarily with events on the scale of months and years, rather than the ultimate geological processes, except as the latter have influenced the present environment.

It is currently debated whether salt marshes are sources or sinks for a variety of substances: organic carbon, inorganic plant nutrients, and trace metals (Nixon, 1980). Because of the great variation in the geological setting of various coastal wetlands, no single answer is applicable. Some wetlands are environments of rapid deposition, perhaps because the area is impounded, as in the Flax Pond marsh of New York (Woodwell et al., 1977), or perhaps because of rapid isostatic sinking as in the Airplane Lake marsh of Louisiana (Delaune et al., 1978). Other marshes, such as some of those on Cape Cod (Valiela et al., 1978), appear to have stronger terrestrial influences and, in addition, are affected by ice formation and movement. In the Duplin River and its marshes we evaluated the net flux of a number of substances, and a successful approach to the problem was developed, an approach applicable to other estuarine systems as well.

2.1. Geomorphology

The Duplin River watershed covers 11 km^2, of which approximately 80% is intertidal marsh and mud flat and the remainder is permanently submerged. It is bounded on the east by Sapelo Island (Figure 1.2) and to the north and west by a ridge of Pleistocene sand, some of which is high enough to support truly terrestrial vegetation. The main stream channel, 12 km in length, opens to the south into Doboy Sound approximately 2 km from the inlet to the ocean. The Duplin River is not a river in the usual sense of the word. Like a number of other tidal streams in coastal Georgia, it has no significant permanent source of fresh water at its head, and the form of the stream channels is completely the result of tidal action. Although most of the drainage area is covered at least fortnightly by estuarine water, the environment has some characteristics of a terrestrial system, notably the soil structure and the flora and fauna.

The salt marsh is dissected by tidal creeks which meander and form a dendritic pattern similar in appearance to a truly riverine stream. Because they are shaped by both flood and ebb currents, which differ in velocity, the larger streams have both flood and ebb channels, with point bars between them (Howard et al., 1975). The position of the major stream channels appears to have changed relatively little within historic time.

Organisms significantly influence the land form of the marsh itself and of the smaller distributary streams. The most influential species is *S. alterniflora* Loisel. The presence of this grass on the streamsides hastens the development of natural levees along the Duplin River and its smaller distributaries. Except on the highest spring tides, the natural levees impede the flow of water across the marsh on the rising and falling tide; the water must flow up the distributaries to their heads and then flow back, inside the levees, to fill the expanse of the marshes. Once the marsh is in steady state with respect to sediment, material carried in suspension is moved directly to the center of expanses of relatively level marsh, apparently producing lateral pressure on the levees. In the marshes of Sapelo Island we have found that levees creep toward the stream channels at a rate of at least a decimeter per year. This, as well as erosional undercurring, leads ultimately to slumping of the levees into the creeks. Blocks of *Spartina* sod may slump slowly down a stream bank; in some instances, masses of sod up to one-half meter in width may tumble into the center of a distributary (Frey and Basan, 1978). During low tide, such precipitous erosion is promoted by rainfall, which probably reduces the ion strength of interstitial water along existing cracks or crab burrows. The resulting loss of binding strength of the clay allows the blocks of sod to fall. However, the entire levee does not slump or break away, so the pattern of water flow is not greatly influenced by levee erosion.

Oysters (*Crassostrea virginica* Gmel.) also influence the morphology of stream channels. In the smaller tributaries, oyster bars develop at intervals, producing a pool and riffle structure much like that of a terrestrial stream. Since oysters grow on any solid object, a log or any stranded object may form a site for a durable clump of oysters. Such a clump may in time catch enough sediment to permit growth of *Spartina*. This sequence of new marsh development is particularly noticeable in areas of deposition of unconsolidated sediment.

Numerous burrowing crabs of several species affect the structure of the marsh in much the same way that burrowing animals affect the structure of terrestrial soils. By bringing up subsurface sediment from the burrows and by reworking the upper 10 to 20 cm of sediments (Chapter 4), they promote erosion, percolation of water, and aeration of subsurface sediments. It is difficult to judge the significance of the activities of crabs relative to the recurrent development of evaporation cracks during daytime low tides in spring. Almost certainly bioturbation, or the mixing of the soils by the activity of animals, has a significant influence on the microbial biochemistry and biogeochemistry of the sediments, and it probably influences the shear strength of the soil. In general, bioturbation of the soils is more extensive in Georgia salt marshes than in marshes in some other parts of the world (Frey and Basan, 1978).

The broad expanses of marsh behind the natural levees contain a perched water table, usually about 10 cm below the surface. Refilled on spring tides and slowly depleted by evaporation and transpiration, this water undergoes little lateral movement. Ground water movement, particularly slow in the central regions of marsh, is not directly significant in the transport of materials (Nestler, 1977a). The perched water table does not have ready egress through the natural levees, although the levees themselves are highly permeable to creek water. During low tide one sees here and there a burrow in the levee from which dark-colored water is trickling, but the burrows usually open to the surface immediately behind the levee and do not drain the perched water table. Remarkably little water comes out through the levees, considering the large number of burrows and evaporation cracks present. The structure of the interior of the levee, which must be responsible for holding the water table in place, has not been described. We can postulate that percolation through the soils is limited by the absence of peat and the presence of ubiquitous fine clay. We know that there is a rather sharp demarcation separating the stagnant perched water table, in which tracer dye moves hardly at all, and the levee, where tracer dye disappears in one or two tidal cycles. These striking differences are also reflected in the salinity of the interstitial water. The salinity of the perched water table is usually higher than that of tidal creek water and changes little over time, while the salinity of interstitial water in the levees closely follows changes in the salinity of tidal creek water.

2.2. Physical Conditions

The sediments of the Duplin River and its marsh soils comprise marine or estuarine clays and sand. The clays are a mixture, principally of kaolinite and montmorillonite. The sands are reworked Pleistocene sands eroded from the margins of the drainage area (Howard et al., 1975). Organic matter in the soils and sediments amounts to 10 to 20% of the dry weight. In the Duplin River and its distributaries the sediments may be unconsolidated or consolidated and bedded (Howard et al., 1975). The marsh sediments have a soil-horizon structure much like that of terrestrial soils, presumably because of the influence of the dense stands of *Spartina* and *Juncus,* and because of the working of the soil by animals, especially crabs.

Water temperature in the Duplin River varies from a winter minimum of around 10°C to a summer maximum of 30°C. Temperature variations in the marsh are more extreme, and this is reflected in water temperature in the headwaters (Pomeroy et al., 1972; see Figure 8.1). Temperature at the surface of sediments exposed to mid-day sun in summer may be as high as 40°C, although 30° is more usual. In winter the sediments, excepting the exposed surface, tend to be a few degrees warmer than the water (Pomeroy, 1959).

2.3. Water Chemistry

Except in periods of substantial local rainfall, salinity of the Duplin River is determined by that of Doboy Sound, which is a part of the delta of the Altamaha River. The Altamaha has a seasonally variable flow, with maximal flow in winter and spring and very low flow in late summer. Salinity in Doboy Sound and the Duplin River varies from a summer maximum of 28 ‰ to a fall minimum of 15 ‰ (Pomeroy et al., 1972). The headwaters, and occasionally the entire Duplin estuary, may be flushed out by a heavy rain (Imberger et al., in ms.). Salinity of interstitial water follows that of the Duplin River in the levees. However, the interstitial water of the sediments in the interior parts of the marsh tends to have higher salinity–35 to 40 ‰ in the short *Spartina* meadows and >40 ‰ in the *Salicornia* zones and barren areas–as a result of evaporation and transpiration (Nestler, 1977a). Fortnightly fluctuations in salinity may occur as a result of the alternation of neap and spring tides (Christian et al., 1978). Near Sapelo Island, where ground water intrudes, interstitial salinity is much reduced, and this reduction is signaled by the presence of *Juncus*. Because of the effects of evapotranspiration, rainfall, and stream influence, salinity is not a conservative property. Nevertheless, salinity is a useful indicator of a number of aspects of the physical regime, and it can be treated as a quasiconservative property for the purpose of estimating flushing on a short-term basis (Section 2.4).

The pH of Duplin River water is around 8, just slightly lower than sea water. The pH of the sediment is lower than that of the water, and while the diurnal cycle of pH in the Duplin River is not more than 0.1 pH unit, the surface sediments may vary by a full pH unit in the course of a day (Pomeroy, 1959). Changes in pH result primarily from the utilization of CO_2 and HCO_3 by the microflora in the water and in the surface sediments (Pomeroy, 1959).

Although the bulk of the soil is anaerobic, water draining through burrows with each tidal cycle probably introduces oxygen to those regions in contact with the percolating water. Oxygen is also introduced into the soil through the roots of *Spartina*, so there are oxidized microzones around the *Spartina* roots and around burrows. This creates a three-dimensional mosaic of oxidized and reduced soil, with gradients between them. Such a configuration of redox gradients promotes variability in the chemistry and particularly the microbial biochemistry of the soil. We do not have absolute values of the redox potential, but Teal and Kanwisher (1961) report relative values. Both color and the presence of hydrogen sulfide indicate a reducing environment below a depth of about 1 cm, except for cracks, burrows, and the microzone around *Spartina* roots. Reducing conditions are maintained by the continuing input of organic matter, primarily from *Spartina* (Chapter 3).

2.3.1 Organic Matter

Compared to the ocean, the Duplin River estuary is highly enriched in both dissolved organic carbon (DOC) and particulate organic carbon (POC) (Figure 2.1). Dissolved organic carbon in the Duplin River is 10 to 20 times that in the ocean (Sottile, 1973). Unfortunately, the term DOC is a catch-all which expresses our analytical ignorance about a complex mixture. Moreover, the usual analytical method (Menzel and Vaccaro, 1964) is not more than 90% efficient when applied to oceanic DOC, and its efficiency when applied to estuarine water is unknown. Therefore, analytical differences in total DOC, while of some interest, are not absolutely accurate. Attempts to fractionate total DOC, attempts which go back many years, are at best qualified successes. Some groups of compounds, such as amino acids, can now be separated in seawater quite successfully. However, most of the compounds in solution have not been identified, and our present knowledge of the Duplin River is limited largely to total DOC. A set of samples taken by Imberger et al. (in ms.) during a rainless period in the summer of 1977 showed little variation in DOC (Section 2.4). Samples taken at the same time in Doboy Sound showed highly variable DOC concentrations, reflecting inputs of water high in DOC from the Altamaha river swamps and water low in DOC from the ocean. The assumption here is that most of the DOC measured analytically is relatively refractory material with a half-life in excess of a month. Particularly in the lower parts of the Duplin River, where substantial DOC is probably received from the Alta-

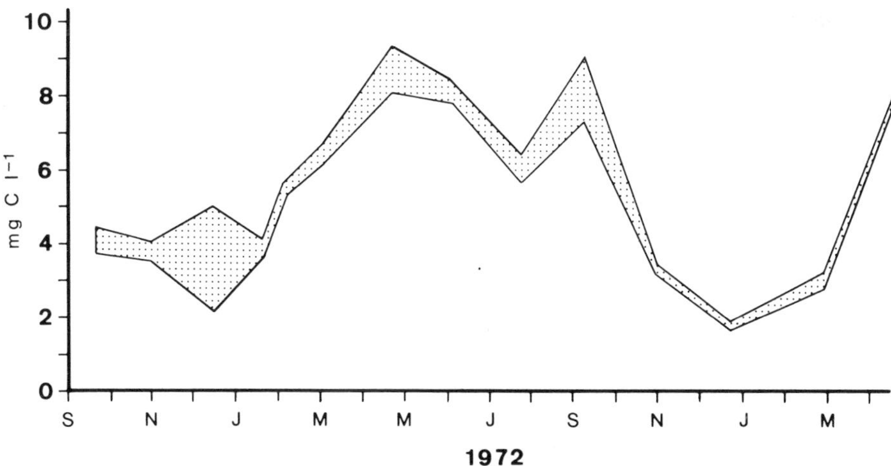

Figure 2.1. Dissolved organic matter in the Duplin River and Study Creek, expressed as the envelope of 95% confidence limits of the mean of all values at the two locations. Milligrams carbon l^{-1}. Adapted from Sottile (1973), with permission of the author.

maha River, labile DOC would have been largely degraded in transit from the river swamps, over 16 river km away. However, there is a significant output of labile DOC from the salt marsh (Hanson and Wiebe, 1977), a fact that is reflected in the higher DOC values in the hydrographically isolated headwaters.

The major components of the refractory DOC are thought to be humates and fulvates. No detailed treatment of the chemistry of that class of complex natural compounds can be included here, but a few generalizations may be useful. Humates and fulvates are complex polyphenolic compounds believed to be degraded little by microorganisms. They may be significant in the chemical environment as reactive compounds and in the biological environment as chelators of otherwise toxic metals (Wood, 1980), but they should not be regarded as a potential microbial substrate. Therefore, the bulk of the standing stock of DOC in the estuarine water is not potentially a part of the food web. Since humates and fulvates originate in the fresh-water river swamps as well as in the salt marsh, the Duplin River probably contains a mixture of materials from those two sources as well as some truly marine humates. The humates from the black-water Altamaha River should be quite different from those originating in the salt marsh. However, once they enter the estuary, they will be greatly modified by increased ion strength and by excess Ca^{++}, and the humate chemistry typical of "black water" cannot be expected to apply. Reactivity of the humates will be reduced, particularly in the presence of relatively large amounts of Ca^{++}, and molecular shape and size may change. Indeed, Wheeler (1976) demonstrated that a high molecular weight fraction of DOC originated in the salt marsh (Figure 2.2). This material, in the 1000-30,000 MW range, was yellow in color and presumably consisted of humates and fulvates, although Wheeler reported that 10% of it was carbohydrate.

Particulate organic carbon (POC) is fully as complex in its chemical composition and distribution as is DOC. Odum and de la Cruz (1967) found a mean standing stock of 8.7 mg l^{-1} (ash-free dry weight) for all tidal stages and seasons in the Duplin River. On spring tides they found a significantly greater concentration in the water on ebbing (10.1 mg l^{-1}) than on flooding tides (6.5 mg l^{-1}). This and other considerations led them to postulate that most of the POC in the Duplin River was derived from *Spartina*. However, recent work with carbon isotopic ratios suggests that most of it is algal, originating from phytoplankton and benthic diatoms (Haines, 1977). Because filter-feeding animals are continually collecting and defecating POC, most of the matter is in fact fecal. Microbial transformation makes more of the organic carbon available to the filter feeders through each successive sequence of collection, grinding, gut passage, and microbial proliferation in the feces. *Spartina* leaves which fall into the water or onto the surface of the marsh appear to be degraded relatively rapidly (Burkholder and Bornside, 1956; Odum and de la Cruz, 1967), and carbon isotope studies suggest that much of the transformation and the associated food web occurs in the marsh, rather than in the water of the tidal creeks and the Duplin River (Section 4.1.3).

The stems of *Spartina* plants are the fraction of the vegetation most resistant to decay. In the absence of snow and ice, most dead stems remain standing throughout the winter. In that position they are probably less subject to mechanical breakup by invertebrates than they would be in the water. However, even fallen stems tend to degrade slowly. They float on the water and accumulate in rafts, sometimes several meters across. Some rafts float over the marsh on spring high tides and become stranded behind the levees where they smother the living *Spartina* shoots and cause bare areas to persist for a year or more. A significant fraction of the rafts flush out of the estuaries into the ocean, and it is not unusual to encounter them tens of kilometers out at sea, particularly after the spring period of high river flow and high tidal amplitudes. Although rafting is highly visible, quantification of the standing stock of rafted material suggests a relatively small possible export, probably less than 1% of net production (P. Wolf, *pers. comm.*).

The marshes of Sapelo Island are underlain by some 10 meters of sediments, mostly clays, which exert a significant influence on estuarine and marsh chemistry. These clays are a nearly equal mixture of hydrated and nonhydrated clays, montmorillonite and kaolinite, making them a versatile, mixed-bed ion-exchange system. Although clays are by definition inorganic, there are undoubtedly electrochemical associations of the clays with dissolved organic substances. As the clays move down an estuary, their electrochemistry supposedly changes with the increasing ion strength of estuarine water, a change which results in decreased effectiveness in sorbing charged ions or molecules (Carritt and Goodgal, 1954). But recently Hunter and Liss (1979) have shown that clays entering estuaries are already coated with organic materials and show little change in charge as a function of salinity. The clays of the Sapelo Island marshes are reworked coastal and estuarine clays, most of which have been in an environment of high ion strength and high concentrations of dissolved organic materials for hundreds, if not thousands, of years (Frey and Basan, 1978).

Although modified by sorbed organic matter, clay minerals exert a significant influence on estuarine and marsh chemistry. Their predominant charge is negative (Hunter and Liss, 1979), and yet they have been shown to be effective in adsorbing $HPO_4^=$, even in aerobic, suspended sediments (Pomeroy et al., 1965). Although the salt marsh contains vast amounts of surface area on its clays, the surface is only effective in sorption reactions in proportion to its contact with estuarine water. Because the surface of the marsh is covered by a layer of living and nonliving organic matter, only a small fraction of the total surface area of clays is in contact with the free water at any instant in time. The organic matter forms an effective barrier to exchanges since suspension of sediments in the water is minimized during high tide. Water currents in the marsh proper are in laminar flow virtually all of the time, further reducing the possibility of clay-water contact. Therefore, the potentially dominant effect of clays is damped to a considerable degree by the physical and biological structure of the system. The sorption reactions lack the instantaneous capacity to buffer biologically induced changes in water chemistry completely. Therefore, we see gradients in dissolved phosphate, for example, which depart from the equilibrium value by an order of magnitude (Pomeroy et al., 1965; 1972).

2.4. Hydrology and the Flux of Materials

An understanding of the hydrology of the Duplin River is essential if we are to interpret ecosystem processes in the estuary and its marshes, and it is also essential if we are to make enlightened comparisons with other marshy estuaries. The Duplin River contains three tidal segments. Few of the tidal creeks along the southeastern coast of

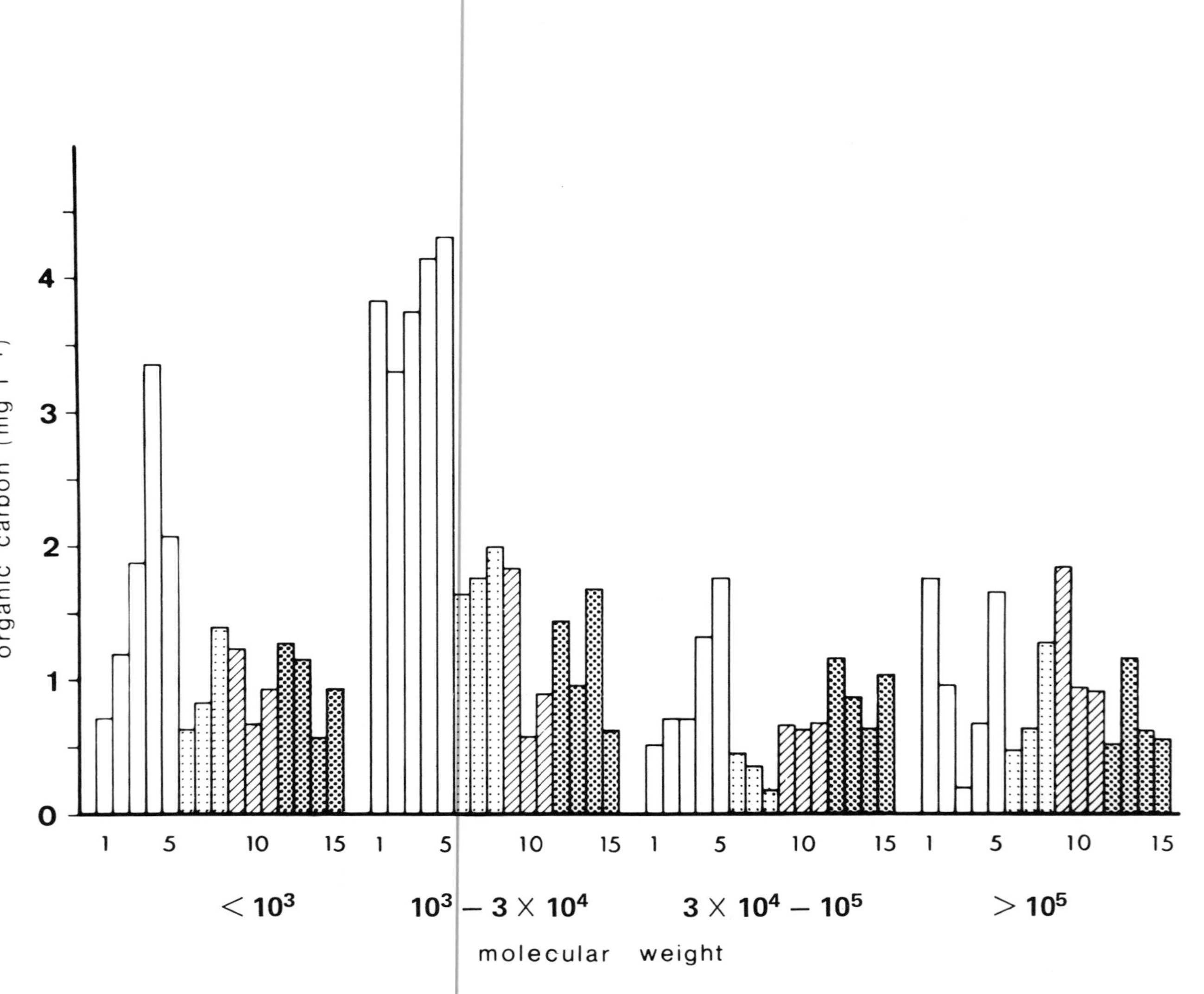
organic carbon (mg l-1)
0
1
2
3
4
1
5
10
15
<10^3
10^3 – 3 × 10^4
3 × 10^4 – 10^5
>10^5
molecular weight

North America are long enough to contain two or more segments, and many contain only one, that one being completely drained of water at low tide. One such creek, a small tributary of the Duplin River called Study Creek, has been investigated by Odum and de la Cruz (1967) and by Sottile (1973).

The more complex Duplin River, with its relatively isolated headwaters, has provided insights into the processes in the marsh which might not have been so readily deduced from the study of a creek with a single tidal segment. In comparing our findings in the Duplin River with the findings of other investigators working in other estuaries, one must keep in mind both the hydrography of the area and the methods used in the study. We found that a lack of a satisfactory hydrographic data base was a major impediment to our ecological work. Because what we finally achieved with a Lagrangian approach we believe to be worth the consideration of others faced with a similar problem, we describe in some detail the methods we adopted.

The currents in the Duplin River estuary are those of a semidiurnal tide (2 cycles per day) which has a mean amplitude of about 2 meters and a spring amplitude of >3 meters. There is little semidiurnal inequality. That is, amplitude varies fortnightly but not greatly from one tide to the next. Tide-gauge records and other observations of anomalous water movements reveal overtides, including some rather high harmonics. Slack water is synchronous with high and low tide everywhere in the Duplin River. The mean tidal excursion is 4.8 km, while the total length of the main channel is 12 km (Ragotzkie and Bryson, 1955). Therefore, the water in the upper part of the estuary is hydrographically isolated from Doboy Sound. The maximal velocity of both flood and ebb tidal currents occurs relatively near high tide, rather than at mid-flood and mid-ebb. Ragotzkie and Bryson (1955) attributed this to the shape of the estuary and found that the maximal currents occur at the time when water is flooding into the marsh, the cross section of the estuary being thus greatly expanded. The shape of the cross section likely promotes the generation of overtides, so it is difficult to say whether what is observed is a slope current or a standing wave, although the presence of visible harmonics on some tide-gauge records suggests the latter. Sinusoidal variations in current velocity within a single tidal cycle (Howard et al., 1975, Figure 5) also suggest standing wave harmonics.

Maximum current velocities in the Duplin River are approximately 1 m sec^{-1}. At the other extreme, water velocity in the *Spartina* stands of the salt marsh rarely exceeds 10 cm sec^{-1}. Therefore, laminar flow is the general case in the salt marsh, and turbulent flow is the general case in the tidal distributaries. This has important implications in erosion, and in transport and deposition of suspended materials. The existence of two types of water movement helps to explain the formation of natural levees, the evidently low rate of deposition of sediments within the salt marshes (Frey and Basan, 1978), and the lack of disturbance of the surface of the sediments in the marshes by tidal inundation.

Figure 2.2. Distribution of dissolved organic carbon by molecular weight. Replicates 1-5 are marsh creek water; 6-8 are Doboy Sound; 9-11 are 3 km offshore; 12-15 are 15-23 km offshore. Adapted from Wheeler (1976) with permission of Limnology and Oceanography. Copyright (1976) by the American Society of Limnology and Oceanography.

Current-meter profiles have been used on several occasions to estimate transport in the Duplin River (Ragotzkie and Bryson, 1955; Kjerfve, 1970; Howard et al., 1975). In a turbulent system it is difficult to achieve a final transport estimate with a precision of better than ±20% (Imberger et al., in ms.), even with a very high density of velocity data. This seriously limits the utility of such studies, since most of the differences in transport which have been found probably fall within the error limits of the method. Certainly, if one is looking for small daily increments in dissolved or suspended materials moving in or out of the system, this Eulerian approach of measuring current velocities at fixed points is futile. The situation in the Duplin River is further complicated by the presence of discrete bodies of water, separated by sharp density gradients, or fronts, which may oscillate with the tide for days (Kjerfve, 1970; Imberger et al., in ms.).

The Lagrangian approach, which takes into account these and other potential variables, has been proposed by Imberger (1977), and the Duplin River has been used as an example of its applicability (Imberger et al., in ms.). This approach involves an evaluation of mass balance within the water body and between defined segments of it. Salinity, or the flux of total salts, forms a convenient basis for evaluation, and naturally occurring pulses of river flow or rainfall can be the medium. By following the flux of fresh water generated by a single thunderstorm limited to the headwaters of the Duplin River, Imberger et al. (in ms.) could evaluate longitudinal mixing in any part of the estuary. The distribution of nonconservative* materials, both in solution and in suspension, could then be compared with that expected from the distribution of salinity, and conclusions could be reached about the transport of materials to and from the marsh, to and from the ocean, and within the estuary. Equations describing the fluxes can be written in the following way.

The net flux of salt, Q, is defined by Imberger as

$$Q = \epsilon A(\delta S/\delta X) \tag{2.1}$$

where

S = mean salinity of a cross section,
X = a defined distance along the estuary,
ϵ = the net longitudinal diffusion coefficient, and
A = the cross-sectional area of the estuary.

A substance, i, has a concentration, $C_i(X, T)$ at any location and time within the defined volume, and the total instantaneous mass, M_i, of the substance in the control volume is the integral of the concentration over the entire volume. Imberger then writes the equation for conservation of mass:

$$dM_i/dt = Q_i + J_i - \sum_{k=1}^{N_1} B_{ik} - \sum_{j=1}^{N_2} F_{ij} \tag{2.2}$$

*Substances which are subject to change as the result of biological activity are termed non-conservative by oceanographers.

where the fluxes have been partitioned so that

Q_i = the longitudinal mixing flux (Equation 2.1) with C_i replacing S,
J_i = the flux of i through the air-water surface,
B_{ik} = the net flux across the bottom-water boundary, including the surface of the marsh,
F_{ij} = the net flux of substance i to another substance in the water and moving with it, while
N_1 and N_2 are constants denoting the number of processes.

In this way the flux of nonconservative materials can be determined, simply by adding to Equation 2.1 terms defining such nonconservative properties as gain or loss from air or sediments or transformation into another compound.

Equation 2.1 is then cast in time-averaged and space-averaged form, written for substance i instead of salinity:

$$Q_i = [\langle \tilde{C} \rangle \langle A \rangle L/T_D]\ a\ [\delta c_i/\delta x] \qquad (2.3)$$

where

$\langle \tilde{C}_i \rangle$ = an estimate of C_i averaged over volume and time,
$\langle A \rangle$ = the area of the mean cross section,
$T_D = L^2/\epsilon$, the mixing time between boundaries of the section, L,
$a = A/\langle A \rangle$, the non-dimensional area,
$\tilde{C}_i$ = the cross-sectional mean concentration of c_i,
$c_i = C_1/\langle C_1 \rangle$, the nondimensional mean concentration, and
$x = X/L$.

When written in this form, the second term $[a\ (\delta c_i/\delta x)]$ approaches 1, so the flux is determined by the first term of Equation 2.3. Other flux terms used in calculating transfer between bottom or marsh and water, or between suspended materials and water, are described in Chapter 9. The flux, J_i, was assumed to be zero for substances studied, so it is deleted from further discussion.

Equation 2.2 can now be written in nondimensional form,

$$dm_i/dt = (\alpha/t_D)\,(\delta c_i/\delta x) - \sum_{k=1}^{N_1} \frac{b_{ik}}{\tau_{ik}} - \sum_{j=1}^{N_2} (f_{ij}/t_{ij}) \qquad (2.4)$$

where

$$\alpha = \langle A \rangle\ L\ a/t_D$$

and the nondimensional time scales, representing time in terms of tidal period, are

$$t_t = T_T/T_T = 1,$$
$$t_D = T_D/T_T,$$
$$\tau_{ik} = T_{ik}/T_T, \text{ and}$$
$$t_{ik} = T_{ij}/T_T,$$

where

T_T = the time for one tidal cycle ($\sim$12 hr),
T_{ik} = the characteristic time of boundary exchange, and
T_{ij} = the characteristic time of internal cycling.

Time scales of the individual processes as the inverse of the nondimensional fluxes could then be ordered from shortest to longest characteristic times. This represents an ordering of fastest to slowest fluxes. Since the nondimensional time for one tidal cycle, $T_T = 1$, the other characteristic times for a given substance should be expressed in terms of that—they are greater than one or less than one tidal cycle. Examples which follow show the application to specific nonconservative dissolved materials. A more detailed discussion of the hydrological background and the field sampling regimes is given in Imberger (1977), Fischer et al. (1979), and Imberger et al. (in ms.).

In the application of these principles to evaluation of the flux of materials in the Duplin River, Imberger et al. (in ms.) measured a number of chemical components, including salinity, at high and low tide over several days following a brief rain at the head of the Duplin River. From those components four were selected as exemplifying different specific regimes of flux which would require different specific regimes of field measurement. The goal was not only to describe the flux of materials in the Duplin River, but also to illustrate how to measure the flux of materials in other situations and how to define the requirements in a given situation. The net flux of salt (Equation 2.1) was used to determine coefficients of longitudinal mixing. This had also been done previously, using growth and longitudinal dispersal of a dinoflagellate bloom (Ragotzkie and Pomeroy, 1957). From the longitudinal mixing coefficients, Imberger et al. (in ms.) estimated that during the period of study a dissolved substance, transferred from the marsh to the headwaters of the Duplin River, would pass through the uppermost tidal segment in about 36 hours ($t_D = 3$). Not all dissolved substances behave like salinity, however. While salinity is a quasiconservative property in the Duplin River, most estuarine chemistry is nonconservative. In other words, most chemical species may be in some way modified by physical or biological uptake, release, transformation, or transport. The net flux of the material in the estuary will be the result of all these factors.

In order to compare the magnitude and importance of the processes at work with respect to specific chemical substances, the data were time-scale ordered, with time in terms of tidal periods. We could then compare the time scales of four processes:

1) t_t, the semi-diurnal tidal period (= 1);

2) t_D, the characteristic time required to mix a substance over length, L, along the estuary;

3) τ_{ik}, the mean reciprocal of the net specific rate constant for the flux from sediments to water; and

4) t_{ij}, the comparable parameter of net flux of the species under study within the water column.

Since all were expressed in terms of tidal period, their magnitude could be compared in the case of the measured flux of any substance.

Applying the principles which have been set forth, a team of investigators examined the distribution and flux of a number of dissolved or suspended materials in the Duplin River. Sampling of ammonium in the Duplin River revealed a patchy, scattered distribution of that substance, with variation increasing from mouth to headwaters. Regeneration, both from the marsh and from heterotrophs in the water, was rapid (Chapter 8). Uptake by photoautotrophs in the water may also be rapid, and sorption by sediments and heterotrophic uptake are possible. Using data from the literature on the turnover time of the instantaneous standing stock of ammonium in the Duplin River (Haines et al., 1977), t_{ij} was estimated to be 0.4; that is, there is a net regeneration of ammonium 2.5 times during a tidal cycle. The ordering of the time scales of water column recycling (t_t), sediment-water flux (τ_{ik}) and tidal transport (t_D), and biological flux (t_{ij}) is: $t_{ij} \cong \tau_{ik} < t_t < t_D$. Fixed point sampling at one station, together with transport measurements, would be a satisfactory method of measuring the flux of a substance such as ammonium, in which the biological turnover times are not only less than the characteristic mixing time but are also less than one tidal cycle. However, a large variance is to be expected. Because $t_{ij} < t_t < t_D$, ammonium is, for all practical purposes, recycled internally, and, in spite of the fact that a concentration gradient exists, export by eddy diffusion is insignificant compared to biological processes. This important finding may not be intuitively obvious.

Dissolved organic carbon (DOC), more chemically complex than ammonia, is composed of a very large number of mostly undefined chemical species. These compounds vary widely in the rapidity with which they are assimilated by heterotrophs in the water, and for the present purpose they were divided into labile ($t_{ij} \leqslant 1$) and refractory ($t_{ij} > 1$) DOC. Much DOC originates in the marsh, albeit from several discrete sources (Sottile, 1973; Hanson and Wiebe, 1977; Gallagher et al., 1976). Most of the DOC produced in the marsh is probably labile and will be utilized rapidly by the microflora within the marsh, while a smaller amount which escapes from the marsh is refractory. Of the total DOC output from the marsh, about 80% is refractory (Sottile, 1973), but because of the diversity of sources, this amount is variable. Measurements of DOC in the Duplin River by the method of Menzel and Vaccaro (1964) revealed little variation in the lower two tidal segments, with a mean of 7.1 ± 1.5 mg C l^{-1} (mean ± one standard error of the mean). In the uppermost tidal segment DOC was 10.8 ± 0.7 mg C l^{-1}. The concentration gradient (Figure 2.3) and the low variability tend to confirm that most DOC is produced in the marsh, but only the refractory DOC enters the lower segments of the Duplin River in significant amounts. Since high and low tide concentrations are similar at any point, the t_{ij} for labile DOC is less than or equal to the tidal period, and the probable ordering of time scales is $\tau_{ik} \sim t_{ij} \sim t_t < t_D$. As in the case of ammonium, little labile DOC appears to be lost by longitudinal mixing, although a concentration gradient of total DOC is present. Because there was a

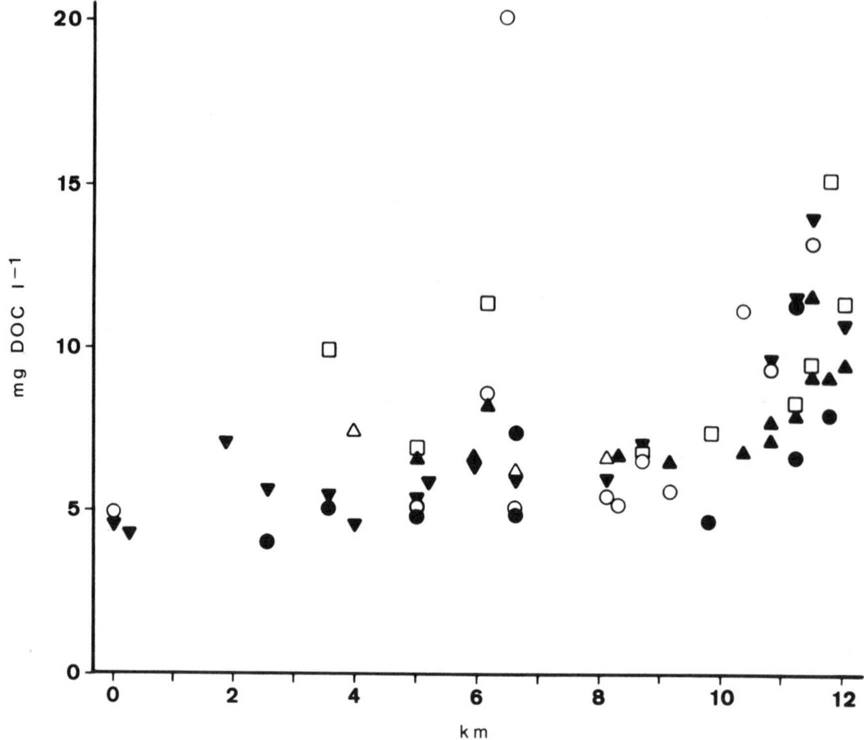

Figure 2.3. Distribution of dissolved organic carbon in the Duplin River. Both high-tide and low-tide values are shown, but high-tide values are translated downstream to correspond to the translation of salinity values on the same day. The mouth of the estuary is on the left. Adapted from Imberger et al. (in ms.)

nearly linear increase in the minimal concentration of DOC with distance up the Duplin River, Imberger et al. (in ms.) concluded that this was largely refractory DOC, with a time ordering of $t_t < \tau_{ik} < t_D < t_{ij}$. As a conservative substance in solution the refractory DOC is essentially lost by longitudinal mixing.

Reactive silicate was found to have a very regular distribution. It appears to originate in the marsh, either from the dissolution of diatom frustules or from the diagenesis of aluminosilicate clays. Concentrations increased from 68 μM at the mouth of the Duplin River to 250 μM in the headwaters, and day-to-day variation at any station or tidal stage was very low (Figure 2.4). The only known significant sink of silicate is uptake by diatoms, both in the water and on the surface of the marsh sediments. Presumably, only the diatoms in the water influenced silicate concentration in the Duplin River, the diatoms in the surface sediments having taken up silicate as it exchanged between sediments and water. To maintain the observed distribution, longitudinal mixing along the concentration gradient must be balanced by uptake of silicate by diatoms in the water. Imberger et al. (in ms.) estimated the uptake to be 0.08 to 0.14 μM h^{-1}. The ordering of time scales of flux of silicate is $t_t < t_D < \tau_{ik} < t_{ij}$. Since the distribution of diatoms in the Duplin River is patchy, the distribution of silicate would be ex-

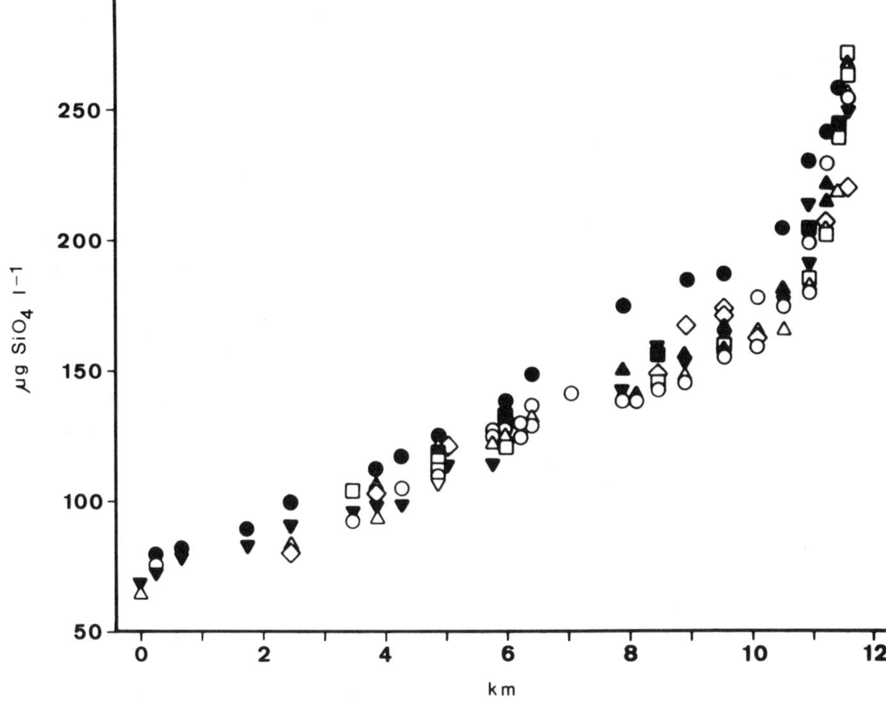

Figure 2.4. Concentration of silicate in the Duplin River. High and low tide values are positioned as in Figure 2.3. Adapted from Imberger et al. (in ms.)

pected to be patchy, if that nutrient were limiting to diatom production. However, silicate does not appear to be limiting to the diatoms, so biological uptake of that element is not sufficient to produce a patchy distribution. Most of the silicate produced in the marsh is lost by longitudinal mixing to Doboy Sound, the net loss being estimated to be 0.3 mMoles m^{-2} day^{-1}.

Because it does not behave like dissolved substances, particulate organic carbon (POC) shows variation in tidal time as well as distance. It is as heterogeneous as DOC, and it includes a living component. However, the nonliving component is always quantitatively dominant, even in phytoplankton blooms. The distribution of POC is made more complicated by the effects of turbulent transport on the suspension of particles. Since much of the POC drops to the bottom during slack water and is resuspended on ebb and flood tides, $\tau_{ik} \sim 1$ for that component. Sources of other labile fractions include production by photosynthesis ($t_{ij} = 0.25$), loss to respiratory processes ($\tau_{ik} = 0.25$), and the uptake and release of POC in the water column ($\tau_{i\ell} \sim 4 \rightarrow 10$, see Chapter 9). The time ordering of fluxes of POC is $t_{ij} < \tau_{ik} < 1 < t_D < t_{i\ell}$. Imberger et al. (in ms.) suggest that, in this case, a sampling regime at several fixed points and at more frequent time intervals would have been more appropriate than the linear series. They also point out that the flux of POC from the marsh through the estuary and into the ocean is highly complex, and that the effects of storm events probably override day-to-day biological and even physical processes. The work of Imberger et al.

(in ms.) shows the difficulties in generalizing about fluxes into or out of salt-marsh estuaries. In the case of the Duplin River the most significant exports appear to be living organisms and refractory DOC, such as humates. That theory is verified on a different time scale by the work of Haines (1976; 1976a; 1977) with $\delta^{13}C$.

When the physical and chemical features of marsh ecosystems are considered, it is apparent that each estuary must be evaluated individually. There are, of course, regional similarities—peat in the marshes of northeastern North America and clay in the marshes of southeastern North America; substantial tidal amplitude on the Atlantic and Pacific coasts and little tidal amplitude in the Gulf of Mexico. Within a single region, the dominant macroorganisms and land forms may vary little, and superficial appearances may suggest that estuaries are all much the same. However, since extensive sets of variables influence estuarine systems, the resulting biological structure and function may actually vary substantially within a region, although these variations may not be apparent on cursory examination. The significance of these variations becomes especially important when generalizations are needed, either for evaluations of human impact on marshes or for interpretations of the basic processes at work (Chapter 10) in these marshes.

Salt-Marsh Populations

3. Primary Production

L.R. Pomeroy, W. M. Darley, E.L. Dunn, J. L. Gallagher, E.B. Haines, and D.M. Whitney

The vast expanses of the cordgrass, *S. alterniflora*, in the marshes of Sapelo Island so impress the scientist as well as the casual viewer, that early workers assumed this species was the only significant primary producer. Indeed, the first studies supported this view. By the early 1960s production of *Spartina* leaves and stems had been measured and found to be among the most productive natural plant populations (Teal, 1962). The golden-brown layer of diatoms on the creek banks and marsh surface was also found to contribute significantly to primary production, albeit its production was just a small fraction of that of the grass (Pomeroy, 1959). However, early measurements indicated that phytoplankton production was trivial (Ragotzkie, 1959).

We began our study with these assumptions, believing that only a few gaps remained in the data on primary production. To fill the gaps we initiated studies on the production of *Spartina* roots and rhizomes, and at the same time, we undertook a comprehensive study of the benthic microflora. From the first, our observations did not agree with the presumed unimportance of the phytoplankton and benthic algae. Early simulations with our preliminary model of carbon flux (Chapter 9) showed that algae were one-tenth to one-fifth as productive as *Spartina*. These findings led to a series of studies which changed our view of the trophic structure of the system. The results of those studies are described in this chapter.

3.1. The Community of Higher Plants

In southeastern North America the gentle gradients in elevation and the tides of 2-meter amplitude produce broad zones of marsh vegetation, while in northeastern North America the typically steeper gradients produce narrow zones. Intrazonal species diversity among plants is low in southeastern, but high in northeastern and Pacific Coast saline marshes. The Duplin River marsh is typical of southeastern marshes in this respect. Here *Spartina* occupies 95% (1045 ha) of the marsh (Reimold et al., 1973), which can be divided into at least four rather distinct types of stands. The stands adjacent to the creeks are tall (>2 m), robust, and have a low density (30 to 50 stems m^{-2}). Figure 3.1 shows a few creeks leading back from the Duplin River which are bordered by tall plants. Creek banks, which occupy about 8% of the marsh surface and compose the zone most frequently inundated, have soil salinities of 25 to 28 $^{0}/_{00}$ (Gallagher et al., 1973). This creekbank zone, along with the levees, is better drained than the back-marsh areas of short *S. alterniflora*. Broad, low-density stands of spindly plants, as seen in the upper left portion of Figure 3.2, are found at the heads of the creeks where drainage is poor and the substrate very soft. At a slightly higher elevation are broad, flat swards of short *S. alterniflora* which contribute approximately 45% of the *S. alterniflora* coverage (Reimold et al., 1973). Density of the short (25 cm) *Spartina* may exceed 300 plants m^{-2}, and soil salinities average 35 to 40 $^{0}/_{00}$. This marsh form is seen as the uniform light grey area in the upper right side of Figure 3.1. In addition, robust plants, 0.5 to 1.0 meters tall, form patches totalling 20% of the area of the marsh (Gallagher et al., 1972). These plants, which may have shoot densities of 80 m^{-2}, can be seen as the dark-colored marsh inside the Y-shaped stream system in the upper center of Figure 3.1.

The distribution of the various growth forms and their associated fauna and microflora appears to be a function of the edaphic features created by the drainage patterns. Thus, *S. alterniflora* forms the communities in the lowest elevations of the marshes of the Duplin River estuary. Most of the remainder of the macrophyte-covered soil is lower (15 to 20 $^{0}/_{00}$) and soils are often sandier than in the *S. alterniflora* zone (Gallagher et al., 1980). The remaining marsh macrophytes occupy the fringe of salt flats and the interface with the high ground, where the wetland species mix with those from the upland. *Fimbristylis castanea* (Michaux) Vahl and patches of *S. patens* characterize the upper boundary where the slope is gradual. Where the slope is steep the transition may be directly from *S. alterniflora* to upland species such as *Serenoa repens* (Bartram) Small and *Pinus spp*. Diversity increases along the upper fringes and on the edges of the high-salinity salt flats. Typically, two species of glasswort, *Salicornia virginica* L. and *Salicornia bigelovii* Torrey, are mixed with *Batis maritima* L., *Sporobolus virginicus* L., very short *Borrichia frutescens* (L.) DC., dwarf *J. roemerianus*, and *Limonium nashii* Small. The scale of the mixing may be on a shoot-to-shoot level or patch-to-patch, the patches being several meters across. With *S. alterniflora* eliminated from competition, the plants with highest salt tolerance dominate this environment of wide temporal variability in salinity. During rainy periods and spring tides, salinity may drop to less than that of seawater. Conversely, salinity may rise to more than double that of seawater during neap tides and drought conditions (Gallagher, 1979).

3.1.1 Physiological Ecology of Salt Marsh Plants

The distribution and productivity of salt-marsh vascular plants results from the interaction of environmental gradients with the physiological capabilities and versatility of the plant species. Even within a species such as *S. alterniflora*, there are large differences in growth forms at different locations in the environmental gradient (Keefe, 1972; Turner, 1976). Plants with C_4 photosynthesis are generally characterized by high rates of net photosynthesis with high temperature optima, high light saturation,

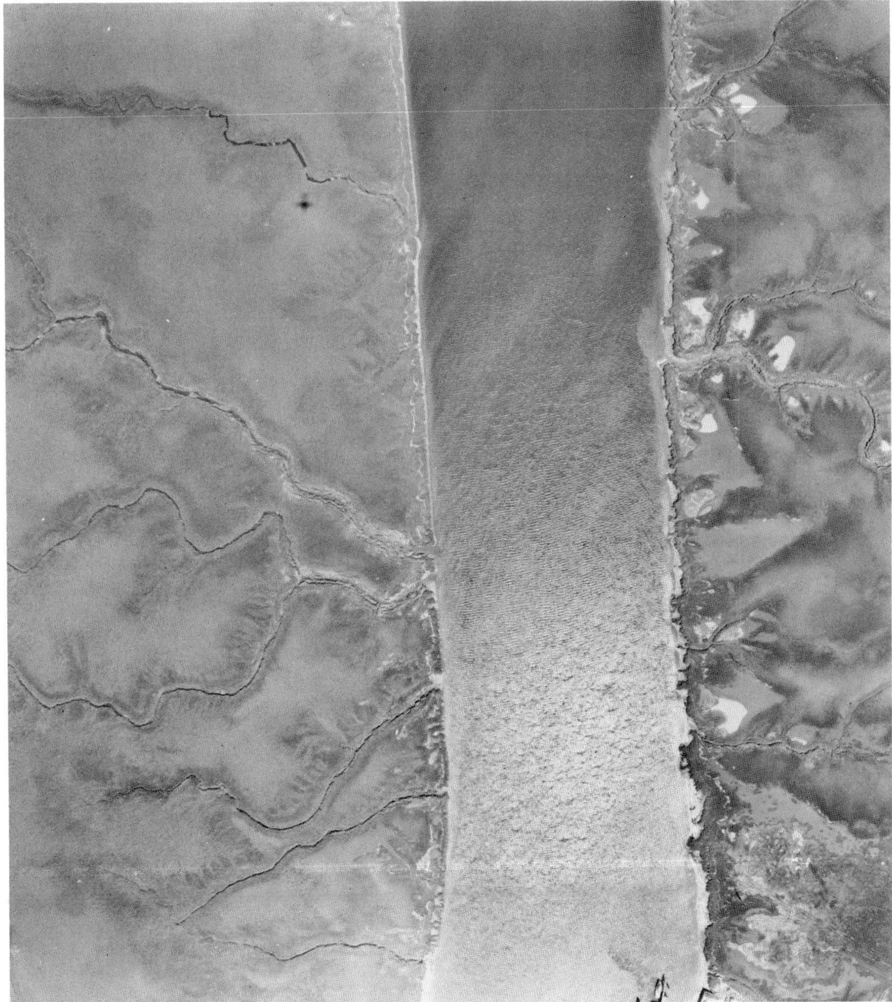

Figure 3.1. Aerial photograph of a portion of the Duplin River near its mouth, together with adjacent marshes. The tide is low. In this region there are broad expanses of short *Spartina* marsh. White areas are former sites of rafted dead stems of *Spartina*.

and lower transpiration rates. *S. alterniflora* is a C_4 species, while *J. roemerianus*, which occupies regions of lower salinity, is a C_3 species. The succulent species which occur in the high-salinity flats and fringing areas include no crassulacean-acid metabolism plants (Antlfinger and Dunn, 1979). In the Georgia salt marshes we are largely concerned with the differences between the tall and short forms of *S. alterniflora* and between the forms of *Spartina* and *J. roemerianus*, since these species occupy most of the area of the marsh.

Because of the subtle interactions between microenvironment and individual plant physiology, it was necessary to observe the physiological responses of intact plants

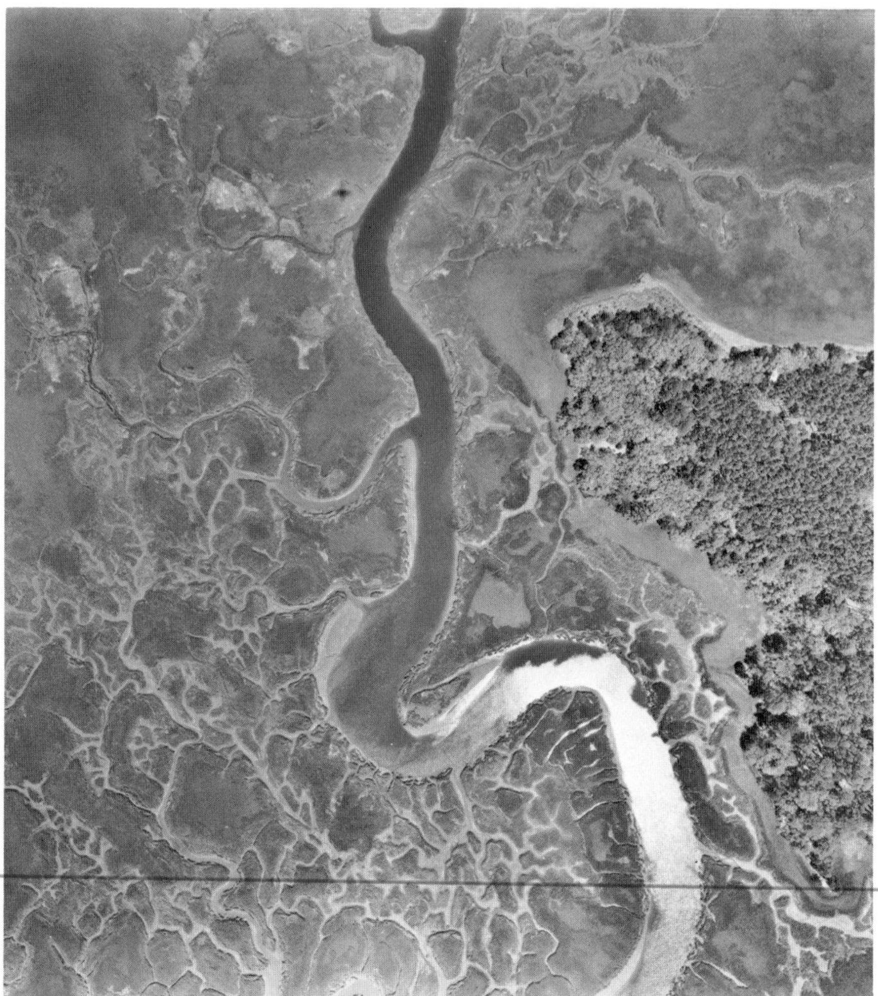

Figure 3.2. Aerial photograph of a portion of the head of the Duplin River, showing extensive development of stream channels in the marsh, with associated low-marsh tall *Spartina.*

growing in place in the salt marsh. In order to do this, a mobile laboratory was brought to sites on Sapelo Island where stands of plants were accessible. Intact, *in situ* plants of all species were enclosed in temperature- and humidity-controlled chambers (Giurgevich and Dunn, 1978, 1979, in press). Placement in chambers always caused some change in orientation of plant parts and therefore resulted in some disturbance of the microenvironment of the natural plant community. Differences in the size and growth form of the species required different degrees of change in plant orientation; tall *S. alterniflora* and *J. roemerianus,* for example, experienced decreases in shading.

With the mobile lab, five types of gas-exchange experiments were conducted on each species:

1) CO_2 and water vapor exchange rates were followed over one day, with radiation and water vapor content of the chambers maintained at ambient values.

2) Light-dependence experiments utilized stepwise increases of radiation with chamber temperature held constant as gas exchange rates were measured.

3) Temperature-dependence experiments utilized a range of temperatures with radiation near full sunlight (200 nEins. cm^{-2} sec^{-1}).

4) CO_2-dependence experiments utilized CO_2 concentrations from ambient to zero in 50 to 75 ppm increments under constant radiation and temperature.

5) Temperature dependence of dark CO_2 exchange was determined by darkening the chamber at dusk.

The great problem for salt marsh plants is to obtain CO_2 without the loss of water vapor through transpiration. There is a close analogy between salt-marsh plants and desert plants, which experience an absolute water shortage, and the physiological strategies for coping with the problem are similar. Although water as such is never lacking in the salt marsh, it must be desalted at significant physiological cost. It is not energetically competitive for a plant to lose water through its stomata, yet the stomata must be open to receive CO_2. The water-use efficiency of photosynthesis (mg CO_2 uptake per g HOH transpired) effectively represents the ability of a plant species to solve these conflicting demands.

Carbon dioxide and water-vapor exchange rates were determined from concentration differentials between inlet and outlet air streams from the chambers. Leaf diffusive resistance, or conductance, to water vapor was calculated using an Ohm's Law analogy of diffusional flux (Gaastra, 1959; Jarvis, 1971). The rates of gas exchange for *B. maritima*, *S. virginica*, and *B. frutescens* were expressed on the basis of tissue dry weight. The responses of *S. alterniflora* and *J. roemerianus* were expressed on the basis of leaf surface area, or were converted to a dry-weight basis using factors provided by Giurgevich and Dunn (1979).

Net primary production was calculated from daily tracking experiments and the temperature response of dark respiration (Table 3.1). Daily totals of CO_2 fixed were obtained by integrating the daily response curve for each plant sampled. Nighttime respiratory CO_2 loss was calculated at the mean night temperature for each season using mean monthly hours of darkness. Net primary production in tall and short *S. alterniflora* and in *J. roemerianus* was calculated for five seasonal periods. Net primary production in *B. frutescens*, *S. virginica*, and *B. maritima* was calculated for 182-day periods corresponding to winter and summer responses (Antlfinger and Dunn, 1979). The responses of all species were presumed to be representative of the time of year at which they were measured. A factor of 0.614 was used to convert CO_2 fixed to grams dry weight of plant tissue for *S. alterniflora* and *J. roemerianus*. The ash contents of *B. maritima, S. virginica*, and *B. frutescens* were generally much higher than those of the other species, so a unique conversion factor was generated for CO_2 exchange rates to tissue dry weight (Antlfinger and Dunn, 1979).

Total net primary production for tall *S. alterniflora*, short *S. alterniflora*, and *J. roemerianus* calculated from the ambient tracking experiments (Figure 3.3) was somewhat higher than harvest estimates (Table 3.2 used in conjunction with area estimates in Table 3.3). Although the magnitude of the differences between estimates varied by species, the relative position of the estimates was the same for both methods. Blum et

Table 3.1. A Comparison of Respiration Rates in the Dark and in the Light Relative to Photosynthesis Rates in Salt-Marsh Macrophytes

Species	Ps[a]	PR[b]	DR[c]	Night DR/Daily Ps[d]	PR/Ps[e]
	mg CO_2[d] m^{-2} h^{-1}			%	%
S. alterniflora-tall[f]	21.3	2.4	1.4	11	11
S. alterniflora-short[f]	14.4	6.5	1.9	26	45
J. roemerianus[g]	16.9	9.1	2.1	14	54
B. frutescens[h]	20.2	14.8	2.2	16	65
B. maritima[h]	14.3	13.6	3.6	17	95
S. virginica[h]	45.5	38.7	4.0	29	85

[a]Photosynthesis calculated from seasonal mean values near optimum temperature (15-30°C)

[b]Photorespiration, means of seasonal values extrapolated from the initial slope of the dependence of photosynthesis to zero CO_2 at the same temperatures as photosynthesis determinations.

[c]Dark respiration, means of seasonal values determined at 25°C.

[d]Mean of seasonal values of total nightime respiration as a % of total daily photosynthesis.

[e]Mean of seasonal PR values calculated as a % of hourly rates at the same temperature.

[f]From Giurgevich (1977).

[g]From Giurgevich and Dunn (1978).

[h]From Antlfinger (1976).

al. (1978) also noted consistently higher productivity estimates from gas exchange data for the same species in a North Carolina marsh. The estimates of net primary production from gas exchange data agreed closely with those from harvest data for short *S. alterniflora* and exceeded those from harvest data for *J. roemerianus* and tall *S. alterniflora* by factors of 1.4 and 2.7 respectively.

Because the entire aboveground portion of short *S. alterniflora* could be enclosed in the chamber, plant position and leaf orientation did not differ greatly from that of the natural communities. Thus the measured responses of both net photosynthesis and dark respiration for short *S. alterniflora* were fairly representative of an undisturbed community. Since whole plants of tall *S. alterniflora* and *J. roemerianus* could not be enclosed in a chamber, there were discrepancies between leaf orientation within the chambers and that in undisturbed communities; for tall *S. alterniflora* and *J. roemeri-*

Table 3.2. Net Annual Aerial Primary Production Estimates for *S. alterniflora* and *Juncus roemerianus*

g dry wt. m^{-2}			
	S. alterniflora		
Streamside	High Marsh	Whole Marsh	*Juncus* roemerianus
3300[a]	2200[a]	2288[a]	1500[c]
3700[b]	1350[b]	1538[b]	2200[b]
2000[c]	400[c]	528[c]	

[a]From Odum and Fanning (1973).

[b]From Gallagher et al. (1980).

[c]From Gallagher et al. (1972).

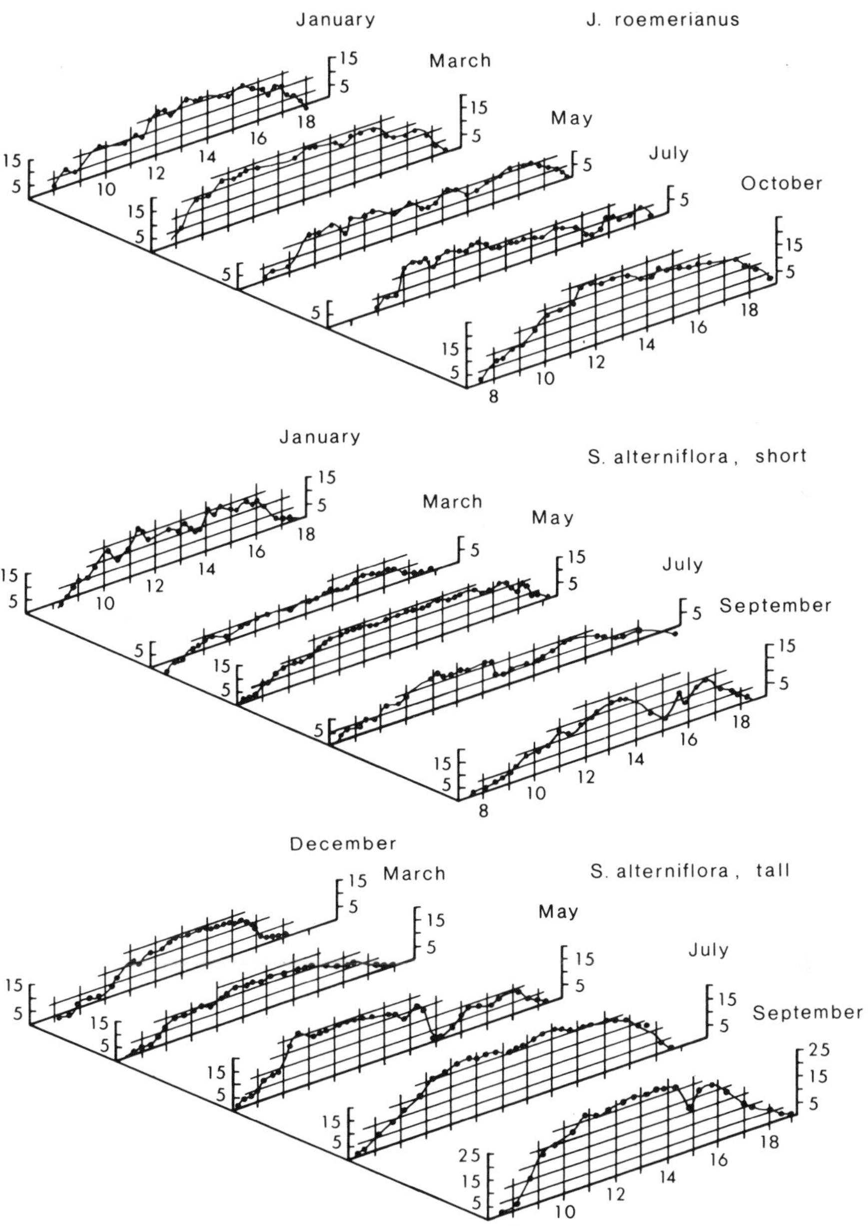

Figure 3.3. Mean daily rate curves of photosynthesis of *S. alterniflora* and *J. roemerianus*, measured *in situ* in the Sapelo Island salt marsh (mg CO_2 · dn^{-2} · hr^{-1}, from 0700 to 1900 hr). Adapted from Giurgevich and Dunn (in press). Copyright Springer-Verlag, Berlin.

anus, the estimates of net primary production from gas exchange techniques were higher than harvest values. This can be attributed to two factors: exposure of leaf tissue to more light and higher physiological activity in younger leaves.

Table 3.3. Distribution of Plant Stand Types at Two Sites[a] in a *S. alterniflora* marsh in Georgia[b]

Type	Distribution % North	South
Streamside (tall)	5	8
Levee top	37	18
Stream head	46	46
Back marsh (short)	12	28

[a]Based on data in Reimold et al., 1973.
[b]In the northern and southern areas of the Duplin River watershed.

While there is potential for strong interaction between solar radiation and photosynthesis, actual relationships in nature may vary with an extensive array of environmental factors. For example, tall *S. alterniflora* increased its photosynthesis with increasing light during all seasons at most temperatures (Figure 3.4). The oldest tissue in December usually exhibited a more flattened response curve, but photosynthesis was light saturated only at high light intensity. Conversely, short *S. alterniflora* had lower rates of photosynthesis and became saturated under specific seasonal-temperature-radiation combinations. Only during July did the light response of the short form of *S. alterniflora* approximate that of the tall form. In both forms the greatest influence of temperature on the light response of photosynthesis was noted during winter on the oldest tissue (Giurgevich and Dunn, 1979). Seasonal rates of photosynthesis of *J. roemerianus* were generally similar to those of tall *S. alterniflora* and higher than those of short *S. alterniflora*. *J. roemerianus* exhibited strongly increased photosynthesis with increasing radiation during all months except September (Giurgevich and Dunn, 1978).

The seasonal patterns of photosynthetic response to light in tall *S. alterniflora* and in *J. roemerianus* suggest light limitation in their dense natural communities. This ability to utilize available radiation to the fullest extent is responsible in part for the relatively high net primary production of these plants. Although the rates of photosynthesis in short *S. alterniflora* were lower, this form also maintained positive rates of CO_2 fixation during all seasons. Because of the shorter height and more upright leaf orientation, available light was higher in communities dominated by the short form. However, short *S. alterniflora* is unable to utilize increased light fully, because of the light-saturation of photosynthesis.

The responses of the marsh plants to temperature vary within and between species. Both forms of *S. alterniflora* maintained net CO_2 fixation over similar temperature ranges and had similar seasonal temperature optima. In short *S. alterniflora* the rate of photosynthesis was significantly lower than in the tall form at all temperatures during all seasons except winter. Temperature optima in *J. roemerianus* did not correspond as closely to prevailing seasonal temperature and were lower than those of *S. alterniflora*. Although photosynthesis of *J. roemerianus* was often similar in magnitude to that of tall *S. alterniflora*, the more flattened seasonal temperature response of *J. roemeranus* may be an expression of lower physiological plasticity. For example, *J. roemerianus* did not acclimate to prevailing summer temperatures, so the total CO_2 fixed was reduced (Figure 3.4).

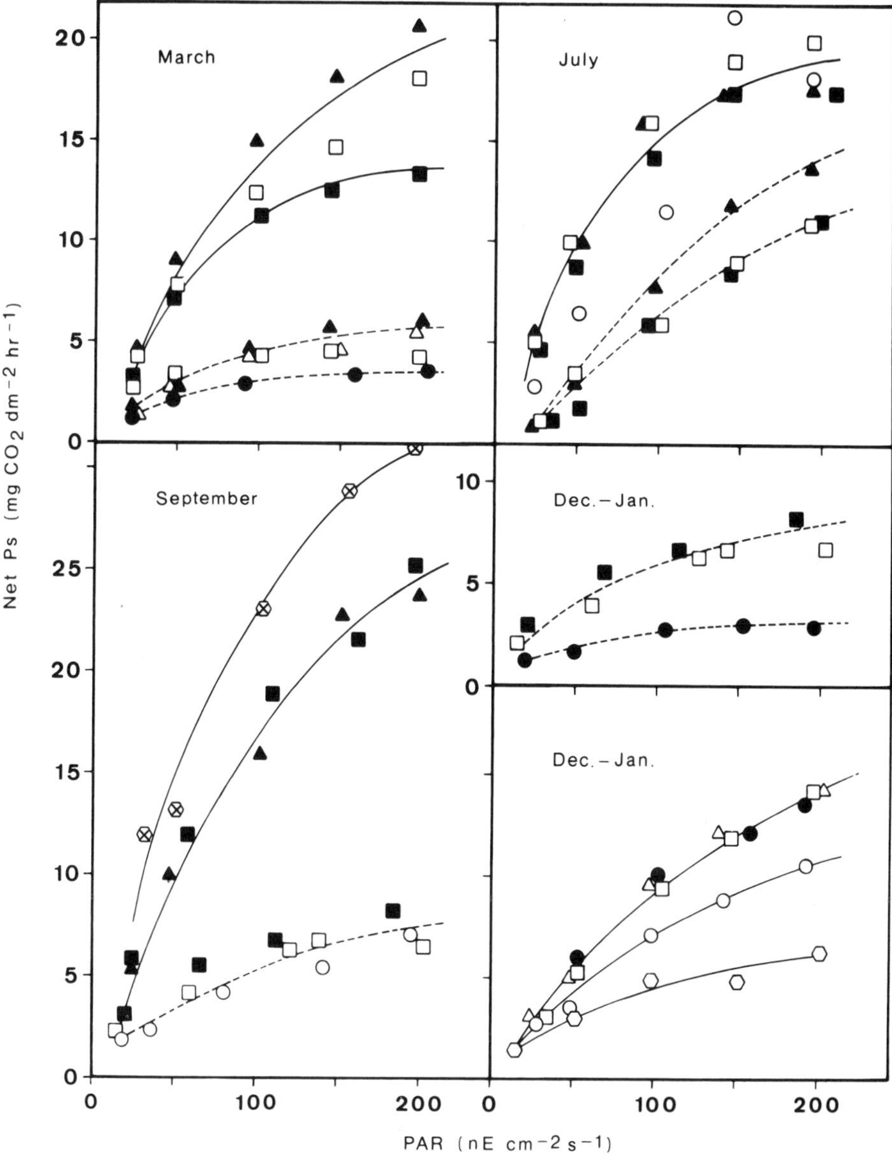

Figure 3.4. Seasonal responses of net photosynthesis to light for the tall (solid line) and short (broken line) forms of *S. alterniflora* at 5°C (⬡), 10°C (○), 15°C (●), 20°C (△), 25°C (□), 30°C (▲), 33°C (⊗), and 35°C (■). Adapted from Giurgevich and Dunn (1979). Copyright Springer-Verlag, Berlin.

Field measurements of the response of photosynthesis to variations in CO_2 revealed the greatest differences in photosynthetic capacity of the different salt marsh species (Figure 3.5). Tall *S. alterniflora* showed typical C_4 responses of a low CO_2 compensation point and a steep initial response of photosynthesis to increasing intercellular CO_2

872615

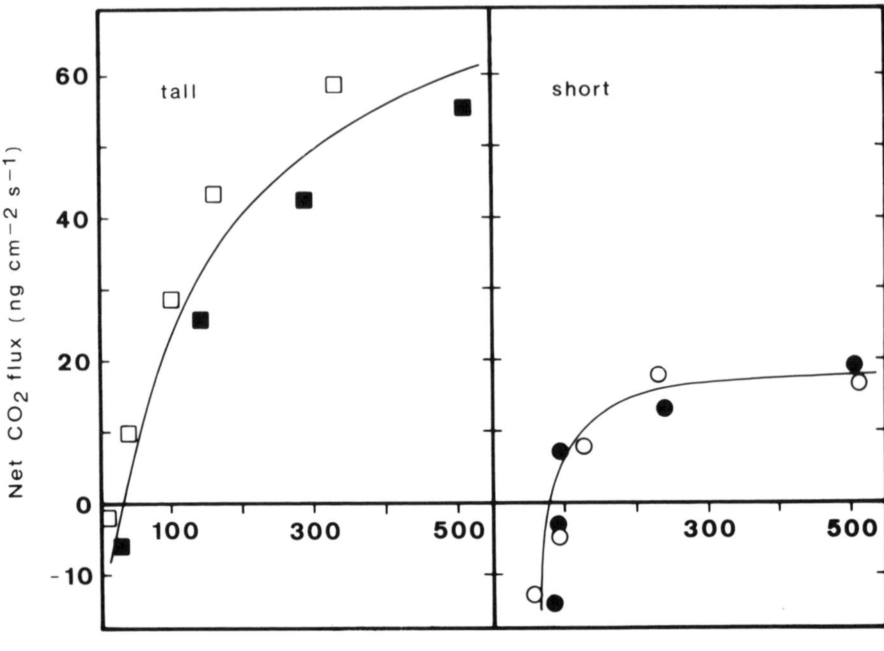

Figure 3.5. An example of the dependence of net photosynthesis on the intercellular CO_2 concentration in tall and short *S. alterniflora.* Measurements were made at 25°C on two samples (open and closed symbols) of each and 200 nEins cm^{-2} s^{-1}. Adapted from Giurgevich and Dunn (1979). Copyright Springer-Verlag, Berlin.

concentration. The short form showed a similar initial response to increasing CO_2, but revealed an elevated compensation point and a distinct CO_2 saturation of photosynthesis at low intercellular CO_2 concentrations. In *J. roemerianus* the photosynthetic responses to intercellular CO_2 were typical of C_3 plants, with a higher CO_2 compensation point and a linear and less steep initial response to increasing CO_2. In all species except short *S. alterniflora* the internal resistances to CO_2 uptake were generally larger than the stomatal resistances. At optimum temperatures in all cases except tall *S. alterniflora*, internal structural and metabolic limitations were more important in controlling photosynthesis than was stomatal resistance. However, in all species temperature increases resulted in increased stomatal resistance and decreased internal resistance (Antlfinger, 1976; Giurgevich and Dunn, 1978, 1979). In addition, in all species except short *S. alterniflora*, because of the response of photosynthesis to increasing intercellular CO_2 concentration, an increase in stomatal resistance did result in a decrease in photosynthesis. Thus, stomatal control of photosynthesis became more important at higher temperatures.

Some of the intrinsic limitations of the C_3 compared to the C_4 photosynthetic pathway are reflected in the magnitude of the internal resistance of C_3 plants. Another significant consequence of the differences in internal versus stomatal diffusive resistances is the water-use efficiency of photosynthesis. Water-use efficiency, which represents the ability of a species to solve the conflicting needs of CO_2 uptake and prevention of H_2O loss, is affected by air temperature and humidity as well as by intrinsic plant characteristics.

The quantity of CO_2 respired by aboveground and belowground tissue is a significant factor in the carbon balance of a plant. Species differences occurred in the rates of aboveground dark respiration and total nighttime CO_2 losses (Table 3.1). Tall *S. alterniflora* had the lowest respiratory rates; the species inhabiting the marsh areas of high or fluctuating salinities had higher respiration rates and nighttime CO_2 losses. Respiration rates of living, belowground parts of the plants are poorly known, but ratios of belowground to aboveground biomass and productivity (Stroud, 1976; Gallagher, 1975; Gallagher and Plumley, 1979) indicate a large quantity of living biomass belowground. The relatively high rates of respiration probably include CO_2 loss from the belowground biomass as well as from the leaves.

Estimates of the rates of CO_2 release in the light, calculated from linear extrapolations of the dependence of photosynthesis on intercellular CO_2 concentration, have provided estimates of daytime respiration which compare to those derived from several other techniques (Chollet and Ogren, 1975). Our measured rates of daytime respiration at seasonal temperatures showed dramatic differences among the salt-marsh macrophytes, with the highest rates produced by succulent species inhabiting areas of highest salinity. Some fraction of our measured CO_2 release in the light may be mitochondrial respiration continuing in the light. Probably some mitochondrial respiration continues in the light, but the relative rates of respiration in the light and dark are still uncertain (Mangat et al., 1974; Chapman and Graham, 1974). Also, some of the CO_2 release in the light, measured at low CO_2 concentration, may originate from respiration of belowground tissues and diffusion of CO_2 through leaves via arenchyma tissue of roots and rhizomes (Anderson, 1974).

Physiological investigation of the vascular plants in the salt marsh helps our understanding of the high and variable productivity which characterizes these saline wetlands. *S. alterniflora* is revealed as a versatile plant, able to cope with rather high salinity at a large but acceptable respiratory cost. At the brackish salinity of the tidal creeks, *S. alterniflora* has a high net productivity. At the highest salinities, where conditions are especially analogous to those in deserts, C_3 succulents compete successfully with *S. alterniflora*, and in soils with nearly fresh interstitial water, *J. roemerianus* competes successfully. While one can walk through the marsh and describe such distributions, only a knowledge of the underlying physiology permits us to define and explain the limits of each species and growth form in terms of specific metabolic adaptations.

3.1.2 Shoot and Root Production of *Spartina* and *Juncus*

Both winter and summer extremes of climate influence the annual cycle of growth of marsh plants. In January, when shoot biomass of streamside *S. alterniflora* is low, there are two types of living tissue. The most obvious is the senescing tissue on large, partially dead shoots that matured during the previous summer. There is also the tissue found in the small shoots that emerged in the fall, but did not elongate. Senescence and death is a continual process, but maximum rates occur from winter through spring (Gallagher et al., 1980). Much of the living biomass of shoots lives through the winter and dies in very early spring. The most rapid increase in biomass occurs in April and May (Gallagher et al., 1980), and a second burst of growth comes with flowering in August and September. Seeds mature in October and November and shattering occurs with winter storms and winds. The pattern for the short-*Spartina* marsh areas is similar to that of the streamside, but moderated, for in those areas biomass is lower, area growth rates slower, and flowering less common.

Because of its role in the detritus food web (Odum and de la Cruz, 1967; Teal, 1962), aboveground productivity in the Sapelo Island marshes has been measured many times. Several studies (Table 3.2) report seasonal production patterns in both high marsh and streamside *S. alterniflora* (Smalley, 1959; Odum and Fanning, 1973; Gallagher et al., 1980). The first two measured changes in living and dead biomass with time, but adjustments were not made for losses which occurred between sampling periods. Gallagher et al. employed the paired-plot technique of Wiegert and Evans (1964) in which plant losses between sampling periods were estimated. They found the shoot productivity of short *S. alterniflora* to be 1300 g dry wt m^{-2} yr^{-1} and that of tall *S. alterniflora* to be 3700 g dry wt m^{-2} yr^{-1}. Because the findings of Gallagher et al. (1980) agree closely with those of Giurgevich and Dunn (in press), even though the latter used a very different approach, we cite the values of Gallagher et al. as the best available data for the region.

Comparing the productivity of Duplin estuary macrophytes with productivity of plants in other marshes depends on the spatial scale within which the comparison is framed. Production varies widely over a given marsh, depending on the duration of tidal inundation, drainage, substratum elevation gradient, age of the marsh, soil fertility, and salinity. Variation in biomass and production in *Spartina* marshes is particularly large. Gallagher et al. (1980) indicated almost a 2.5-fold difference in production between tall and short *Spartina*. The tall *Spartina* had a mean biomass of 1966 g m^{-2} while the shortest stands back from the streams contained 397 g m^{-2} (Gallagher et al., 1972). The proportion of various types of stands within the marsh becomes an important factor when assessing the productivity of the marsh as a whole. Reimold et al. (1973), using aerial photography and extensive sampling of plants of the Sapelo Island marsh, compared the distribution of the various types of stands of *S. alterniflora* from two areas (Table 3.3) and found great variation between sites.

Production of *Juncus roemerianus* is estimated at 2200 g dry wt. m^{-2} yr^{-1} (Gallagher et al., 1980), but because this species occupies only 6% of the marsh (Gallagher et al., 1972), the overall contribution is small. The net aerial productivity of the minor species found in the Sapelo marshes varies widely (Table 3.4). Individually these species may be more productive than the major species. The production of *S. cynosuroides*, for example, exceeds the highest production estimates for creekbank *S. alterniflora*. But in the Sapelo Island marshes the minor species occupy such a very small percentage of the total area that their contribution to the total primary production is indeed negligible. Gallagher et al. concluded that, while there is considerable uncertainty in the existing estimates of photosynthesis, greater precision could be achieved only with a multi-year, multi-point study which would probably yield values within the range of those in hand. Therefore, the available data, not only for Sapelo Island

Table 3.4. Aerial Net Primary Productivity of Plant Species Which Make Up a Minor Part of the Total Plant Cover[a]

Species	Annual Production g C M^2 yr^1
Distichlis spicata	4214
Spartina cynosuroides	5996
Spartina patens	3824
Sporobolus virginicus	1372

[a]From Linthurst and Reimold (1978).

marshes but all other sites, indicate only the order of magnitude of production, and not much more can be expected for the present.

The annual net aboveground primary production figures (Table 3.2) do not include organic matter leached by the tidal water or removed by grazing herbivores. Many plants lose soluble photosynthate in these ways, and *Spartina* is no exception. Leaching losses from *S. alterniflora* are about 6.1 g C m^{-2} yr^{-1} (Gallagher et al., 1976), a small amount compared to shoot production. The organic matter lost may, however, be especially important because it becomes readily available for heterotrophic metabolism. Turner (1978), in another study of loss of soluble photosynthate, reported release of 400 mg C m^{-2} day^{-1}. The disparity between the two studies has not been resolved, and it is of sufficient magnitude to be of some import in attempting to understand or to model the food web. After *Spartina* plants senesce and die the leaves are quickly colonized by bacteria and fungi. Leaching of organic matter from the dead plant community was found to be slightly higher than the estimate of Gallagher et al. (1976) for leaching from the living plants, but much lower than Turner's estimate of leaching from the living *Spartina*.

The underground biomass of many of the marsh plants exceeds the aerial biomass (Gallagher, 1974), and the amount of photosynthate allocated to underground parts varies widely both between and within species (Table 3.5). For example, both tall,

Table 3.5. Estimates of Underground Production by Marsh Plants near Sapelo Island[a]

Plant	Annual Production g C m^{-2}
S. alterniflora	
Streamside	770
Back-marsh	770
Spartina patens	120
Sporobolus virginicus	220
Distichlis spicata	420
J. roemerianus	1340
Salicornia virginica	140
Borrichia frutescens	320

[a]From Gallagher and Plumley (1979).

streamside and short, high-marsh *S. alterniflora* produced the same amount of underground material per year (2000 g dry wt m^{-2} or 770 g C m^{-2}) but the net aerial primary production was 2.5 times greater in the tall plants (Gallagher and Plumley, 1979). The standing crop of underground biomass, both living and dead, is greater in the high marsh, where drainage is reduced and salinity is higher.

3.2. The Epibenthic Algal Community

The epibenthic and epiphytic algae of salt marshes are poorly known compared to the more obvious vascular plants, although they contribute significantly to the total primary production of the marshes. The microalgal flora of the Georgia salt marshes includes a diverse assemblage of several hundred species of pennate diatoms that compose 75 to 93% of the total algal biomass (Williams, 1962). Filamentous blue-green algae (*Anabaena oscillarioides, Microcoleus lyngbyaceous, Schizothrix calcicola*) and a single species of *Euglena* constitute most of the remainder of the microalgal community. Macroscopic algae are poorly represented in Georgia salt marshes. The small (1 cm in height) red algae, *Caloglossa leprieurii* and *Bostrychia radicans*, are found on standing dead *Spartina* culms during the summer months (Chapman, 1971). A border of *Ectocarpus confervoides* develops on the stems of streamside *Spartina* in mid-winter (Pomeroy et al., 1972). Although blue-green algae occasionally form small, conspicuous patches on the mud, the large mats of filamentous green and blue-green algae or the populations of fucoid seaweeds which have been described for salt marshes in Delaware, New York, and Massachusetts (Brinkhuis, 1976; Sullivan, 1975; Van Raalte et al., 1976a; Whitney et al., 1975) do not develop in the Georgia salt marshes. A few genera of macroscopic algae, such as *Rhizoclonium, Ulva, Enteromorpha* and *Vaucheria* are found only infrequently. The paucity of macroscopic algae is likely caused by the high turbidity of the estuarine water, the rapid sedimentation which occurs in the lower areas of the marsh (Williams, 1962), and the extremes of temperature and desiccation. The motility of the pennate diatoms, filamentous blue-green algae, and euglenoids enables them to remain in or near the euphotic zone at the surface of the shifting sediments.

In an intensive study of the standing crop of the diatom assemblage in Sapelo Island salt marshes, Williams (1962) found that an average of 90% of the cells belonged to one of four genera: *Cylindrotheca, Gyrosigma, Navicula* or *Nitzschia*. Diatom cell numbers and total cell volume tended to be highest, 3100 to 6400 cells mm^{-2} and 7.4 to 22.7 ml m^{-2} respectively, in the tall *Spartina* near the creeks, decreasing toward the creek bottom and also toward the high marsh. In the vegetated portion of the marsh, diatom cell numbers fluctuated annually with the winter values which averaged about 10 times higher than the summer values. Total diatom volume also exhibited an annual cycle, but variation in volume was less pronounced than variation in numbers because the small diatoms ($< 3000\ \mu m^3$), which made up the majority of the assemblage, were those most affected by the summer decline in numbers. Week-to-week changes in standing crop in the vegetated portion of the marsh were slight, while the creekbank assemblage showed irregular, short-term variation with no obvious seasonal cycle. Cell numbers on the creek banks were about twice those in the vegetated portion of the marsh. Rain during ebb tide, which has a noticeable scouring action on the surface of exposed salt marsh sediments, was responsible for many of the sudden decreases in diatom standing crop (Williams, 1962).

Whereas Williams enumerated and measured living cells in his estimation of standing crop, other workers in Atlantic coast salt marshes (Sullivan, 1975, 1976; Van Raalte et al., 1976) have emphasized diatom species diversity and community structure rather than standing crop. In his study of the diatom assemblage in a Delaware salt marsh, Sullivan (1975) found that each of several marsh habitats had a unique diatom community, although 32 of the 104 species encountered were found throughout the marsh. Decreases in species diversity of the salt-marsh diatom assemblage have been noted following long-term experimental manipulation of nutrient levels and light intensity (Sullivan, 1976; Van Raalte et al., 1976).

3.2.1 Conceptual and Technical Considerations

Conceptual and technical considerations are important when working with salt-marsh epibenthic algae, because epibenthic algal production differs from higher-plant production in several fundamental ways. The algal standing stock is much smaller than that of the higher plants in the marsh, but the algae have a much higher turnover rate and respond to environmental changes more quickly than do the higher plants. Much of the algal biomass is probably consumed directly by herbivores rather than indirectly through the detritus food web, as occurs in most of the higher-plant production in Georgia salt marshes (Reimold et al., 1975). The seasonal variation in algal production also differs from that of the higher plants. Much of this algal production occurs when the higher plants are dormant, thus increasing the relative contribution of the algae to the total energy flow during the winter months (Gallagher and Daiber, 1974; Van Raalte et al., 1976a).

The consequences of the intimate and complex relationship of algae to their substrate becomes an important consideration when we are attempting to define the functional role of the epibenthic algae in the salt-marsh ecosystem. The microalgae live in and on the top few millimeters of sediment, a habitat whose microenvironment is very difficult to describe. The interface represents a boundary between a dark, nutrient-rich, anaerobic sediment and either an illuminated, aerobic, comparatively nutrient-poor water column or, at ebb tide, the atmosphere. This microenvironment is extremely patchy and is subject to rapid and extreme variation, being directly affected by many factors. It is influenced by variations in tidal exposure, sedimentation, higher-plant cover, and surface and subsurface herbivores and detritivores. These factors, in turn, affect light intensity, temperature, pH, salinity, levels of organic and inorganic nutrients, intensity of grazing, and the stability of the sediment surface. The habitat of the epibenthic algae is virtually impossible to define or to reproduce adequately; thus, when attempting to measure the performance of the algae in their native habitat, we must maintain the integrity of the natural relationships in the surface layer of sediment.

Another interesting aspect of the algal-substrate relationship is the well-documented vertical-migration rhythm exhibited by many benthic diatoms. In some cases the migration appears to have characteristics of an endogenous circadian rhythm (Brown et al., 1972; Palmer and Round, 1967). In the creekbank regions, where the sediments are exposed to strong tidal currents, diatoms move to the surface during daytime low tides, but at high tide and during nocturnal low tides, they retreat beneath the surface, remaining there within the top mm. One may take advantage of this migration rhythm to obtain clean suspensions of epibenthic diatoms for various physiological experiments (Eaton and Moss, 1966; Williams, 1963). Motile diatoms readily migrate into the top of a double layer of fine netting or lens tissue spread on the mud surface soon after the tide has ebbed. After an hour or two, the top layer is recovered and cells are washed off with clean sea water. The technique does not recover all of the cells from the mud and undoubtedly selects for certain species in the assemblage, but it is a useful way to obtain sediment-free, field-grown material.

Because of strong light attenuation in muddy sediments (99% in 0.2 to 1 mm), the migration rhythm has a pronounced effect on photosynthesis (Perkins, 1963; Williams, 1962). Diurnal changes in algal productivity have been described on intact salt-marsh sediments (Darley et al., 1976; Gallagher and Daiber, 1973), but it is possible that these changes reflect a composite effect of decreasing light intensity that results from the effects of migration superimposed upon the effects of a metabolic rhythm in photosynthetic capacity. The relative impact of these two effects can be established by comparing daily photosynthetic rates of algae obtained with netting and maintained in clean cell suspensions, with daily photosynthetic rates of algae residing in intact sediments. Since the clean-cell suspension will experience a constant light intensity during experimental measurements, variations in photosynthetic activity of the cell suspension will indicate a metabolic rhythm. The result of such an experiment with creekbank algae (Figure 3.6) showed a slight rhythm, with an early morning increase in the photosynthetic capacity of the cells in suspension. In contrast, the photosynthetic rate of algae on the sediment shows a much larger diel change, with peaks occurring during daytime low tides. Although both migration and metabolic rhythms appear to affect productivity, the two rhythms are not in phase. The effect of migration in reducing light intensity appears to have a much more important impact on productivity measurements than does the effect of metabolic rhythm. Therefore, if the results are to be representative of field conditions, productivity measurements on creekbank sediments, carried out in the laboratory, must be conducted during the actual period of low tide, when the diatoms have migrated to the top of the sediments.

Intact cores from the short *Spartina* marsh (Figure 3.6) show no apparent rhythmic change in potential productivity over a 24-hour period. Gallagher and Daiber (1973) also noticed a reduced amplitude to the rhythm in the higher portions of the marsh, and Pomeroy (1959) reported that algae in this marsh area remain at the surface at all times. The comparison of intact sediments and algal suspension was not repeated in the high marsh since it is not possible to obtain a large enough sample from that area with the netting technique.

Measurements of epibenthic algal biomass present problems not encountered in the study of phytoplankton or higher plants, and the measurement of chlorophyll content is still the best available method. Sediments, however, contain high concentrations of chlorophyll degradation products which cause serious errors in data obtained through standard methods for measuring chlorophyll. One recently developed solution to this problem (Whitney and Darley, 1979) is to partition the acetone extract with hexane. Chlorophyllides and pheophorbides remain in the acetone phase, while chlorophylls *a*

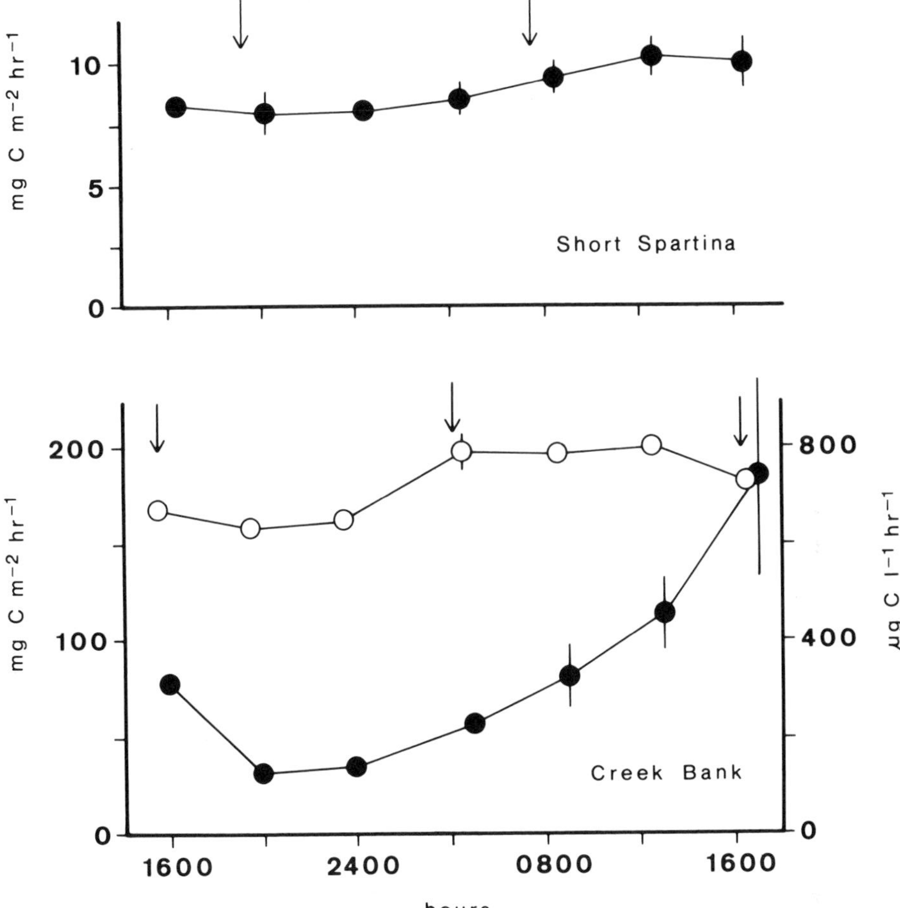

Figure 3.6. Diurnal variation in photosynthesis in intact sediments (•) and cell suspensions (○) in continuous dim light (20 μEin m^{-2} s^{-1}) in the laboratory at 29°C. Bars = one standard error of the mean. Arrows indicate the time of low tide.

and *b* and pheophytins *a* and *b* are quantitatively transferred to the hexane phase. With slight modifications of the standard acidification procedures, the latter pigments can be measured free of interference from the other degradation products. It is also difficult to decide how much of the measured chlorophyll is actually in the euphotic zone in sediments, or even to obtain that layer if the euphotic zone could be identified. We have succeeded in this regard by flattening the surface of sediment cores against polyethylene film, freezing the sample, and then slicing off 1 to 2 mm sections with a specially designed microtome. The interpretation of assimilation ratios (mg C fixed per mg chlorophyll *a* per hour) or turnover rates derived from these chlorophyll measurements is problematical, however, because it is not known how much of the measured chlorophyll has been exposed to optimum light conditions.

3.2.2 Benthic Algal Productivity

Comparing results of the studies of benthic algal productivity in different Atlantic coast salt marshes is difficult because of differences in study techniques and in environments, but epibenthic algae clearly make significant contributions to total salt-marsh primary productivity.

Pomeroy's (1959) study in the Sapelo Island salt marshes was the first attempt at a description of the seasonal primary productivity of benthic microalgae and of the factors which influence it. He measured productivity at low tide, using a flowing-air system with CO_2 absorption columns. High tide was simulated by placing bell jars filled with seawater over the exposed sediments and monitoring dissolved oxygen. Since the abundant crab burrows found in most zones interfered with the latter technique by permitting water to drain from the bell jars, high-tide observations were limited to the bare creekbank zone. In that zone hourly rates of photosynthesis during low-tide conditions were highest during the winter. Based on these studies, Pomeroy estimated annual gross algal production at 200 g C m^{-2} and net algal production at not less than 90% of this value. Most other investigators have only measured epibenthic algal production under water, even though much of the marsh surface is exposed more than half of the time. Gallagher and Daiber (1974) used dissolved oxygen changes over flooded cores in the laboratory to measure epibenthic algal productivity in a Delaware salt marsh. They estimated annual gross production at about 80 g C m^{-2}, which is about one-third of net angiosperm aboveground production in that particular marsh. On a unit area basis, the bare creek bank was the least productive of the several areas studied, including a salt panne. In the vegetated portion of the marsh, productivity was highest in the spring, or spring and summer, but no seasonal pattern was seen in creek-bank and salt-panne areas.

Van Raalte et al. (1976a) measured epibenthic algal productivity in the vegetated portions of a Massachusetts salt marsh by incubating the top 0.5 cm of sediment cores with $^{14}C\text{-}NaHCO_3$ in the field. Brief spring and fall peaks in productivity coincided with blooms of filamentous green algae. Annual algal production was 105 g C m^{-2}, or about 25% of the aboveground grass production for this marsh.

More recently, a detailed study of *in situ* epibenthic algal productivity has been carried out in the marshes near Sapelo Island (Whitney and Darley, unpubl.). Biweekly measurements with ^{14}C were carried out in plexiglass chambers enclosing otherwise undisturbed areas of the marsh surface. Incubation took place under both exposed (low tide) and submerged (high tide) conditions at the time when these conditions occurred naturally in the marsh. Preliminary analysis of the results has yielded an overall annual estimate of net productivity of approximately 190 g C m^{-2}, a figure which is close to Pomeroy's (1959) estimate and indicates that epibenthic algal productivity is nearly 25% of aerial *Spartina* productivity (Gallagher et al., 1980). More than 75% of this algal production occurs when the marsh is exposed at ebb tide, and during this time the bare creek bank is the most productive area of the marsh (Table 3.6). Conversely, it is the least productive when submerged at high tide. Migration of the diatoms into the sediment and the turbidity of the estuarine water probably account for low rates at high tide. The tall-*Spartina* levee area is almost twice as productive when submerged as when exposed, although the reason for the difference is not readily ap-

Table 3.6. Average Hourly Epibenthic Algal Productivity in Certain Salt Marsh Areas, Sapelo Island, Georgia (mg C m^{-2} hr^{-1})

Area	Exposed	(mg C m^{-2} hr^{-1})	Submerged
Bare creekbank	132		5
Levee (tall *Spartina*)	11		23
High marsh (short *Spartina*)	7		6

parent. In the high marsh, algal production is nearly equal under high- and low-tide conditions, with no obvious seasonal trends. The findings of Darley and Whitney demonstrate the need to study ecosystem components under natural conditions, particularly in the creekbank area. For example, the high rate of total carbon fixation by the creekbank algae would have gone unnoticed if only underwater productivity had been measured. The converse, of course, holds for the algal populations of the levee.

Under laboratory conditions many diatom species utilize dissolved organic carbon to support heterotrophic growth. Under conditions of limiting light intensities and relatively high levels of dissolved organic substrates, conditions which often exist in the habitat of the epibenthic algae, diatoms and other algae might also be using DOC to supplement their normal autotrophic nutrition. To test this hypothesis, Darley et al. (1979) collected natural populations of creekbank algae with the netting technique, then incubated them with ^{14}C-labeled glucose, acetate, lactate, and glycerol at 10^{-6} M. The measured assimilation rates of these compounds per unit of algal biomass suggest that the algae were obtaining less than 1% of their carbon heterotrophically, even when several substrates were each present in the surface sediment layers at concentrations of 10^{-6} M. Although this study suggests that heterotrophy is not significant to the carbon budget of autotrophically growing cells, it could be important for maintenance metabolism in cells which are buried by sediment.

3.2.3 Controlling Environmental Factors

Many factors could limit the standing crop biomass and productivity level of salt-marsh epibenthic algae. For example, we might assume that exposure at low tide would be a stressful condition for a single-celled alga, but a comparison of productivity rates under exposed and submerged conditions (Table 3.6) indicates that production is not adversely affected during low tide. Higher areas of the marsh may dry out considerably when a neap tide cycle coincides with hot, dry weather, and algal productivity would presumably decline during such periods, but this problem has not yet been investigated. The tidal regime may also affect productivity indirectly through its influence on other parameters, including salinity, pH, temperature, light intensity, and nutrients.

The *salinity* of the marsh surface varies from moderate at flood tide, to high following evaporation at low tide, to low during rains at ebb tide. Salt-marsh and estuarine diatoms appear very tolerant of salinities normally encountered in their habitat; salinity is not likely to be an important controlling parameter. Fourteen species of diatoms isolated from Sapelo Island salt marshes grew well in salinities of 10 to 30 ‰; several grew well over a salinity range of 1 to 68 ‰ (Williams, 1964). Cultures and mixed natural populations of benthic estuarine diatoms from the Netherlands maintained photosynthetic rates of not less than 70% of their initial rate following a six-hour exposure to salinities of 4 to 60 ‰ (Admiraal, 1977). It is likely, however, that the occasional extreme salinities of over 100 ‰ which have been reported in creekbank and panne-marsh zones (Gallagher and Daiber, 1974; Sullivan, 1975) have a short-term deleterious effect on productivity.

Although the *pH* of the marsh surface generally remains between 7 and 8, algal photosynthesis during low tide can increase it to over 9 (Pomeroy, 1959; Sullivan, 1975). Inadequate supplies of CO_2 and HCO_3^- under these conditions may limit photosynthesis, but more definitive information is needed on this point.

Seasonal variation in *temperature* has less influence upon epibenthic algal productivity than it does upon *Spartina* productivity, although Williams (1962) reported that the average warm-weather division rates of diatoms were twice those of cool weather. Pomeroy (1959) and Van Raalte et al. (1976) point out that photosynthetic rates are independent of temperature at sub-optimum light intensities. Tides have an ameliorating effect on sediment temperature, warming in the winter and cooling in the summer, and this mechanism might be responsible for the higher productivity on the submerged levee (Table 3.6). Temperature may also indirectly affect the algae by affecting activity of grazers. Williams (1962) attributed the warm-weather decrease in diatom standing crop to increased activity of grazers rather than to any direct effect of temperature upon the algae itself.

Light intensity has been implicated by several workers as the most important factor influencing epibenthic algal productivity in Atlantic coast salt marshes. While it is well known that full sunlight inhibits photosynthesis in phytoplankton (Belay and Fogg, 1978), there is little direct evidence of severe photoinhibition in epibenthic algae. Using cell suspensions collected with the netting technique from the creekbank area at Sapelo Island, Williams (1962) found maximum photosynthetic rates at about one-half of full sunlight, with only moderate inhibition at full sunlight. In similar experiments with diatoms from a Massachusetts intertidal sand flat, Taylor (1964) reported that photosynthesis was saturated at about 16% full sunlight with only a 10% decrease from maximum rates at full sunlight. Recent *in situ* experiments (Whitney and Darly, unpubl.) with creekbank algae at Sapelo Island also show a broad plateau in photosynthesis vs light intensity curves, with only slight inhibition at high light intensities (Figure 3.7). Epibenthic algae which experience high light intensities in their natural habitat are well adapted to these conditions by both their photophysiology and their migratory behavior.

In the vegetated portions of a marsh, it is possible that too little light penetrates the grass canopy to saturate algal photosynthesis. In a Massachusetts salt marsh an average of approximately 20% of the available summer light reaches the marsh surface (Estrada et al., 1974), although the intensity varies greatly due to the uneven canopy. Based on a comparison of photosynthetic rates in naturally or experimentally shaded and unshaded sediments, light is considered to be of greatest importance in determining primary productivity in these marshes (Van Raalte et al., 1976a). However, in a year-long study in a Delaware salt marsh, reduction of the light intensity penetrating the grass canopy by 30% or even 60% had no significant effect on the chlorophyll concentrations

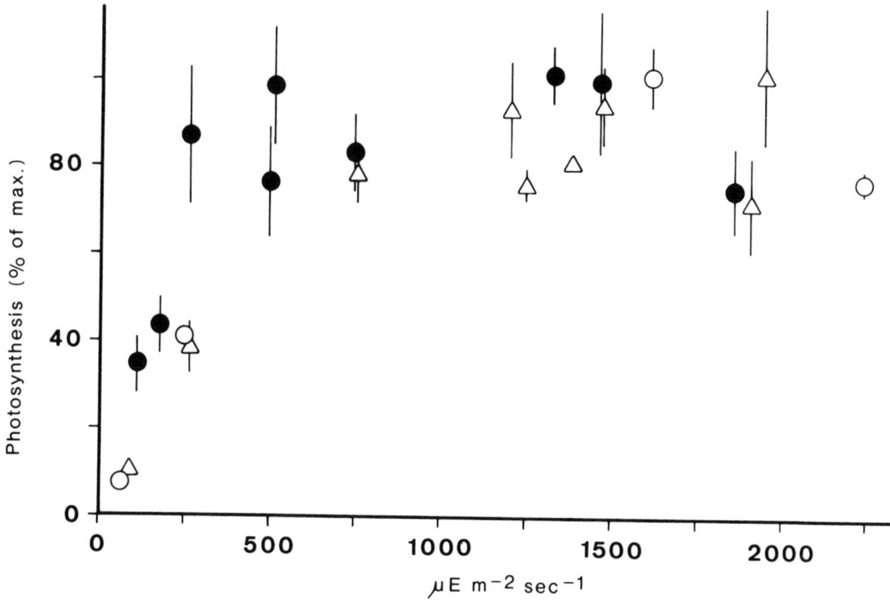

Figure 3.7. Response of photosynthesis to light intensity in epibenthic creekbank algae at Sapelo Island. Measured in the field during low tide on three days during summer, 1976. Bars = one standard error of the mean.

○ = June 23, 1976
● = June 24, 1976
△ = August 23, 1976

in the surface sediment, suggesting that the epibenthic algae were not light-limited under the natural grass canopy (Sullivan and Daiber, 1975). But removing the grass canopy and exposing the sediment to full sunlight caused a significant increase in soil chlorophyll and led to the appearance of mats of filamentous algae instead of the normal algal assemblage of diatoms. However, since the increase in chlorophyll can be attributed to the change in community structure and to the absence of vascular plant competition for nutrients, it does not necessarily indicate that the original diatom assemblage was light-limited.

In recent field experiments at Sapelo Island (Darley et al., 1981), sediment cores were incubated at various light intensities in fiddler-crab exclosures. In the summer experiment (Figure 3.8), algal biomass, as chlorophyll *a*, increased in sediment cores exposed to intensities ranging from full light to 25% of full light and remained constant or decreased in sediment cores exposed to lower light intensities. In the winter season, there were no significant differences among the increases in algal biomass which occurred in cores exposed to light intensities ranging from 6 to 100% of the light intensity available on a very clear day. Thus, these algae seem not to be limited by light at the average intensity–about 25%–that penetrates the canopy in these marshes, although limiting conditions are approached during the summer months. The winter algal assemblage appears to be better shade adapted than the summer assemblage. However, changes in community structure of the algal assemblage were not studied.

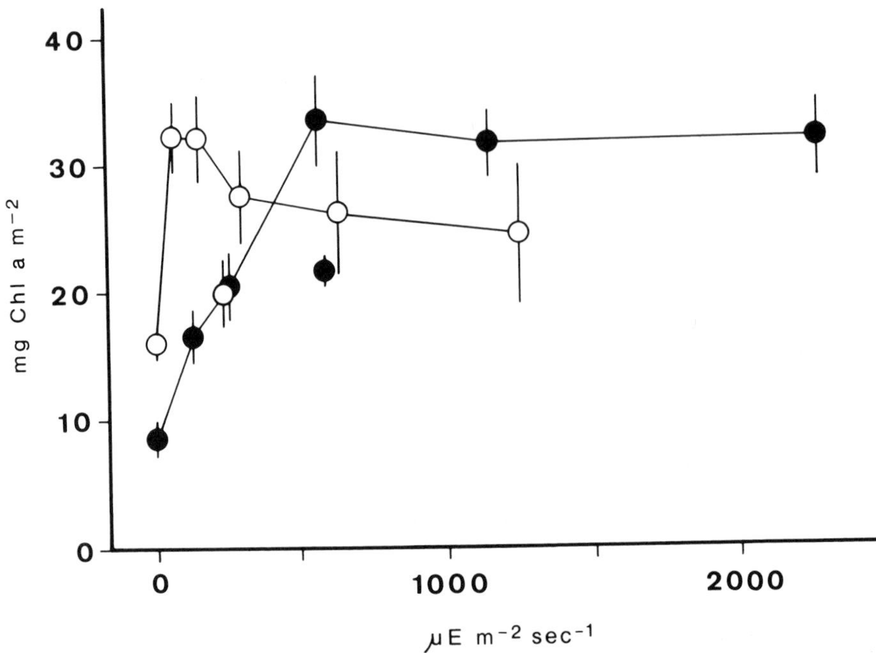

Figure 3.8. Algal biomass as a function of light intensity. Chlorophyll *a* in the top 2 mm of sediment cores (3 cm diameter, from short *Spartina* marsh) incubated in the field for 2 weeks in summer (●) or 1 week in winter (○) in fiddler crab exclosures. Ambient light intensity was modified with neutral density screening. Initial chlorophyll concentrations were 20 mg m^{-2} in summer and winter. Bars = 95% confidence limits.

Limitation by nutrients has also been investigated in the epibenthic algae of Atlantic coast marshes. Although algae in the vegetated portions of Massachusetts marshes are thought to be light-limited (Van Raalte et al., 1976a), they may also be nutrient-limited, since nitrogen enrichment stimulated productivity. Algal biomass, as chlorophyll *a*, measured in late summer was not affected by nutrient enrichments, however, because fertilization had stimulated grass production which, in turn, decreased the light intensity reaching the sediment surface and confounded the results of the experiment (Estrada et al., 1974). Similar problems were encountered in a study in a Delaware marsh (Sullivan and Daiber, 1975). Removing the vascular plant cover increased edaphic chlorophyll concentration. Nutrient enrichment of clipped plots did not result in any further significant increases in chlorophyll, except for those plots enriched with nitrogen during the summer months. These experiments were long-term studies in which nutrients were applied in pulses, biweekly in Massachusetts and monthly in Delaware, and the results were evaluated after several months to a year.

In further short-term experiments at Sapelo Island (Darley et al., 1981), sediment cores were incubated in the field in fiddler-crab exclosures and fertilized daily with nutrient solutions. In the short-*Spartina* marsh (Figure 3.9), the algal standing crop, as chlorophyll *a*, and productivity of the algae both increased significantly when the cores were fertilized daily with nitrogen or with a complete nutrient solution con-

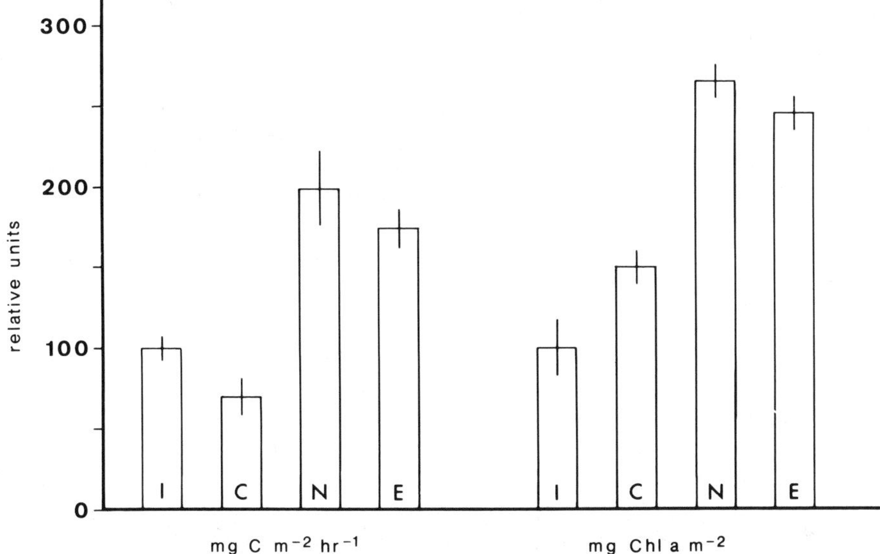

Figure 3.9. Summer nutrient enrichment experiment in short *Spartina* marsh at Sapelo Island. Initial (I) values are 14.9 mg C m^{-2} h^{-1} and 33.5 mg chlorophyll *a* m^{-2}. Sediment cores were incubated for 8 days in fiddler crab exclosures at 50% ambient light intensity with nutrient solution added at low tide daily. Controls (C) received seawater. Nitrogen-enriched cores (N) received 3 μM N as NH_4Cl in seawater. Completely enriched cores (E) received a complete nutrient solution. Bars = 95% confidence limits.

taining nitrogen. Similar results were obtained in winter experiments, suggesting that the algae are nitrogen-limited in both summer and winter. In the bare-creekbank marsh zone, similar experiments indicated that the algae there are limited by the grazing activity of the snails and fiddler crabs. Nutrient enrichment only stimulated algal production following an algal bloom which was the result of grazer removal, and which had depleted the relatively high standing stock of nutrients in the surface creekbank sediments. Short-term experiments such as these reflect the actual status of the algae in the marsh, whereas the long-term experiments indicate the response that would result from recurrent perturbation of the marsh ecosystem.

3.2.4 The Fate of Algal Production

There is very little quantitative information on the fate of the carbon fixed by epibenthic algae or on the effects of such losses on subsequent algal production. Qualitative observations suggest that fiddler crabs (Williams, 1962; Shanholtzer, 1973; Haines and Montague, 1979), snails (Kraeuter and Wolf, 1974; Wetzel, 1975), and herbivorous fish such as mullet (Hughes, 1980) graze on epibenthic algae. This grazing is thought to be a major route for loss of algal biomass. Removing mud snails from experimental plots of creekbank sediments resulted in highly significant increases in chlorophyll *a* and in productivity (Pace et al., 1979). The effect of meiofaunal grazing on epibenthic algae is unknown.

Excretion of organic matter by epibenthic algae represents another potential loss of fixed carbon. This loss has proved difficult to measure because of the intimate association of an active bacterial flora with the algae. The bacteria would be expected to assimilate any labile organic matter nearly as fast as it was released. The only available excretion data (Darley et al., 1976) suggest that the amount lost is very small, approximately 1% of the total fixation. This measurement, however, represents only the excreted matter that remains unassimilated in the sediment for more than 15 minutes.

Gallagher (1975a) found that epibenthic pennate diatoms were lifted from the marsh as a surface film by the flooding tide. Over one-third of the net photosynthesis of the water column occurred in the surface film. Quantitative estimates of these losses have been measured only on the bare creekbank area of the Sapelo marshes. The algal patches lifted from the sediment surface during each high tide constitute about 10 to 15% of the algal biomass (Whitney, unpubl.). This loss will be less in the vegetated marsh areas because water velocities there are much lower. Although a significant portion of the algal biomass is removed this way from a given area of bare creek bank or mud flat, many of the unsettled algae do not remain in the water column. The benthic algae rapidly leave the floating patch of surface film, and at least a portion of them are deposited upon the marsh surface further upstream where current velocities rapidly decrease (Whitney, unpubl.). Rain storms during the time of ebb tides erode the top few millimeters of sediment from creek banks and mud flats, resulting in a drastic reduction in the diatom standing stocks (Williams, 1962). This loss is not as great in the more protected, vegetated portion of the marsh.

Algal biomass may also be lost to deeper sediments when cells are unable to regain the surface following the rapid sedimentation that occurs in the lower marsh areas. As much as 5 to 6 mm of sediment may be laid down on mud flats and creek banks during the course of one day (Shimmel, 1979). The amount of biomass lost by this mechanism has not been measured, although the small but nearly constant amount of "live" chlorophyll *a* found at depths of several centimeters (Whitney, unpubl.), is presumably that of buried algae cells.

3.3. Phytoplankton

Early workers in turbid, shallow southeastern estuaries regarded photosynthesis by phytoplankton as a minor source of fixed carbon when compared to photosynthesis by the marsh grass. Ragotzkie (1959) found that the critical depth (that depth at which the integrated mean light intensity reaches the compensation point) was usually above the bottom of the estuary and so concluded that phytoplankton production was negative. His sampling regime comprised several stations, all of them located in deep, central portions of the Duplin River. Although the effect of water turbidity on available light is probably the most severe limiting factor in phytoplankton photosynthesis, Ragotzkie's results were likely biased by his sampling locations. Several recent studies in estuaries of the southeastern United States, including the Duplin River, found phytoplankton to be a significant source of organic matter for the estuarine food web, if not for the salt-marsh food web as well (Sellner and Zingmark, 1976; Thomas, 1966; Whitney et al., in ms.).

The species composition of the phytoplankton communities in the estuaries of Georgia is similar to that described for North and South Carolina estuaries (Hustedt, 1955; Zingmark, 1978). Pelagic diatoms, such as *Skeletonema costatum, Rhizosolenia sp., Asterionella sp.*, and *Coscinodiscus sp.* are dominant. Benthic pennate diatoms, such as *Nitzchia* and *Pleurosigma*, are always present in small numbers. Fresh-water and brackish-water species are intermittently present, depending upon the inputs of fresh water to the estuary. Several species of dinoflagellates in the 15 to 25 μm range, as well as several green flagellates in the 5 to 15 μm range, are present and may at times dominate the community (Whitney et al., in ms.). Blooms of dinoflagellates which frequently occur in the upper tidal segment of the Duplin River, were erroneously reported to be *Gymnodinium* (Pomeroy et al., 1956; Ragotzkie and Pomeroy, 1957), but are actually *Kryptoperidinium*.

3.3.1 Methods

Early estimates of photosynthesis by phytoplankton in the Duplin River and other estuaries of the southeastern United States were obtained by measuring changes in oxygen in light and dark bottles. This is a satisfactory method, because the absolute rate of photosynthesis in estuaries is high enough so that incubation need last less than 12 hours. Because most of the early studies were either 12 or 24 hours in duration, the major potential source of error probably was exhaustion of some essential element, even carbon, at the high rates of photosynthesis sometimes reported. That the oxygen method is a measure of gross photosynthesis is not a serious problem. Other sources of error almost certainly exceed the difference between net and gross production in such highly productive water.

More recent work employed the ^{14}C method (Steeman-Nielsen, 1952; Schindler and Holmgren, 1971), and in productive waters incubation usually lasted not more than six hours. This method is believed to approximate net photosynthesis. However, it has recently been questioned by critics (Sieburth, 1977; Joiris, 1977; Gieskes et al., 1979) who point out that the amount of community respiration apparently exceeds photosynthesis. Because there are many potential sources of this sort of error, the ^{14}C method has not yet been proven to be at fault. Recent work has produced higher estimates of photosynthesis than the earlier studies did, however, and if the present criticisms have merit, future estimates will be still higher. Thus, we can probably say that the present estimates of phytoplankton photosynthesis are still conservative figures.

3.3.2 Production of Phytoplankton

Estimates of net production by phytoplankton in salt-marsh estuaries in the southeastern United States have varied from <100 to nearly 400 g C m^{-2} yr^{-1} (Table 3.7). In every case these were values for the subtidal portions of the estuaries. Whitney et al. (in ms.) found that the highest photosynthetic rates for phytoplankton occurred in the water over the marsh on spring high tide. However, the number of daytime hours per year when the marshes were inundated was so small that the increase did not greatly influence the estimate of total photosynthesis within the system. The annual produc-

Table 3.7. Summary of Annual Phytoplankton Production in Southeastern Estuaries and Coastal Waters

Study	g C m^{-2}	Location
Ragotzkie (1959)	248[a]	Duplin River, Ga.
Williams and Murdoch (1966)	225[a]	Beaufort Channel, N.C.
Williams (1966)	100[a]	Bogue Sound, Newport River, North River, and Core Sound, N. C.
Thayer (1971)	67	
Sellner and Zingmark (1976)	346	North Inlet, S. C.
Whitney et al. (in ms.)	375	Doboy Sound-Duplin River, Ga.

[a]Represents gross production.

tion of phytoplankton in the Duplin River and adjacent Doboy Sound is estimated to be 375 g C m^{-2} yr^{-1} (Whitney et al., in ms.). On the basis of equivalent areas this is equal to half the rate of production of shoots of *Spartina*. However, since the area of standing water–exclusive of spring high tides–is 21% of the total area of the estuary, the other 79% being intertidal marsh, the contribution of phytoplankton is actually one-eighth of the aboveground production of *Spartina* (Table 3.8).

3.3.3 Environmental Factors

The major factor limiting photosynthesis in the Duplin River and other estuaries of the coast of Georgia is light. Because of high turbidity, and at times high concentrations of humic substances as well, the euphotic zone is limited to the upper 1 to 2.5 meters (Figure 3.10). In the deeper portions of the Duplin River and much of Doboy Sound, the critical depth is above the bottom, and the entire water column is usually mixed by tides and winds (Ragotzkie, 1959). Highest photosynthetic rates are in the upper 0.5 meters of the water column, but lower rates do continue to compensation depth, suggesting that some part of the phytoplankton community may be shade adapted (Whitney et al., in ms.).

Table 3.8. Production for the Duplin River Marsh and Estuary as a Whole, Prorated on the Basis of 21% Subtidal and 79% Intertidal Area

Producer Population	g C m^{-2} yr^{-1}	% Aboveground	% Total
Spartina whole plants	1216		84
Spartina roots	608		42
Spartina shoots	608	73	42
Benthic algae	150	18	10
Phytoplankton	79	9	6
Total	1445		
Total aboveground	758		

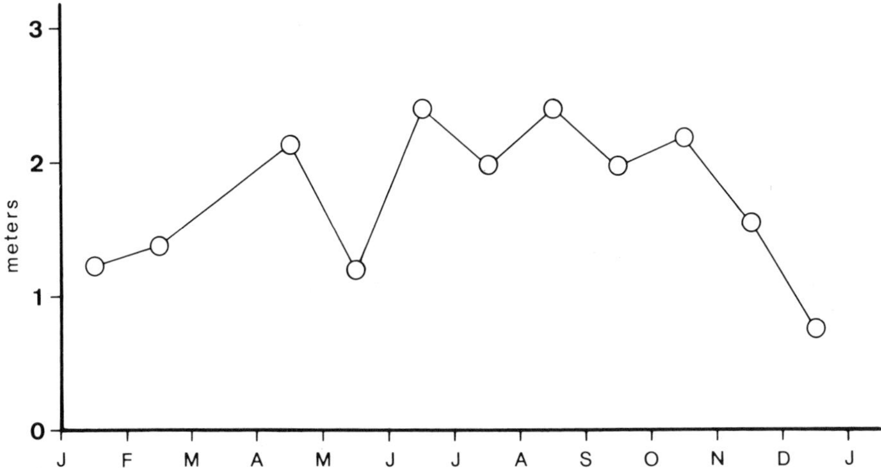

Figure 3.10. Depth of compensation light intensity in the Duplin River at the time of monthly photosynthesis studies.

There is little evidence that nitrogen and phosphorus limit phytoplankton photosynthesis in estuaries of the southeastern United States (Williams, 1972). Physical and biological processes from both fresh and coastal waters tend to concentrate nutrient elements in estuaries, with sediments and salt marshes acting as temporary storage facilities or sinks (Pomeroy et al., 1965; Pomeroy, 1970; Woodwell and Whitney, 1977; Woodwell et al., 1979). Even when standing stocks of nitrogen and phosphorus in the water column are low, rapid biological and physiocochemical processes maintain a continuous supply sufficiently adequate for phytoplankton growth (Pomeroy et al., 1972; Haines, 1979).

The annual 25°C range of water temperature in the salt-marsh estuaries is greater than that in coastal ocean waters, and this factor could affect phytoplankton photosynthesis. Indeed, as Figure 3.11 indicates, the lowest rates of photosynthesis occur during the winter (Thomas, 1966; Sellner and Zingmark, 1976). However, this is also the season most given to stormy weather, which enhances water turbidity. No one appears to have separated experimentally the effects of winter temperatures from the effects of seasonal turbidity. Moreover, at sub-optimal light intensity, temperature effects may be less pronounced (Pomeroy, 1959; Van Raalte et al., 1976a).

The effect of grazing on phytoplankton productivity, or even on our ability to measure production rates in the Duplin River, is not well understood and constitutes a significant residual problem. The activities of consumers can stimulate or depress algal productivity, depending on the relative densities and the factors controlling the grazers and herbivores (Brenner et al., 1976; Fenchel and Kofoed, 1976; Pace, 1977; Pace et al., 1979). Use of the marsh model to predict the effects of grazing is discussed in Chapter 9.

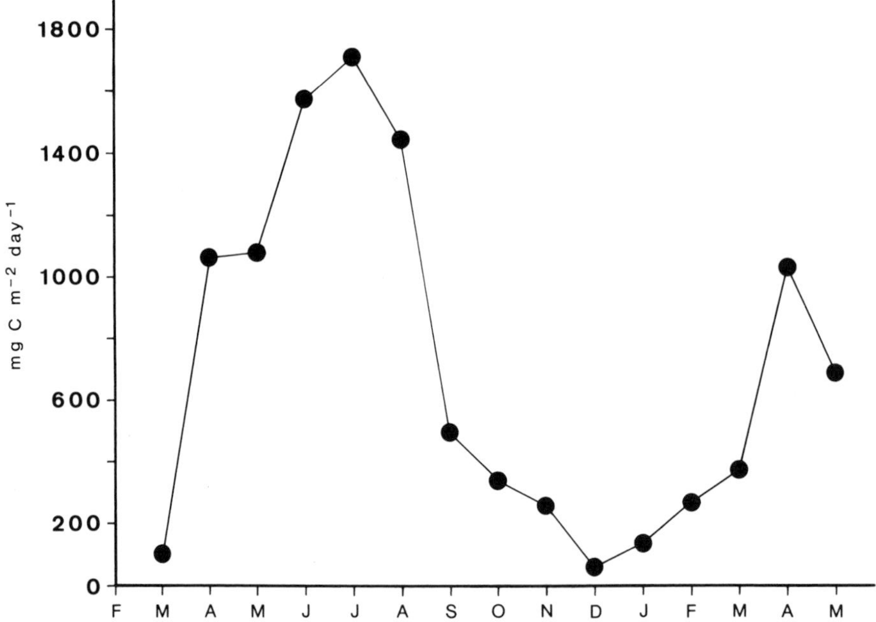

Figure 3.11. Rate of phytoplankton photosynthesis in the Duplin River and Doboy Sound, based on means of measurements taken monthly at a series of stations.

3.4. Total Plant Production

The attitude of investigators toward the relative significance of the several populations of primary producers in salt-marsh estuaries has shown a continuing shift over three decades. While early workers viewed the marsh grass as virtually the sole source of primary production, we no longer consider it the only significant producer. In the marshes of Sapelo Island, production by algal populations, when prorated according to the area occupied, is more than one-third of the aboveground production by *Spartina* (Table 3.8). This fact takes on greater significance when we recognize that nearly all of the *Spartina* shoot production becomes detritus, a relatively refractory food source, while most of the algal production, going directly to grazing consumers, is a relatively high-quality food source.

The fate of carbon fixed by the primary producers of the salt marsh is of paramount importance to any explanation of how this ecosystem operates. Just as our perception of where the organic carbon is produced has changed, so too our perception of its transfer and ultimate transformation has altered. Teal (1962) estimated that roughly half of the net production by *Spartina* is removed from the marsh as POC or DOC, becoming in that form an important energy input to the estuarine and nearshore food chains. Unfortunately, direct measurements in support of this hypothesis were never obtained (Wiegert, 1979; Chapter 9). Furthermore, recent work with stable carbon isotope ratios by Haines (1977) failed to show significant amounts of carbon with the isotopic ratios characteristic of *Spartina* in the waters of the nearshore or estuarine zones.

Even the fauna of the estuary and larger tidal creeks had $^{12}C/^{13}C$ ratios, ratios which suggest carbon resources not wholly traceable to *Spartina* detritus (Haines and Montague, 1979).

Clearly, the earlier, simpler concept of the marsh-estuary-nearshore interaction needed to be reexamined and revised. In view of recent findings and given our continued belief in the marsh as a net source of fixed carbon, two explanations—which are not mutually exclusive—seem possible.

1) We recognize two quantitatively major food chains in the aerobic waters of the tidal creeks, one based on *Spartina* detritus and the other on microalgae in the water and on the surface of the intertidal sediments. Further, we postulate a large amount of total detrital breakdown occurring in the water covering the marsh at high tide and in the upper portions of the tidal creeks. This breakdown results in a transfer of major amounts of organic carbon from both algae and *Spartina* to the fauna, thus accounting for the faunal $^{12}C/^{13}C$ ratios and for the absence of large amounts of *Spartina* detritus in the lower tidal creeks and the estuary.

2) The anaerobic transformation of carbon in the marsh soil results in the export of large amounts of energy to the water in the form of reduced sulfur compounds, the end products of sulfate reduction (Chapter 7). When these reduced compounds are oxidized by chemoautotrophs in the water, carbon of low $^{12}C/^{13}C$ ratio is fixed and transferred to the fauna. Little detritus of *Spartina* origin is exported from the marsh and its interior tidal creeks (Howarth and Teal 1979, 1980; Peterson et al., 1980).

The relative validity of these two explanations of the data of Haines (1977) and of Haines and Montague (1979) will be evaluated in Chapters 4, 7, 9, and 10. Neither of these explanations eliminates the possibility that a significant fraction of the production in the salt marsh is transported to the estuary or even to the coastal waters. That possibility is considered further in Chapter 9.

4. Aquatic Macroconsumers

C.L. Montague, S.M. Bunker, E.B. Haines, M.L. Pace, and R.L. Wetzel

The productive plant populations of the salt-marsh estuary support abundant populations of animals. Ranging in size from small copepods, polychaete worms, and snails to large fishes, birds, and mammals, these macroconsumers share the need to obtain sufficient energy-rich organic carbon to meet the demands of maintenance and reproduction. These demands and the manner in which they are met form the subject of this chapter and the next. In this chapter we consider the aquatic macroconsumers and in Chapter 5 the terrestrial ones. Because the marsh is alternately flooded and drained, certain problems of classification and definition inevitably arise. The distinction is somewhat contrived, but convenient. In general, "aquatic" denotes those organisms that live in the intertidal area or in the tidal creeks and are constantly or periodically submerged. Their trophic relationships generally begin with algae or *Spartina* detritus. Fiddler crabs (*Uca*), mussels (*Geukensia*), mullet (*Mugil*), blue crab (*Callinectes*), and shrimp (*Penaeus, Palaemonetes*) are all examples of aquatic macroconsumers found in the Sapelo Island ecosystem.

Population sizes of individual species are often large, but species diversity of the Sapelo Island salt marsh is low, perhaps because of the stressful environment and the dearth of niches due to the structural and productive dominance of *S. alterniflora* (Figure 4.1). Nonetheless, the secondary production of this salt marsh is among the highest in the world (Wiegert and Evans, 1967; Odum, 1961).

In the intertidal zone, biomass of aquatic macroconsumers may exceed 15 g C m^{-2}. Eighty to 200 mud fiddler crabs (*Uca pugnax*), 400 to 700 marsh snails (*Littorina irrorata*) or mud snails (*Ilyanassa* [*Nassarius*] *obsoleta*), and seven to eight ribbed mussels (*Geukensia demissa*) per square meter of marsh are not unusual populations. (Teal, 1958, 1962; Wolf et al., 1975; Odum and Smalley, 1959; Kuenzler, 1961). In addition, quantities of the pulmonate snail, *Melampus bidentatus*, the marsh clam, *Polynesoda caroliniana*, and several polychaete worms are abundant. On mud flats, densities of 500 to 1600 mud snails (*Ilyanassa obsoleta*) are found per square meter, along with other macroconsumers (Pace et al., 1979). Estuarine-dependent commercial

Figure 4.1. Photograph of the marshes at Sapelo Island showing the structural uniformity imposed by the dominance of *S. alterniflora.*

species such as white shrimp (*Penaeus setiferus*), brown shrimp (*P. aztecus*), blue crabs (*Callinectes sapidus*), and oysters (*Crassostrea virginica*) provide a multimillion dollar seasonal income for coastal Georgia. In recent years, an average of 2 to 4 $\times$ 10^6 kg (wet weight) of shrimp and 3 to 4 $\times$ 10^6 kg (wet weight) of crabs have been removed annually by commercial fishermen in Georgia (Anderson, 1970; Palmer, 1974). The marsh supports a productive sports fishery as well. Many furbearing mammals–raccoon, muskrat, mink, and otter–inhabit salt marshes. Shorebirds, game waterfowl, and endangered species of birds, reptiles and mammals depend on Georgia marshes for food and shelter. Alligators, porpoises, and manatees inhabit tidal creeks.

Interest in the role of aquatic macroconsumers in Georgia salt-marsh ecosystems arose from fundamental questions concerning primary production, decomposition, and mineral cycling, which were posed by ecologists in the late 1950s and early 1960s. The first research on macroconsumers was descriptive and emphasized fluxes of material or energy. Studies on the marsh snail *Littorina irrorata*, the salt marsh grasshopper, *Orchelimum fidicinium* (Smalley, 1959a; Odum and Smalley, 1959), fiddler crabs and other marsh crabs (Teal, 1958, 1959), and the ribbed mussel (Kuenzler, 1961, 1961a) provided some of the data used in the energy-flow budgets first drafted for the salt-marsh ecosystems (Teal, 1962; Figure 4.2). But in these early studies many of the energy-flow rates, population figures, and food-item quantifications were only estimates or were based on very limited data.

Because Teal's summarization indicated that macroconsumers accounted for little of the total ecosystem production and energy flow, the thrust of the research in the late 1960s and early 1970s shifted to *Spartina* and to the decomposers through which most of the energy flowed. This emphasis on magnitude ignored the factors controlling the flows of material and energy. We know that much of this control is provided by macroconsumers. Studies of this group have enhanced our understanding of salt marsh-estuarine carbon flow and of macroconsumer foods, feeding, and energetics. These organisms are involved in the food chain arising from the decay of dead *Spartina*. However, stable carbon isotopic analysis has recently changed our thinking about the utilization and availability to macroconsumers of organic carbon produced by *Spartina* (Haines, 1977) as compared to the carbon fixed by the algae. A section of this chapter is devoted to isotopic analysis. General accounts of macroconsumers and lists of Georgia salt-marsh species are available in the references indicated in Table 4.1 and in Teal (1959a, 1962) and Teal and Teal (1969).

4.1. Foods and Feeding Categories

Knowledge of the foods and feeding rates of salt marsh and estuarine animals is fundamental to understanding carbon flow from *Spartina* and algae to top carnivores and commercial species. This plant-animal transformation also is one way in which carbon can get out of the marsh and into the consumers in the estuary. Extensive studies of gut contents (Darnell, 1961; Odum and Heald, 1975) have identified the foods of macroorganisms in other coastal ecosystems, but specific information on animal feeding rates is scarce. Ingestion and assimilation rates have been estimated for the mud snail, *Ilyanassa* (*Nassarius*) *obsoleta* (Wetzel, 1975, 1976), the mud fiddler crab, *Uca pugnax* (Shanholtzer, 1973), and the grass shrimp, *Palaemonetes pugio* (Johannes and Satomi, 1967). Rates determined by measuring uptake and retention of radioisotopes (Wetzel, 1975) are preferred, since radioisotopes can establish the relative use of dead plant matter, associated microbes, and microalgae. Here we describe the food available to macroconsumers and several methods used by the consumer to obtain these foods. The recent studies by Martin and Martin (1979) used enzyme analyses to identify specific foods potentially digestible by consumers. Although these techniques have not yet been applied to marsh-estuarine species, they should be useful in a number of cases.

4.1.1 Foods

Carbon is available to macroconsumers in the salt marsh in the form of live vascular plants, dead plants, microbes (both heterotrophs and chemoautotrophs), algae (benthic and planktonic), live and dead animal tissue, and feces. It had been assumed that all of these except living *S. alterniflora* are commonly ingested by the aquatic macroconsumers,* but evidence is accumulating against the direct assimilation of dead *Spartina*. Though only a few animals have been tested, none has demonstrated the ability to

*The crab, *Sesarma*, is considered as a grazer in Chapter 5.

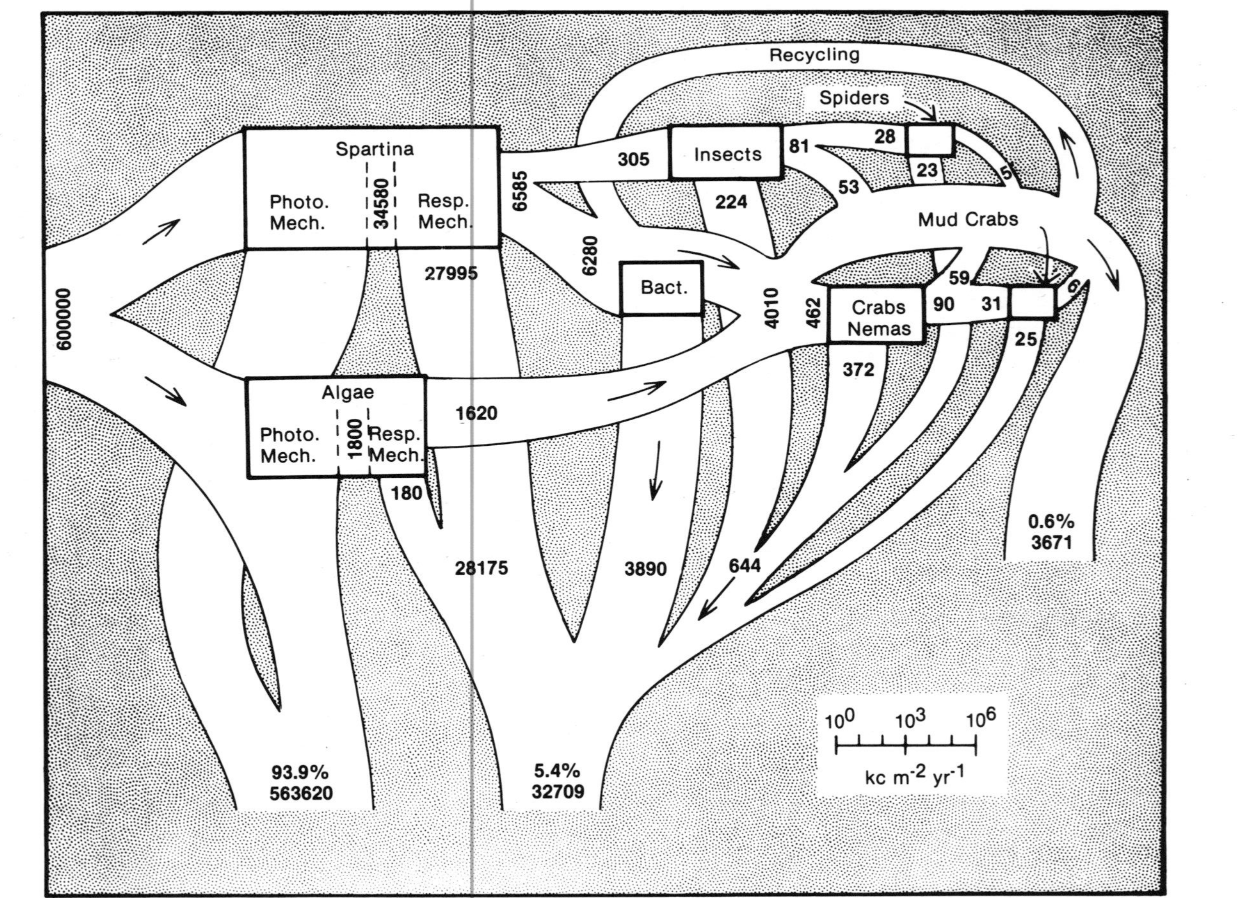

Figure 4.2. Box model of energy flow in the Georgia salt marshes. The two outward flows at the bottom are losses as heat, while the outward flow at the right is net secondary production exported from the marsh. Adapted from Teal (1962).

Table 4.1. Selected References from the University of Georgia Marine Institute in Some General Categories of Macroconsumer Research[a]

Author(s)	Research Categories	Organism(s) Studied
Bahr (1976)	1,2	*Crassostrea virginica* (oyster)
Basan and Frey (1977)	5	invertebrate traces
Bunker (1979)	1,3,4	*Mugil cephalus* (mullet)
Dahlberg (1975)	6	fish
Dahlberg and Heard (1969)	3,6	sharks and rays
Dahlberg and Odum (1970)	2	fish
Edwards and Frey (1977)	5	invertebrate traces
Frankenberg and Burbanck (1963)	1,2	*Cyathura polita* (isopod)
Frankenberg and Smith (1967)	3	numerous macroconsumers
Haines (1976a)	3	macroinvertebrates
Haines (1976b)	3	*Uca spp.* (fiddler crabs)
Haines and Montague (1979)	3	macroinvertebrates
Heard (1975)	3	*Ictalurus catus* (catfish)
Hoese (1971)	3	bottlenosed dolphin
Jacobs (1968)	2	copepods
Johannes et al. (1969)	3	*Bdelloura candida* (turbellarian)
Johannes and Satomi (1966)	5	*Palaemonetes pugio* (grass shrimp)
Johannes and Satomi (1967)	1,3	*Palaemonetes pugio* (grass shrimp)
Kale (1964)	3	*Telmatodytes palustris* (marsh wren)
Kale (1965)	1,2,3	*Telmatodytes palustris* (marsh wren)
Kale and Hyppio (1966)	6	birds
Kraeuter (1973)	6	pycnogonids
Kraeuter (1976)	5	marsh macroinvertebrates
Kraeuter and Setzler (1975)	2,6	jellyfish
Kraeuter and Wolf (1974)	5,6	macroinvertebrates
Marcus and Marcus (1967)	6	opisthobranchs
Martof (1963)	6	herpetofauna
Menzies and Frankenberg (1966)	6	isopods
Mishima and Odum (1963)	1	*Littorina irrorata* (marsh snail)
Odum, W. (1968)	3,4	*Mugil cephalus* (mullet)
Odum, W. (1968a)	3	*Mugil cephalus* (mullet)
Pace, et al. (1979)	5	*Ilyanassa* (*Nassarius*) *obsoleta* (mud snail)
Pomeroy, et al. (1969)	3	7 macroconsumers
Pomeroy, et al. (1963)	5	zooplankton
Pomeroy, et al. (1966)	3	5 macroconsumers
Rickards (1968)	3	*Megalops atlanticus* (tarpon)

[a]Note: 1 = energetic rates, 2 = population estimates, 3 = foods and feeding, 4 = assimilation of dead *Spartina*, 5 = impact of macroconsumers on the marsh-estuarine ecosystem, and 6 = species lists and general accounts of macroconsumers.

assimilate dead cordgrass. *Spartina* is particularly refractory because of its chemical composition (Valiela et al., 1979). Some of the *Spartina* is degraded anaerobically by fermenters and sulfate reducers, and the resulting reduced sulfur compounds may be used by aerobic chemoautotrophs to fix CO_2, thus providing an additional producer base for the aquatic macrofaunal food web (see Section 4.1.3 and Chapter 7).

The concept of detritus as substrate for microbes that in turn are food for detritivores was reviewed by Darnell (1967). Experiments have been performed with the mud snail, the grass shrimp (Wetzel, 1975, 1976), the marsh snail (Wetzel, 1976), and the striped mullet, *Mugil cephalus* (Bunker, 1979; Odum, 1968, 1970). These detritivores assimilate the decomposers associated with plant detritus, egesting the dead plant material, which is then reinvaded by the decomposer organisms. Coprophagy by marine invertebrates under laboratory conditions is common (Frankenberg and Smith, 1967), but the contribution of feces to the diet of salt marsh macroconsumers is unknown. Fecal pellets are often enriched in organic carbon owing to sediment-sorting by macroconsumers (Kraeuter, 1976; Section 4.2.), so that after reinvasion by decomposer microbes, microfauna, and meiofauna, feces should be a rich and renewable food resource for marsh animals (Newell, 1965; Hargrave, 1976; Levinton and Lopez, 1977).

Algae, both benthic and planktonic, are an important food source for many salt marsh and estuarine animals. When available, algae may be a more readily utilized food than decomposer microbes. The diet of the mud snail, *Ilyanassa*, which feeds on the algal-rich mud flats, consists of 75% algae and 25% decomposer microbes (Wetzel, 1975, 1976). Fiddler crabs readily assimilate benthic diatoms (Shanholtzer, 1973; Montague, unpubl.). Grass shrimp and mullet graze phytoplankton and benthic algae (Welsh, 1975; Odum, 1968, 1968a). Oysters, even those in small tidal creeks surrounded by *Spartina* marsh, may subsist entirely on phytoplankton (Haines, 1976, 1976a; Haines and Montague, 1979). Algae may, in fact, be preferred foods of these aquatic macroconsumer rather than only dietary supplements ingested to compensate for variations in the supply of detritus. The "detritus-algae" feeders in the energy budget of Teal (1962) might be more accurately named "algae-detritus" feeders.

As crabbers and sports fishermen know, animal tissue, both live and dead, is also a common food for marsh and estuarine macroconsumers. Predator-scavenger food chains have been known to fisheries science for years (Cushing, 1975). The magnitude of this trophic pathway is far smaller than the producer-consumer flow of carbon, and its role as a control of the larger flow is still poorly known.

Based on the observations summarized above, grazers on benthic algae and phytoplankton and decomposers of *Spartina* detritus must be considered the major initial links in the salt marsh-estuarine aquatic food web. Decomposers on or in plant detritus initiate the transfer of carbon fixed by *Spartina* to forms utilizable by the fauna of the marsh and estuary. Fecal pellets serve to renew food resources by providing a concentrated, in some cases, pulverized, nutrient-enriched substrate for microbial decomposers. Thus, the plant-microbe-animal transformation is the key to understanding how salt-marsh primary production feeds the secondary production that, in turn, supports the coastal fisheries.

4.1.2 Feeding Categories

Most aquatic macroconsumers are omnivores. Fiddler crabs, for example, ingest algae, detritus, foraminiferans, nematodes, inorganic particles, and sometimes carrion (Teal, 1958, 1962; Shanholtzer, 1973). Many different organisms find food on the same decaying plant biomass. For example, detritus is a source of microfauna for the killifish (Jeffries, 1972), the grass shrimp, (Welsh, 1975), and many other salt-marsh macrofauna. Yet, because they feed in different areas, at different times, on different particle sizes, or use different searching and sorting techniques, these omnivores seldom compete directly for any particular food. Fiddler crab species, for example, live and feed in different zones of the marsh. There are virtually as many different feeding behaviors as there are species of macroconsumer in salt marshes.

Macrofaunal feeders may be arbitrarily grouped into five general categories: deposit feeders, suspension feeders, deposit-suspension feeders, predator-scavengers, and strict predators. This categorization is used purely for ease of discussion and, although each category has typical examples among the dominant animals, many species combine or modify these methods to create unique, competitive, and species-specific feeding strategies.

Deposit feeders, which constitute a major part of the macroconsumer biomass, assimilate organic matter contained in the salt-marsh soils and sediments. The foods available in these intertidal and subtidal zones include microalgae (mostly diatoms), decomposer microbes, microfauna, and meiofauna. Fiddler crabs, snails and the polychaete worms are typical examples of deposit feeders. Fiddler crabs sort sediments and thus increase the organic content of ingested material. The organically enriched rejecta and feces of the deposit feeders serve as food for other organisms.

Suspension feeders, typically bivalves, assimilate organic matter filtered from the water. Tidal water contains phytoplankton, suspended benthic diatoms, microfauna, microbes associated with particles, and free microbes. Suspension-feeding species found in the marshes of Sapelo Island include the ribbed mussel, *Geukensia demissa*, the marsh clam, *Polynesoda caroliniana*, the edible clam, *Mercenaria mercenaria*, and the oyster, *Crassostrea virginica*. Except for the ribbed mussel and the marsh clam, which occur in the marsh proper, these bivalves are found in tidal saline creeks, rivers, and sounds. Most bivalves sort inorganic from organic particles, egesting the former as pseudofeces which form sediment nearby. The animal also produces organically enriched feces by this sorting process.

Deposit-suspension feeders include the many salt-marsh and estuarine animals that obtain food by filtering organic matter from water and by sorting sediments. Mullet routinely sort sediments and are also known to graze dense patches of phytoplankton (Odum, 1968, 1968a, 1970). Grass shrimp consume deposits as well as suspended material (Johannes and Satomi, 1966; Welsh, 1975), as do commercial brown and white shrimp.

Some prominent estuarine carnivores, commonly thought to be scavengers, include the blue crab, the white catfish, and the shark. Most estuarine carnivores will consume carrion when it is available, but because of high demand, carrion is never available for long. Since a diet restricted to carrion probably cannot be maintained in salt marshes and estuaries, these predators scavenge only opportunely. The blue crab, for example, is known to capture other crabs, grass shrimp, minnows, and snails, to dig for infauna, and to consume bivalves, submerged aquatic plants, macroalgae, and organic detritus (Van Engel, 1958; Darnell, 1961; Virnstein, 1977). Sharks and rays subsist to a large degree on live crustacea and bottom fishes (Dahlberg and Heard, 1969; Dahlberg, 1975). The white catfish and juvenile Atlantic tarpon are also primarily predators but occasionally scavengers (Heard, 1975; Rickards, 1968).

Strict predators, carnivorous animals that capture only living prey, are less common than predator-scavengers in salt marshes and estuaries, but they do occur. Most crustacean larvae and some adult copepods are strict predators (Marshall and Orr, 1960). At the other extreme in size, the Atlantic bottlenosed dolphin, *Tursiops truncatus*, consumes principally live fish and squid (Ridgeway, 1972).

The more omnivorous salt-marsh and estuarine macroconsumers defy even broad classification in the trophic scheme. For example, while the killifish, *Fundulus heteroclitus*, is generally a predator, consuming snails, grass shrimp, and other crustaceans, it also filters detritus particles from the water, probably in search of associated fauna, and feeds readily on carrion (Jeffries, 1972; Vince et al., 1976; Valiela et al., 1977). Menhaden (*Brevoortia tyrannus*), which could be considered a suspension feeder since it strains particles from the water, is also a predator on zooplankton (Jeffries, 1975). Some copepods and pelagic larvae capture diatoms and ciliates as well as removing nonliving particles from the water; in this sense they are both predator-scavengers and filter feeders (Berk, et al., 1977; Poulet, 1976).

Omnivory seems to be the rule for salt-marsh and estuarine macroconsumers. Some animals may have diets specialized in food size and location; some, such as the blue crab larvae which require yellow dinoflagellates (Van Engel, 1958), may have developed very specific food requirements; but all these animals will supplement their special requirements with a wide range of other foods. Why are there so many omnivores? Scarcity of food, variability in food type and quantity from place to place or through time, and nutrient imbalances can all produce food generalists in an ecosystem (Emlen, 1973). The diffuse nature of the food sources may also play a role. The macroconsumer may take his nourishment in algae and detritus suspended in water or spread over the surface of the marsh or may find it in the discrete structural or compartmentalized vascular plants. Filter and suspension feeders have much to gain by assimilating all useable organic matter ingested. Organisms that find their food in large packets can discriminate easily and thus have at least the potential to specialize.

The copious secondary production in the algae-detritus food chain of the salt marsh may be food limited. The presence and extent of food limitation can be established by comparing the supply rates of food to the use rates and manipulating food-supply rates in the field. Limiting factors for macroconsumers are considered in more detail in Section 4.2.

4.1.3 The Food Web Analyzed with $^{13}C/^{12}C$ Ratios

Stable carbon isotope ratio analysis of food sources and marsh fauna has helped delineate the structure of salt-marsh food webs. During photosynthesis, plants discriminate against ^{13}C, the heavier stable isotope of carbon, to a varying degree, depending on the photosynthetic pathway and source of inorganic carbon. The depletion of ^{13}C

in organic matter that results from this discrimination is expressed as $\delta^{13}C$, the parts per thousand difference from the $^{13}C/^{12}C$ ratio of a carbonate standard.

Delta ^{13}C values are distinctly different for several groups of primary producers (Haines, 1976a). Marsh grasses, which have the C_4 pathway of photosynthesis, incorporate atmospheric CO_2 with less discrimination against the heavier carbon isotope than C_3 plants. The latter lack the undiscriminating enzyme, PEP carboxylase, for retrieving atmospheric CO_2. *S. alterniflora, Spartina cynosuroides, Distichlis spicata*, and *Sporobolus virginicus* are C_4 marsh grasses with $\delta^{13}C$ values of -12 to -14 ‰. *Juncus roemerianus, Salicornia virginica, Batis maritima*, and *Borrichia frutescens* are C_3 marsh plants with $\delta^{13}C$ values of -22 to -25 ‰. Benthic algae, phytoplankton, and organic carbon seston in the estuary have $\delta^{13}C$ values of -16 to -26 ‰, with the $\delta^{13}C$ of benthic algae averaging 17 ‰, and that of phytoplankton and seston averaging 21 ‰ (Haines and Montague, 1979).

Once fixed during photosynthesis, the carbon isotopes show little further fractionation during death and decay of plant material or during assimilation of organic matter by animals (DeNiro and Epstein, 1978; Haines, 1977; Haines and Montague, 1979). However, no data exist on the correspondence between the $\delta^{13}C$ value of detritus and that of the decomposers which degrade and assimilate the detritus. A definitive experiment would have to separate the microbial component from the decomposer-detritus complex. With this cautionary note in mind, if location and feeding habits of consumers are carefully considered, together with available carbon sources, the distinctive $\delta^{13}C$ values of particular autotrophs can be traced through the estuarine food web. But note that the success of this method also depends on the presence of no more than two distinct sources of autotroph carbon, carbon with significantly different $\delta^{13}C$ values. If more than two sources are possible, combining of categories to reduce the number to two may be possible. Two potential problems of interpretation of the isotopic ratios present themselves. The first involves detritus from the C_3 marsh plant, *J. roemerianus*; the second involves the possibility of a contribution of carbon from microbial chemoautotrophs utilizing carbon dioxide from the water. Since *Juncus* is a minor contributor of organic carbon in the marshes of Sapelo Island, the first problem can be minimized by taking samples from creeks far from stands of that species. The second problem, raised by Peterson et al. (1980) will be considered later in this section.

Small animals or small quantities of animal tissue (as little as 1 or 2 mg DW organic matter) can be analyzed accurately, since the analytical error of the carbon isotopic analysis method is only 0.2 to 0.4 ‰ ($\delta^{13}C$ units). The isotopic composition of an animal reflects the $\delta^{13}C$ value of the primary producer or producers at the origin of the food chain, but this technique does not distinguish between direct assimilation of plant material or assimilation of primary microbial consumers of that plant material. Delta ^{13}C values of animals without food in their guts actually represent carbon assimilated over a long period, a period roughly equal to the turnover time of body carbon in the animal, but the value is weighted for the recent past, since more of the earlier assimilated carbon already will have been respired. Gut-content analyses represent carbon or foods ingested, but not assimilated, over a short period, a period roughly equal to the turnover time of gut contents.

Grazers, which consume live *Spartina* exclusively, have $\delta^{13}C$ values close to that of their food plant (Chapter 5). The aquatic macrofauna of the marsh shows somewhat more variation. Many deposit feeders, or detritivores, mirror the $\delta^{13}C$ values of the living vascular plants nearby (Figure 4.3), presumably a function of the higher proportion of *Juncus* detritus in a *Juncus* stand and of *Spartina* detritus in a *Spartina* stand. But the identity is not so close or uniform as in the case of the more specialized grazers; $\delta^{13}C$ values of fiddler crabs, for example, are skewed in the direction of benthic algae, on which they are known to feed.

Fiddler crabs collected in pocket stands of C_3 plants within the *Spartina* marsh have higher $\delta^{13}C$ values than can be explained just by assimilation of benthic algae along with the C_3 plant detritus; these crabs probably get significant *Spartina* carbon by tidal transport or by out-of-stand foraging. *Uca minax* collected in freshwater tidal marshes dominated by southern wild rice (*Zizaniopsis*), a C_3 marsh macrophyte, and away from the influence of other C_3 plants and of *Spartina*, have $\delta^{13}C$ values close to that of *Zizaniopsis* (Haines and Montague, 1979).

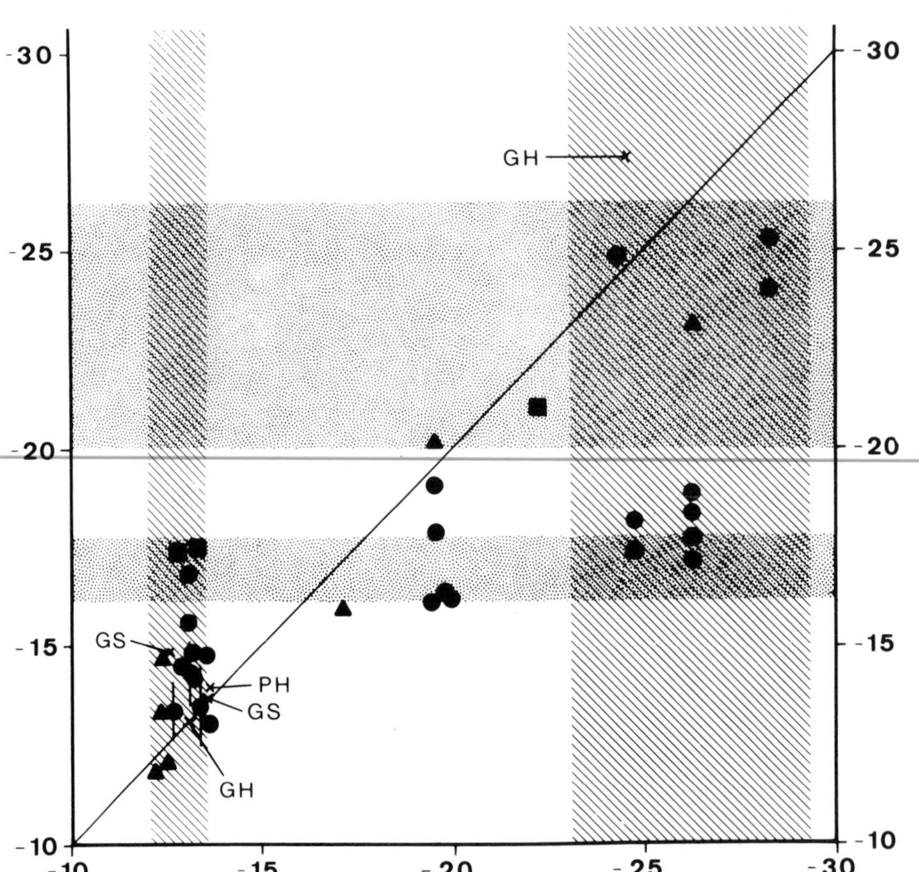

Figure 4.3. Ratios of $^{13}C/^{12}C$ in estuarine fauna in relation to potential food sources in the Sapelo Island marshes. Diagonal line shows equivalent plant/animal $\delta^{13}C$ values. Crabs (●); gastropods (▲); pelecypods (■); grasshoppers (GH); planthoppers (PH); grass shrimp (GS). Adapted from Haines and Montague (1979) with permission of Duke University Press. Copyright (1979) by the Ecological Society of America.

In contrast to other marsh animals, filter-feeding mussels, clams, and oysters have $\delta^{13}C$ values in the lower range characteristic of algal and chemoautotrophic microorganisms. Different species of bivalves occupying different marsh zones have isotopic compositions which suggest intermediate to no influence of *Spartina* carbon. Ribbed mussels, which occur in the salt marsh itself, have the highest values, averaging −17 ‰. Edible clams that are embedded in tidal creek bottoms have $\delta^{13}C$ values of −18 to −19 ‰. The oyster, which forms banks along the creek sides, has isotopic compositions of −21 to −24 ‰, suggesting marked dependence on carbon from microorganisms other than those associated with *Spartina* detritus.

A further result of the isotopic carbon study emerged from comparison of the $\delta^{13}C$ values and size. In seven out of nine species segregated into size classes, the $\delta^{13}C$ value of individuals decreased as the size of those individuals decreased. Smaller members of a species seem more dependent on phytoplankton or chemoautotrophic bacteria, while the larger individuals assimilate organisms degrading the abundant *Spartina* detritus in the upper reaches of the tidal creeks.

The export of energy-containing reduced sulfates from marsh soils to water, to the extent that it does occur, makes the interpretation of low $\delta^{13}C$ values more difficult; the ingestion of more phytoplankton than *Spartina* is then no longer the only possible explanation for low $\delta^{13}C$ values. Peterson et al. (1980), in proposing an explanation for these low faunal values, cite the utilization of reduced sulfur compounds by chemoautotrophs. Sulfate reduction appears far less important as a pathway of energy transfer between marsh soil and water in the Sapelo Island marshes than in the Great Sippewissett marsh as described by Howarth and Teal (1979). Nevertheless, sulfate reduction does occur in the Sapelo Island marshes, particularly in the levee soils, where considerable water exchange takes place (Nestler, 1977, 1977a). Peterson et al. (1980) report $\delta^{13}C$ values of −19 ‰ to −38 ‰ for bacteria using chemical energy and fixing carbon from seawater. They calculate that seston containing a mixture of 30% of this bacterial carbon ($\delta^{13}C$ values of −35 ‰) and 70% *Spartina* detritus ($\delta^{13}C$ value of −12 ‰) would have a mean value of −19 ‰, well within the range for salt marsh seston and soils. The extent to which this confounding factor operates in the marshes of Sapelo Island remains a pressing question. However, existing data on the amount of dark fixation of carbon in the creeks and the lack of appreciable movement of interstitial water from high marsh argue against copious chemoautotrophic carbon fixation in the Sapelo Island marshes.

4.1.4 Intraspecific Variation of $\delta^{13}C$ Values

Individual variation in isotopic composition within a consumer species reflects the variation in foods assimilated by members of that species during the carbon turnover time. Plants themselves, because of the well-mixed CO_2 in air, show relatively little variation in isotopic carbon composition from individual to individual. Minimum variability should result from extreme specialization and uniformity in feeding—for example, in restriction in diet to a single species of plant. The specialist grazers mentioned earlier not only reflect the isotopic composition of their food plant, but vary little from individual to individual. However, most salt-marsh consumers are generalist feeders, and individual variation among members of the same species reflects the circumstances of time and place of feeding. If the generalist individuals assimilate average amounts of all or most of their possible foods during the carbon turnover time, then all individuals within a species will tend to reflect the average isotopic composition of all foods available to that species. If, on the other hand, some individuals specialize over the short run, then variability in the isotopic carbon composition of the consumers can be substantial.

Distinguishing between these two alternatives depends on the relative mobility of the generalist individuals during the carbon turnover time and on the relative availability of various foods. If animals travel rapidly and widely or if foods are well mixed, the generalist consumer will reflect a mean value with lower variance. Thus, the number of factors that must be considered when interpreting differences in carbon isotopic variability include motility, turnover time of body carbon, temporal and spatial variability of food within a given habitat or feeding area, and genetic variance with respect to food preference. If several of these factors can be ruled out, then the variance of isotopic composition may be compared and conclusions may be drawn about species differences, taking into consideration only the remaining factors.

Within each of three species of crab in the marsh–the fiddler crab, *Uca pugnax*, a deposit feeder, the squareback crab *Sesarma reticulatum*, which is believed to eat live *Spartina* and deposits, and the marsh crab, *Eurytium limosum*, a predator–isotopic carbon values varied with standard deviations of only 1 ‰ (Haines and Montague, 1979). For $\delta^{13}C$ values of clams (*Mercenaria mercenaria*) collected from six different sites, there were standard deviations of 0.6 ‰ (Haines and Humphries, unpubl.). Mullet (*Mulgil cephalus*) and killifish (*Fundulus heteroclitus*), on the other hand, are much more variable in their isotopic compositions with standard deviations of 3 ‰.

Fiddler crabs, mud crabs, and clams should not vary significantly in individual food preferences. These are comparatively slow moving or sedentary animals existing in an environment in which food concentrations on the macro-scale are relatively constant, either by virtue of uniform distribution of plants (*Spartina*, benthic algae) or because of turbulent mixing in the creeks, though not on the marsh (see Chapter 2). In contrast, mullet and killifish, although quite mobile, live in an environment with patchy food distributions, and they show short-term preferences for habitats and feeding behaviors that lead to increased individual variability in isotopic composition.

In summary, stable carbon isotopes can help ecologists gain a better understanding of the food-web complexity and carbon transformation processes of the salt marsh. However, since the isotopic composition of macroconsumers is influenced by a large number of factors, many of which tend to produce similar results, the interpretation and explanation of observed ratios must be done carefully and must take into consideration other essential ecological information. Currently, the greatest needs are for measurements of different species in different places and for experimental validation of the use of whole-animal ratios.

4.2. Impact of Macroconsumers on Salt-Marsh Metabolism

Macroscopic marsh animals degrade much less energy than do microbes (Teal, 1962; Pomeroy et al., 1977), but the behavior and byproducts of macroconsumers may influence ecosystem energy flow far more than their energetics would imply. For example, the ribbed mussel is more important as a builder of phosphorus-rich surface marsh sediments than as a consumer of energy (Kuenzler, 1961a). This filter feeder daily removes one-third of the phosphorus in the overlying water and deposits most of it as pseudofeces. Unfortunately, many of the earlier studies of salt-marsh macrofauna (Odum and Smalley, 1959; Smalley, 1960; Teal, 1962) did not address the role of macroconsumers as controllers of energy flow. Other reviews treat aspects of animal controls on marine environments (Carriker, 1967; Kraeuter and Wolf, 1974; Shanholtzer, 1974; Hargrave, 1976).

Macroconsumers can prey on, and thus control, standing stocks of certain resources. Standing stocks of benthic algae, as chlorophyll *a* m^{-2}, and their production, as carbon fixed m^{-2} hr^{-1}, increased when mud snails were removed from experimental plots on an intertidal mud flat (Pace, 1977; Pace et al., 1979). Conversely, feeding on overcrowded prey populations can increase the production rate of the remaining organisms by temporarily reducing intraspecific competition for available space and nutrients (Hargrave, 1976; Wiegert, 1975). Although grazing by mud snails does not increase the specific rate of productivity by the benthic algae, grazing by the salt-marsh amphipod, *Orchestia grillus*, did stimulate the metabolic activity of microbial decomposers, the principal food of *O. grillus* (Lopez et al., 1977). Similar responses were elicited by Brenner et al. (1976) from another amphipod, *Talorchestia longicornis*, grazing on blue-green algae. However, in Brenner's study the increase in specific rate was realized at the expense of a lowered productivity per unit area.

Decomposer activity can be stimulated by grinding ingested particles. Crabs and shrimp grind with a gastric mill (Kaestner, 1970), mullet use a gizzard-like organ (Thomson, 1966; Odum, 1968) and snails use their radula. Grinding not only aids digestion but also accelerates the decomposition of egested material by increasing the surface area available for microbial attack.

Many salt marsh macroconsumers concentrate organic matter in their feces or pseudofeces. This provides decomposers with environments rich in organic carbon. Oysters, clams, and ribbed mussels remove organic matter from the water column and concentrate it on the sediment surface (Bahr, 1974, 1976; Haven and Morales-Alamo, 1966; Kraeuter, 1976). Grass shrimp and mullet ingest organic matter from the water and the sediments and deposit feces which settle in the creeks (Johannes and Satomi, 1966; Welsh, 1975; Odum, 1970). Fiddler crabs and marsh snails deposit feces on the marsh and mud flats (Valiela et al., 1974; Kraeuter, 1976; Lopez and Levinton, 1978; Levinton and Lopez, 1977; Wetzel, 1976). Enhancing the carbon available to decomposers in this manner not only makes an important contribution to the energy flow of the salt marsh (Welsh, 1975; Darnell, 1967; Odum and de la Cruz, 1967), but it also provides a renewable resource for the marsh macroconsumers themselves (Levinton and Lopez, 1977), since they are generally consumers of microbes attached to organic particles (Hargrave, 1976).

In addition to accelerating decomposition, the conditioning and packaging of detritus in macroconsumer feces facilitates the regeneration of nutrients by microbes (Johannes and Satomi, 1966; Welsh, 1975; Hargrave, 1976). Sediment reworking by macroconsumers should increase the turnover of algae and nutrients by bringing limiting nutrients into contact with benthic microalgae. Ribbed mussels and oysters remove nutrient elements and organic matter from water and deposit them on sediments as feces and pseudofeces (Kuenzler, 1961a).

Marsh macroconsumers influence significantly the physical nature of marsh sediments, especially at the soil-water interface. Sediment reworking, or bioturbation, provides habitat for other organisms, alters water turbidity, and affects primary production and detrital decomposition. Oyster reefs support a large community of organisms, are responsible for considerable sedimentation by feces and pseudofeces, and create dams and pools in tidal creeks (Bahr, 1974, 1976; Haven and Morales-Alamo, 1966). Marsh crabs burrow frequently in some areas, perhaps once per day per crab, and cause turnover and local aeration of intertidal sediments to depths of 10 to 100 cm (Wiedemann, 1972; Basan and Frey, 1977; Edwards and Frey, 1977). Burrows, therefore, may stimulate the growth of *Spartina*, and crab burrows provide a habitat for a variety of symbiotic organisms. More importantly, fiddler crabs turn over and aerate the top half-centimeter of marsh when feeding at every low tide (Kraeuter, 1976; Edwards and Frey, 1977). Such bioturbation may stimulate primary production by bringing diatoms into closer contact with light and nutrients. Detrital decomposition and microbial nutrient recycling may be stimulated in the aerobic zone where fiddler crabs have grazed frequently. Mullet rework great quantities of surface sediment while grazing soft creek sediments. A single mullet 20 cm in length, can rework 45 m^2 of sediment, ½ cm deep, per year (Odum, 1970). This disruptive activity aerates sediments, which can accelerate decomposition, and clouds the water, which can retard phytoplankton production. Water turbidity is also altered by filter feeders such as oysters, clams, and mussels, which, when submerged, are constantly removing suspended material.

Marsh snails and crabs can physically affect processes in the intertidal marsh. Marsh crabs transport belowground root material to the marsh surface by burrowing, thus providing additional organic substrate to decomposers (de la Cruz and Hackney, 1977).

The rate of recycling of essential elements by excretion, and incidental to defecation, is inversely proportional to body size (Johannes, 1964). Therefore, the smaller organisms, despite low biomass, accomplish a large fraction of the nutrient regeneration (Pomeroy and Bush, 1959). Protozoa may recycle as much phosphate and ammonia as do all larger organisms (Buechler and Dillon, 1974). Hence salt-marsh macroconsumers influence nutrient cycles not only by direct regeneration but also by mastication, enrichment, packaging of detritus, and reworking of sediments. The relative importance of these varied activities of macroconsumers may be evaluated by manipulating the density of organisms or their byproducts—feces, burrows, physical disruption—and measuring the response of important energetic variables, such as primary production and decomposition. This type of experiment can be performed by modeling, as we did in this salt-marsh study (Chapter 9), and it can also be extended to field manipulations. Examples of field experimentation include the works of Pace (1977) and Stiven and Kuenzler (1979). A complete assessment of the roles of macroconsumers in ecosystems such as the salt marsh should include not only trophic dynamics and energy budgets, but also the impact of macroconsumer behavior and byproducts on ecosystem processes (Welsh, 1975).

4.3. Regulation of Salt-Marsh Macroconsumers

Physical and biological conditions clearly exclude certain species from the salt marshes. But what limits the density and biomass of those populations which are physiologically adapted to this habitat? Tidal inundation, temperature and salinity extremes, and time of day all interact with other environmental conditions to produce physical constraints that limit both the time and space for feeding and the habitat in the intertidal zone. Fiddler crabs, for example, are generally active at low tide, but return to their burrows during high tide. Their activity is further influenced by temperature, time of low tide, and gender, as males will often feed during a portion of high tide (Teal, 1958; Barnwell, 1966, Valiela, et al. 1974). Mud snails (*Ilyanassa obsoleta*) are active during high tide when submerged, but become less active, thus reducing moisture loss, as the mud flat begins to dry during low tide (Schaeffer et al., 1968). Such periods of forced inactivity limit the time available for feeding. Although energy budgets have been developed for some salt-marsh macroconsumers, detailed population models that would allow an evaluation of the constraints imposed by the interactions of various physical conditions do not yet exist.

Predation, resource supply, and competition for resources, both material and spatial, interact with the physical constraints to limit density of salt-marsh macroconsumer populations. Space can limit both the size of populations and the number of species in salt marshes. Individuals must have enough space to insure 1) adequate food or nutrients, 2) tolerable physical conditions or habitat, 3) suitable mating territory, 4) refuge from predators, 5) combinations of these and other requirements. Many species of salt-marsh macrofauna are spatially specialized. Distribution of the clam, *Mercenaria mercenaria*, is determined largely by settling and survival of the larvae (Wells, 1957). Three species of fiddler crab live in the marsh, but there is little overlap in their distribution (Teal, 1959; Miller, 1965). *Uca pugilator* is found primarily on sandy substrates, while *Uca pugnax* occurs on the muddy substrates beneath *Spartina alterniflora*. *Uca minax* prefers lower salinities and thus exists in marshes with some freshwater input. *Littorina irrorata* is found on the marsh surface during winter and early spring, but it is on the stalks and stems of *Spartina* in summer [Figure 4.4] (Smalley, 1959a). Thus, there is little overlap on the marsh surface between these organisms and *Uca pugnax*, which is most active in summer. A similar sharp demarcation occurs between the mud snail, *Ilyanassa obsoleta*, which occupies the lower muddy zone of the creek bank and *Uca pugilator* which occupies sandy areas high on the creek bank. Other deposit feeders, principally *Palaemonetes pugio* and *Mugil cephalus*, live in the water and enter the marsh only to feed at high tide. Spatial specialization may be the result of competitive exclusion. For example, the three fiddler crab species appear capable of survival in each other's habitat (Teal, 1958). Spatial distributions of species reflect adaption to the intertidal zone, where gradients of temperature, moisture, and tidal inundation interact with competition for space to promote specialization in narrow zones (Newell, 1976).

Within the zone of distribution of a population, control of population density may result from interference competition for space. Manipulation of densities of the periwinkle, *Littorina*, and the ribbed mussel, *Geukensia*, significantly affected growth and mortality (Stiven and Kuenzler, 1979). The value of space for these mollusks may or may not have been the amount of food available per individual; when detritus was added, growth and survivorship increased in some experiments but decreased in others. Van Dolah (1978) found intraspecific competition between adult and juvenile amphipods, *Gammarus palustris*, in Maryland salt marshes. These crustaceans compete for *Spartina* culms, which are used as refuges from the predatory killifish, *Fundulus*

Figure 4.4. *Littorina irrorata*, the marsh periwinkle, spends the summer on stems of *Spartina*, above the level of spring high tides.

heteroclitus. Adults displace juveniles from preferred sites, a behavior which increases the exposure of juveniles to predation but protects adults. This interaction between intraspecific competition and predation controls the amphipod population.

Food supply can be one reason why a particular space is valuable to an organism, but in this sense the population is not truly space-limited; the competition takes the form of exploitation of a shared resource rather than intraspecific interference. Salt-marsh primary production is large, and it is tempting to assume animals in the salt marsh are not food limited. However, there is little experimental evidence at present to support or refute such an assumption. Furthermore, available food is difficult to define. Although carbon, as a resource per individual consumer, may be relatively high in the salt marsh compared to other ecosystems, much of this carbon is refractory or is buried in anaerobic soils. Indeed, the assimilable nutrition per individual may be near the threshold of reproductive survival. Before it is utilizable, *Spartina* detrital carbon must be transformed by a decomposer, with an accompanying loss of energy. The age of detritus, the presence of ciliates and meiofauna, and the nitrogen content of detritus influence its value as food for macroconsumers (Tenore, 1977, 1977a; Tenore et al., 1977; Briggs et al., 1979). Widespread omnivory, together with specialization of feeding strategies between species, is evidence for food limitation in salt marshes and the surrounding estuaries. Low abundance of food favors animals with feeding alternatives and produces competition for food that ultimately results in the development of unique feeding behaviors (Emlen, 1973; Levinton, 1972). Different salt-marsh species consume many of the same foods, but do so in different areas, at different times, and using different foraging methods.

Predation is a significant regulator of the macroconsumers in the water column, such as the grass shrimp, *Palaemonetes pugio* (Welsh, 1975), and the amphipod, *Gammarus palustris* (Van Dolah, 1978). Although energy budgets have been developed for several invertebrates (Odum and Smalley, 1959; Kuenzler, 1959; Shanholtzer, 1973), little is known about how much of the production goes to predators. In addition, since these budgets only consider post larval populations, the role of predation in limiting larval recruitment is unknown. Time spent avoiding predators reduces feeding time and consumes energy. For example, the movement of periwinkles (*Littorina*) above the water line during high tide may be related to avoidance of blue crab and other aquatic predators (Hamilton, 1976). Predation also influences the size structure of some macroconsumer populations. Killifish consume the larger individuals of the amphipod, *Orchestia grillus*, and the smaller individuals of the snail, *Melampus bidentatus* (Vince et al., 1976). In low marsh zones where this predation is most efficient, the favored prey were virtually eliminated from enclosures in a Massachusetts high marsh. The dense stands of *Spartina patens* in Massachusetts high marsh reduce predation by killifish and provide refuges for the prey. Thus predation can influence the behavior and distribution of macroconsumer populations.

Several mechanisms may act successively within or between generations, or simultaneously, to control the animal numbers. When the available resource of a population increases, or competition for the resource decreases, or predation decreases, equilibrium densities of the population increase in the models of Wiegert (1979b). Sometimes predation will limit biomass but increase production (see 4.3). Space is valuable for individuals and for species providing substrate for attachment (Connell, 1961), refuge from environment and enemies feeding and mating territory (Crane, 1975), and all other manner of biotic conditions.

The salt marsh is an excellent ecosystem in which to evaluate control mechanisms and test theories of consumer regulation, because diversity is low, species often occupy distinct zones, and macroconsumer production is very high. Experiments can be performed by manipulating densities of suspected resources, competitors, or predators and measuring the impact on population growth (Stiven and Kuenzler, 1979). As more of these experimental studies are completed, our present indistinct view of the regulation of both macroconsumer populations and of ecosystem processes in the salt marsh will come into sharper focus.

5. Grazers on *Spartina* and Their Predators

W.J. PFEIFFER and R.G. WIEGERT

Subsequent to the studies of Smalley (1959, 1959a, 1960), Kale (1964, 1965), and Marples (1966), a hiatus occurred in research published on the grazing food web associated with the *Spartina* plant of the Sapelo Island salt marshes. This relative dearth of interest in that portion of the salt-marsh ecosystem resulted perhaps from Smalley's conclusion that only a relatively small percentage of the energy flow was transferred from the primary producer to the herbivore community. Hence, more recent research centered around those functional groups, in particular microbial populations, that initiate the larger part of the flow of energy and elements in *Spartina* salt marshes.

The lack of interest in the herbivorous organisms feeding on *Spartina* may also have resulted from their inconspicuous nature, the various barriers to the accurate estimation of their abundances and production, and the difficulties posed in assessing the impact of these herbivores on the production and fitness of the host plant. Smalley concentrated his efforts on the grasshopper, *Orchelimum fidicinium*, which, while presenting problems with regard to accurate density estimates, is one of the most conspicuous grazing insects and has a simple life history with one generation per year. For the minute, extremely abundant, and multivoltine planthopper, *Prokelisia marginata*, only a tentative energy budget was possible, given the sampling techniques available in the late 1950s.

From what we know of the grazing insects, the vast majority of the energy flowing from *Spartina* to the grazing community is apportioned to *Prokelisia* and *Orchelimum*. If Smalley's tentative energy budget parameters for *Prokelisia* are reasonably accurate, comparison of the salt marsh with other terrestrial grasslands (Table 5.1) leads to the surprising discovery that *Spartina* grazers greatly surpass the secondary production attributed to grazers in other grasslands. Much of this high secondary production associated with the *Spartina* herbivore community is the result of high primary production of *Spartina* and the domination of the grazing community by poikilothermic organisms, which possess much lower maintenance costs than homeothermic organisms (Humphreys, 1979).

In discussing the grazing food web in the salt marsh we are handicapped by the limited information published, particularly with regard to experimental studies. We have sought here to flesh out the structure of this food web and to suggest insights into some of the functions these organisms may have in the system.

5.1. Vertebrate Herbivory

Bison and other ungulates may have browsed Georgia salt marshes during the Pleistocene (Edwards and Frey, 1977). Since the immigration of Europeans into North America, cattle and other domestic ungulates have been grazed intermittently in the salt marshes of eastern North America (Teal and Teal, 1969; Reimold et al., 1975a). But vertebrate herbivory in Georgia cordgrass marshes, if it was ever important, is scant at present. White-tailed deer (*Odocoileus virginianus*) are occasionally sighted along the upper fringes of Georgia marshes, but little grazing on the shoots is evident. Evidence of grazing by West Indian manatees, *Trichechus manatus*, on new cordgrass communities growing on a dredged-materials spoil area near the mouth of the Altamaha River has been noted recently (M. Hardisky, pers. comm.). Because such observations have not been made on older, established marshes, the manatees may be restricted by physical barriers such as oyster reefs and natural levees.

The greater snow goose, *Anser caerulescens atlantica*, appears to be the only native vertebrate herbivore with an appreciable impact upon Atlantic *S. alterniflora* marshes. During its winter residence from New Jersey southward to North Carolina this species has a severe but highly localized effect on marshland swards. Flocks of geese may denude large tracts of *Spartina* marsh by digging up rhizomes (Howard, 1940; Meanley, 1975; Smith and Odum, 1981) that are presumably laden with stored nutrients.

Seaside sparrows (*Ammospiza maritima*) and sharp-tailed sparrows (*A. caudacuta*) eat cordgrass seeds, but they probably also consume more animal matter than is characteristic of fringillids (Bent, 1968). Red-winged blackbirds (*Agelaius phoenicius*), primarily insectivorous around salt marshes, may also supplement their diets with cordgrass seeds during the autumn and winter months. Cordgrass seeds were rarely found by Sharp (1967) in the stomachs of rice rats (*Oryzomys palustris*) in Duplin River marshes. This limited amount of seed predation is unlikely to have any direct impact on *Spartina* populations because of the large seed crop produced and the prevalence of reproduction by rhizomatous sprouting.

5.2. Primary Production and Herbivory in Grasslands

The energy budget of *Prokelisia*, the dominant herbivore of living *Spartina* in the Georgia marshes, is known only approximately (Smalley, 1959; Wiegert and Evans, 1967). For the moment we assume that the partial energy budget constructed by Smalley (1959) is accurate, although this budget has considerable shortcomings (Section 5.4.3). Table 5.1 shows the aboveground net primary production, herbivore consumption and herbivore secondary production estimates for a variety of grassland ecosystems. Despite a high absolute value, when shown as a percentage of the very large annual net primary production, primary consumption in the *Spartina* salt marsh is lower than in many of the terrestrial ecosystems. In this respect the *Spartina* marsh is similar to the bulrush (*Scirpus*) marshes and sugarcane (*Saccharum officinarum*) plantations, some of the most productive ecosystems in the world (Westlake, 1963). Table 5.1 shows a relationship between the annual net primary production and total herbivore consumption. A regression of the latter on primary production gave a significant correlation: $r = 0.633$, d.f. = 13, $0.01 < p < 0.02$ for systems lacking grazing by domestic ungulates.

Secondary production in the Georgia salt marshes is also extremely high, being almost twice as large as that of any of the other ecosystems in Table 5.1 and 80 times greater than that of the Michigan old field. In the case of those areas, such as the Serengeti plain, with a large ingestion by grazing mammals, the explanation of the relatively low secondary production lies in the differing production/respiration ratios of homeotherms and poikilotherms. In the former this ratio is much lower. The pooled value of the P/R ratio for mammal species is 0.03, whereas the value for all non-social herbivorous insects is 0.39 (Humphreys, 1979). Thus, the high herbivore consumption but low herbivore secondary production of the ungulate-dominated Serengeti grasslands are not surprising. Indeed, in the Serengeti, grasshopper consumption, expressed as percent of the total herbivore consumption, accounted for only 28% in the long grass area, 11% in the short grass area, and 56% in the Kopjes area. Grasshoppers accounted for 82%, 59%, and 94% of the secondary production of all herbivores in the respective areas.

5.3. Effects of Herbivory

Herbivores, by removing part of the machinery of production as well as the product, have the potential for a direct effect on the plant that is not shared by detritivores. Clearly, the removal of large quantities of the plant photosynthetic apparatus, particularly early in the growing season, could have a significant effect on net production. The direction of this effect will, of course, depend on the extent to which the plant has an excess of net production and on the fate of the energy removed. Similarly, seed predators, by removing potential new individuals, can have a direct impact on the primary production, although the impact is delayed at least one generation.

When the amount of energy removed is relatively small compared to the total annual net primary production, as it is in the case in the salt marsh, the direct effects of herbivory, whether harmful or beneficial, become much more subtle. There are many potential effects of herbivores that are only indirectly, if at all, related to the energy and material losses to the plant (Varley, 1967). Because these more subtle effects are difficult to measure, most studies of the flow of plant-herbivore energy and matter have either ignored them or have treated them in a very cursory manner. Some of the potential general effects relevant to the *Spartina*-herbivore interaction are listed in Section 5.4 and specific examples are discussed.

Herbivorous insects, particularly the sap-sucking species, can transmit fungal, bacterial and viral pathogens to plants (Carter, 1973). Because *S. alterniflora* is not regarded as a commercially important species, the lack of evidence for grazing insects as vectors of disease in this species could be a result of no one looking for such inter-

Table 5.1. Comparison of the Aboveground Net Primary Production, Herbivore Consumption, and Herbivore Production in Various Grassland Ecosystems[a]

	Aerial Net Primary Production	Aerial Herbivore Consumption	Aerial Herbivore Secondary Production
	kcal m^{-2} yr^{-1}		
S. alterniflora salt marsh (Georgia)[b]	6260	574	81
Old field (South Carolina)[b]	1240	165	4.7
Old field (Tennessee)[c]	1328	166	29
Old field (Michigan)[b]	1380	20	0.9
Shortgrass prairie (Colorado and Texas)[d]	571	17	1.9
Desert grassland (New Mexico)[d]	794	32	3.3
Mixed grass prairie (South Dakota)[d]	1219	89	6.6
Mixed grass prairie (Saskatchewan)[e]	2132	102	7.3
Tallgrass prairie (Oklahoma)[d]	1292	99	6.4
Stellario-Deschampsietum wet meadow (Poland)[f]	904	172	30
Caricetum elatae tall sedge meadow (Poland)[f]	2692	260	42
Serengeti grassland (Tanzania)[g]			
Long grass area	2473	681	31
Short grass area	1946	743	18
Kopjes area	2473	358	29
Shortgrass prairie (Colorado)[h]			
Ungrazed by cattle	562	11	1.6
Lightly grazed by cattle	958	54	4.5
Heavily grazed by cattle	728	113	7.9
Moorland (Great Britain) grazed by sheep[i]			
Limestone grassland			
Including sheep	1067	285	13
Excluding sheep		1.3	0.3

Alluvial grassland			
Including sheep	915	133	6
Excluding sheep		0.8	0.2
Juncus squarrosus area			
Including sheep	1720	70	3.9
Excluding sheep		4.3	1.0
Montane grassland (North Wales) grazed by sheep[j]	5053	957	48

[a]The section numbers below correspond to the superscripts following the location designations given in Table 5.1. Aerial net primary production estimates obtained from the various sources are corrected for grazing losses (herbivore consumption) where necessary. Consumption is defined as the sum of ingestion and wastage. Wastage is interpreted as plant material detached but not ingested by herbivores; for sap-sucking insects wastage is assumed to equal zero. Where appropriate, production/consumption ratios established by Scott et al. (1979), representing means calculated from data for four grasslands, are employed: tissue-feeding mammals = 0.0112; tissue-feeding arthropods = 0.1336; and sap-feeding arthropods = 0.2339. Except for sap-feeding arthropods, previous ratios correspond to production/(consumption + wastage) values of Scott et al. (1979), Tables 5.2, 5.3 and 5.4, respectively.

[b]From Wiegert and Evans (1967). Herbivore consumption estimates derived from ingestion estimates by assuming consumption/ingestion ratios of tissue-feeding insects and mammals = 1.50; sap-sucking insects, other herbivorous insects and granivorous birds = 1.00. Primary production estimate derived from Gallagher et al. (1980) prorated value (see Table 3.2) and Smalley's (1959a) caloric equivalent for live grass.

[c]From Van Hook (1971). Herbivore consumption derived from ingestion estimate by assuming a consumption/ingestion ratio of 1.50.

[d]From Scott et al. (1979). Consumption and production by dead-plant litter-feeding arthropods not included in present estimates.

[e]From Coupland and Van Dyne (1979). All annual energy flow to aboveground primary consumers attributed to herbivores. Secondary production calculated from Table 8.1 and tissue production/mean standing crop ratios given by Scott et al. (1979).

[f]Primary production of *Carex* tall sedge meadow based on data from on Traczyk (1968) and caloric conversion factor of 4.229 kcal g^{-1} dry weight (Golley, 1961). Primary production of *Stellario-Deschampsietum* wet meadow based on mean of a two-year study by Andrezejewska and Wojcik (1970) and the caloric conversion factor applied above to *Carex. Stellario-Deschampsietum* and *Carex* consumption by grasshoppers obtained from Andrezejewska and Wojcik, (1970), using same caloric conversion factor. Consumption by Homoptera in both meadows based upon the mean value of a three-year study by Andrezejewska (1967). Herbivore production estimates for these two meadows derived from the production/consumption ratios based upon Scott et al. (1979).

[g]From Sinclair (1975). Caloric conversion factor for live grass, employed by Sinclair (4.137 kcal g dry weight^{-1}), derived from Golley (1961). Secondary production estimates arrived at on basis of production/consumption ratios derived from Scott et al. (1979).

[h]From Andrews et al. (1974). Consumption estimates given this paper were assumed to include wastage factors.

[i]From Coulson and Whittaker (1978). Herbivore consumption estimate based on production/consumption ratio of 0.044 for sheep (Perkins, 1978) and similar ratio for Hemiptera of 0.2339 based on Scott et al. (1979).

[j]From Perkins (1978). Sheep consumption of standing dead grass (896 kJ) was not included in herbivore consumption estimate. Hence, sheep production estimate presented by Perkins is proportionately reduced and contribution of standing dead grass to sheep production not included.

actions. The salivary secretions of sap-sucking insects can damage plant tissue adjacent to feeding punctures and, as a result of translocation, at a distance from punctures (Miles, 1968; Dixon, 1971; Carter, 1973). Some investigators have attributed premature senescence of foliage, or hopperburn, to mechanical plugging of xylem and phloem by the abandoned stylet sheaths of sap-sucking insects (DeLong, 1965; Miles, 1968). Consumption of nitrogenous compounds by sap-sucking insects early in the life history of their host plants may depress potential primary production to a degree greater than would be expected from the direct energetic loss (Wiegert, 1964). The timing and the location of herbivory by tissue-chewing lepidopteran larvae, grasshoppers, and beetles can also result in damage to the plant greatly in excess of damage due to the loss of the material ingested (Varley, 1967). Insertion of eggs into plants by insects can also cause premature senescence.

Herbivory need not be viewed merely as a parasitic pathway of energy and material flow nor as a solely detrimental process (Chew, 1974; Mattson and Addy, 1975; Owen and Wiegert, 1976, 1981; McNaughton, 1979; Scott et al., 1979). Mattson and Addy regard grazing insects as cybernetic regulators, in the sense that these herbivores may maintain consistent and optimal primary production. One important impact of such regulators could be to dampen potential asynchrony between nutrient demand by primary producers and nutrient supply from decompositional processes (Lee and Inman, 1975; Mattson and Addy, 1975; Owen and Wiegert, 1976). The rapid response times and the sensitivity of phytophagous insects to the physiological status of their host are traits which suit them well for the role of regulator (Mattson and Addy, 1975). Thus, some of the classic short-term "detrimental" effects of herbivores can be viewed alternatively as being of ultimate benefit to the primary producer.

The regulatory roles of herbivorous insects have usually been discussed with reference to long-lived trees (Mattson and Addy, 1975; Owen and Wiegert, 1976). But both semelparous (single reproduction) species and perennial (or iteroparous) grasses such as *Spartina* may also exhibit coevolved, mutualistic relationships with their herbivores. The potential longevity of a cordgrass genet (a genetically identical clone, *sensu* Harper, 1977) is a barrier to assessment of the influence of herbivorous arthropods. Short-term field of laboratory manipulations of the relationship between grazer and *Spartina* may be misleading. Field experiments are further complicated by the extreme difficulty involved in delineating the entire structure of a genetic individual at a given point in time, as well as in attempting to trace the structural development of a genet during its temporal existence.

5.4. Arthropod Primary Consumers

5.4.1 Species Richness

A list of the arthropod families associated with the grazing food web of *Spartina alterniflora* marshes and the number of species contained within each family is presented in Table 5.2. The primary source of this table is a comprehensive annotated list, compiled by Davis (1978), of insect species (except Lepidoptera) occurring in the coastal zone of North Carolina, South Carolina, and Georgia. We extracted from Davis's

Table 5.2. Number of Likely Herbivore Species from Arthropod Orders Collected in the *Spartina* Marshes of North Carolina, South Carolina and Georgia

Class	Order	Family	Common name	Species
Crustacea	Decapoda	Grapsidae	Grapsid crabs	1
Insecta	Orthoptera	Acrididae	Short-horned grasshoppers	1
		Tettigoniidae	Long-horned grasshoppers	5
	Thysanoptera	Thripidae	Common thrips	2
		Phloeothripidae	Phloeoid thrips	1
	Hemiptera	Miridae	Leaf (plant) bugs	4
		Lygaeidae	Lygaeid bugs	4
		Coreidae	Leaf-footed bugs	1
		Pentatomidae	Stink bugs	2
	Homoptera	Cicadellidae	Leafhoppers	6
		Delphacidae	Planthoppers	4
		Aphidae	Aphids (plant lice)	1
		Diaspididae	Armored scales	4
	Coleoptera	Melyridae	Short-winged flower beetles	3
		Languriidae	Lizard beetles	2
		Phalacridae	Shining flower beetles	1
		Mordellidae	Tumbling flower beetles	2
		Chrysomelidae	Leaf beetles	7
		Anthribidae	Fungus beetles	1
		Curculionidae	Snout beetles (weevils)	3
	Lepidoptera	Noctuidae	Noctuid moths	2
		Geometridae	Inchworms	1
		Pyralidae	Pyralid moths	2
	Diptera	Culicidae	Mosquitoes	2
		Otitidae	Picture-winged flies	3
		Chloropidae	Frit (stem) flies	14
		Anthomyiidae	Anthomyiid flies	1
	Hymenoptera	Braconidae	Braconid wasps	4
		Ichneumonidae	Ichneumonid wasps	6
		Eulophidae	Eulophid wasps	1
		Encyrtidae	Encyrtid wasps	1
		Eupelmidae	Eupelmid wasps	1
		Pteromalidae	Pteromalid wasps	3
		Eurytomidae	Seed chalcid wasps	2
		Scelionidae	Scelionid wasps	2
		Formicidae	Ants	1
		Vespidae	Paper (potter) wasps	3
		Pompilidae	Spider wasps	1
		Sphecidae	Sphecid wasps	2
		Halictidae	Mining bees	1
		Apidae	Honey bees, etc.	1
	9 orders	41 families		109 species

list those species collected in *S. alterniflora* stands, and, using Borror and DeLong (1964) as a reference, we removed species belonging to insect families that are not characteristically herbivorous. Species belonging to genera with trophic characteristics contrary to that of their family as a whole were included or excluded as necessary. For example, Evans and Murdoch (1968) found that adults of all Hymenoptera species inhabiting the field layer of a Michigan old field utilized flowers as a source of energy. Hence, all species from wasp and bee families occurring in *Spartina* stands were incorporated into Table 5.2. This may be too liberal an inclusion, considering the short flowering period of *Spartina.* The marsh ant, *Crematogaster clara*, has been observed to cluster around plant sap exuding from wounded tissue adjacent to portions of leaves fed upon by *Orchelimum*. Hence, this species is included in Table 5.2 although it is not primarily herbivorous. Additionally, adult mosquitoes, such as *Aedes sollicitans* and *A. taeniorhynchus*, associated with salt marshes along the Atlantic Coast of North America, may exploit *Spartina* floral nectar as a carbohydrate source.

Information regarding lepidopteran families and species was derived from Kale's analyses (1964) of stomach contents of the marsh wren. A few families and species absent from Davis's list, but present in Kale's, were added. The squareback crab, *Sesarma reticulatum*, was reported by Jackewicz (1973) to feed on live *Spartina* leaves in Delaware marshes and appears to be relatively abundant in the low marshes of Sapelo Island (Teal, 1958).

The salt-marsh grazing community is species-poor when considered on an areal basis, for most terrestrial grasslands are far richer in grazing insects. The Michigan old field studied by Evans and Murdoch (1968) had, at last count, well in excess of 1500 species of insect associated with the aboveground grazing food web. Viewed in the context of a single host plant, however, the *Spartina* grazing fauna seems surprisingly rich. For instance, the 108 grazing insect species associated with *S. alterniflora* constitute a number greater than the number of species predicted for a similar area of sugarcane. The latter shares certain properties with *Spartina*; both are highly productive, perennial grasses existing in large, and essentially monospecific, stands. If one assumes that *S. alterniflora* dominates 60% of the 443,000 hectares of salt marsh in the coastal zone of Georgia and the Carolinas and then interpolates from the species-area curve in Strong et al. (1977), the number of herbivorous insect species predicted to occupy the equivalent area of sugarcane would be approximately 65.

Table 5.3 compares the families of insect herbivores associated with *S. alterniflora* and *S. patens* to those found with *S. foliosa*, the Pacific coast representative of the genus. The insect fauna of *S. foliosa* was studied in a California salt marsh by Cameron (1972). Although direct comparison of the species richness of herbivorous insects between Pacific coast and Atlantic coast *Spartina* is not totally appropriate, since Cameron examined only a very small area, still, notable disparities are suggested for the Orthoptera, Hemiptera, Homoptera and Hymenoptera.

5.4.2 The Leaf-Chewing Guild

Spartina probably provides at best a mediocre food resource for leaf-chewing herbivorous arthropods. Of nine amino acids examined by Burkholder (1956), *S. alterniflora* leaves were 64% lower in eight of them than the average values for Gramineae. Caswell

Table 5.3. Comparison of the Number of Herbivorous Species Occupying Various Insect Orders Associated with Three Intertidal Species of *Spartina*

	S. alterniflora	*S. patens*	*S. foliosa*
Orthoptera	6	10	0
Thysanoptera	3	8	5
Hemiptera	11	11	1
Homoptera	15	22	7
Coleoptera	19	18	10
Lepidoptera	5	–	2
Diptera	20	16	11
Hymenoptera	29	22	6
Total Herbivorous Insect Species	108	107	42

et al. (1973) regarded C_4 plants, of which *S. alterniflora* is an example, as nutritionally inferior to C_3 plants, partially because of their leaf morphology. The 27% assimilation efficiency estimated for *Orchelimum* by Smalley (1960) is low compared to values reported by Waldbauer (1968) for grasshoppers feeding on C_3 plants. Burkholder (1956) found that creekbank and back-levee *Spartina* leaves were superior to leaves from the high marsh in protein, amino acid, and vitamin concentrations. This phenomenon may in part explain why the high marsh populations of the insect are relatively sparse, although *Orchelimum* is widespread over the marsh (Smalley, 1959a).

The quality of leaf tissue is further reduced by *Trigonotylus uhleri*. This mirid bug empties the contents of mesophyll cells, creating chlorotic patches. Such mesophyll-feeding is concentrated towards the leaf tip and is most common in dense patches of high marsh grass. *Orchelimum* tends to remove strips of leaf tissue from the lower half of the leaf. This spatial separation of feeding suggests avoidance of competition.

The only important members of the leaf-chewing guild in Sapelo Island marshes appear to be the grasshopper, *Orchelimum*, the squareback crab, *Sesarma*, and adults of the weevil, *Lissorhoptrus chapini insularis*. *Orchelimum* possesses the broadest range, the other two species being generally restricted to the low marsh. The genus *Orchelimum* is restricted to North America, and most of its members inhabit freshwater marshes and ditches. But *Orchelimum fidicinium* is restricted to Atlantic and Gulf coast salt marshes, with an apparent northern limit near Delaware (Morris and Walker, 1976). Davis (1978) characterized *O. agile* as an occasional resident of cordgrass marshes in North Carolina, being more commonly found together with two other members of the genus, in more highly elevated tidal marshes along the Atlantic coast.

In the Sapelo Island marshes the salt-marsh grasshopper ingests only 2% of the net annual aboveground primary production (Smalley, 1960). Wastage was not measured. During the active period of approximately 100 days, from early May to middle August, the *Orchelimum* population ingested 107 kcal m^{-2}, assimilated 29.4, respired 18.6, and produced 10.8 kcal m^{-2} of grasshopper biomass. Only one species of Orthoptera in Humphreys's (1979) review had a higher annual production, and that was the 26.2 kcal m^{-2} of *Melanoplus sanguinipes* (Van Hook, 1971). The 2% of net primary production ingested by *Orchelimum* conveys a deceptive impression of the impact of this grasshopper. Mitchell and Pfadt (1974) estimated that prairie grasshoppers waste–

Figure 5.1. *Orchelimum fidicinium*, the salt-marsh grasshopper, in a characteristic cryptic posture with hind legs extended and body close to a blade of grass.

detaching but not ingesting–a quantity of grass equal to 50 to 100% of that ingested. *Orchelimum* actively feeds for only one-third of the growing season of *Spartina* and does not utilize stems and leaf sheaths. Hence, during their active growth period, *Orchelimum* may destroy as much as 5% of net primary production of *Spartina,* with this destruction concentrated upon the most photosynthetically active tissue. Furthermore, *Orchelimum* scrapes leaf tissue deeply from the adaxial surfaces, so translocation from the upper portion of leaves may be interrupted and senescence hastened. Leaching of organic and inorganic compounds from such scarred leaves may be increased. Whether such leaf destruction, premature senescence, and subsequent litter input to the marsh surface has any potentially beneficial effects on cordgrass is unknown. In the tidally influenced salt marsh, the role of consumer organisms in the regulation of nutrient cycling to the benefit of the host plant may be slight.

Among arthropod predators, only lycosid and araneid spiders appear capable of any significant consumption of *Orchelimum*. Kale (1965) reported that *Orchelimum* composed only a small portion of the diet of the long-billed marsh wren, *Telmatodytes palustris griseus*. But in recent years cattle egrets (*Bubulcus ibis*) have foraged in Sapelo Island salt marshes during the summer (Pfeiffer, 1974), and grasshoppers form a dominant portion of this egret's diet in other habitats (Siegfried, 1971).

The species of *Lissorhoptrus* found in Sapelo Island salt marshes represents the only cordgrass grazer for which there is any indication of potential belowground herbivory. Adult *Lissorphoptrus* weevils scrape out short strips of leaf tissue with their mandibles, leaving behind rectangular feeding scars which can be counted. Adults are active from

September through April. Other species of this genus insert eggs into leaf sheaths. After hatching, the larvae mine the leaf sheaths of their semiaquatic hosts for a short period before migrating to the roots, on which they feed until the pupal stage (Webb, 1914; Kuschel, 1951; Grigarick and Beards, 1965; Bowling, 1972; Sooksai and Tugwell, 1978).

Densities of the square-backed crab, *Sesarma reticulatum*, decline from substantial numbers in the tall and intermediate stands of cordgrass growing in the vicinity of the levee to two or three individuals m^{-2} in short *Spartina*. This crab is most active at night, during high tides, and during overcast weather (Teal, 1959; Palmer, 1967). The crab often clips the live leaves off near their base, thus wasting a substantial amount. Although areas of creekbank *S. alterniflora* in Delaware marshes are reduced to stubble by *Sesarma* herbivory (Jackewicz, 1973; Kraeuter and Wolf, 1974), such devastation is not found in Georgia marshes. Ingestion and assimilation by *Sesarma* for the six-month active period, May to October, in Delaware marshes were estimated from laboratory measurements (Jackewicz, 1973). Ingestion of 116 kcal m^{-2} and assimilation of 68 kcal m^{-2}, combined with the 20.8% production efficiency for noninsect invertebrate herbivores (Humphreys, 1979) yields an annual production by this crab of 14 kcal m^{-2}. *Sesarma* densities prorated over the entire marsh may be higher in Georgia than in Delaware as a result of the broader zone of creekbank vegetation in Georgia. The active foraging season in the Georgia marshes is also longer. When studied sufficiently in Georgia, *Sesarma* may prove to be the major consumer of *Spartina* in the leaf-chewing guild.

5.4.3 The Sap-Sucking Guild

The sap-sucking insects feeding on cordgrass are more diverse and more productive than the leaf-chewers. Five species are particularly abundant: the delphacid planthoppers, *Prokelisia marginata* and *Delphacodes detecta*, the mirid (leaf) bug, *Trigonotylus uhleri*, the lygaeid bug, *Ischnodemus badius*, and the armored scale, *Haliaspis spartinae*. These insects either ingest material being translocated through vascular tissue, in particular the phloem vessels, or suck the contents from mesophyll cells. Generally, phloem-feeders have a potentially richer but more variable resource, whereas cell-feeders face a more constant resource, but one yielding moderate rewards per effort expended.

Unquestionably the most abundant and productive insect herbivore in Georgia marshes, the planthopper *Prokelisia* produces five or six generations per year. It achieves its highest population densities, prorated over the entire marsh, during the late fall and winter months. At this time the population density of *Prokelisia* in the high marsh usually fluctuates between 5 to 40 $\times$ 10^3 m^{-2}, but may surge to well above 5 $\times$ 10^4 m^{-2} immediately following large-scale and roughly synchronized hatching of eggs. From November into March the standing stock in the high marsh generally ranges from 0.5 to 4 g dry weight m^{-2}. The dynamics for two generations of *Prokelisia* over such a time interval is illustrated in Figure 5.2. for a high-marsh sampling site near the Duplin River. *Prokelisia* eggs are generally inserted into the ridges occurring on the corrugated adaxial surface of live leaves of short *Spartina*. In tall *Spartina*, as well as in particularly robust leaves of the short-form *Spartina* found in plots enriched with

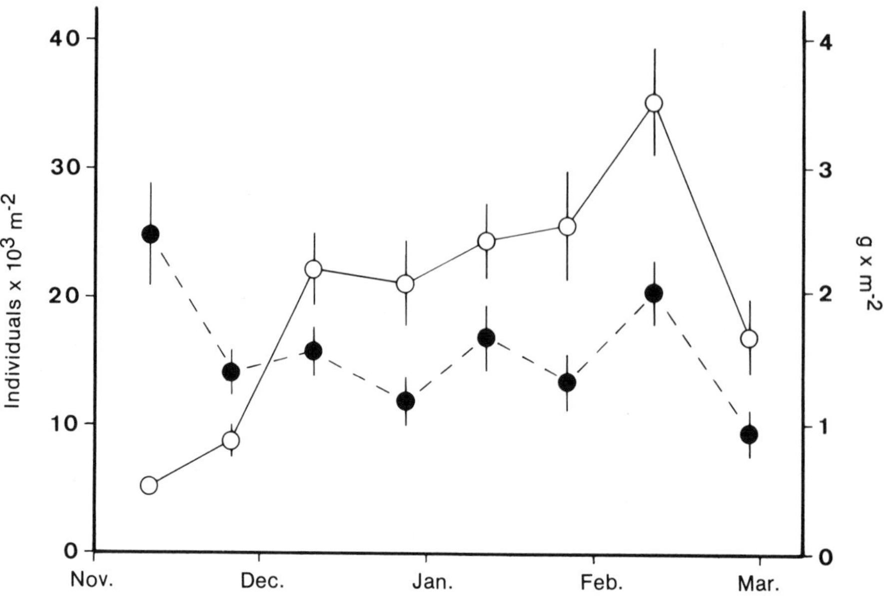

Figure 5.2. Population changes over two generations of *Prokelisia marginata* in short *Spartina* marsh. Population density (●), biomass (dry weight) (○). Values uncorrected for the efficiency of the sampling device (D-VAC), preliminary mean estimate for which ranges between 80-90%.

nitrogen, the leaf tissue underlying the corrugated adaxial surface is sufficiently thick to allow eggs to be inserted through the base of the furrows. For much of the thicker central portion of the leaf blades of tall *Spartina*, vertical deposition of eggs occurs, but insertion parallel to the long axis of the leaf becomes necessary towards the peripheries. The eggs are deposited on the uppermost, unfurled leaves of creekbank plants. Distance above the marsh surface does not appear to be an important influence upon the site of egg deposition in the high marsh, for very young shoots, often with only one or two leaves expanded, frequently bear a proportionately large number of eggs.

Heavy concentrations of *Prokelisia* eggs cause roughly elliptical discolorations near the bases of the leaves, and these discolored areas of leaf tissue are prone to senesce prematurely. Such leaf damage could well account for a greater destruction of *Spartina* than does consumption by *Orchelimum*, although some energy or material may be conserved by the plant through translocation from the senescing tissue. Indeed, *Prokelisia* may benefit a cordgrass genet by hastening the senescence of mature leaves and thereby stimulating the reallocation of resources to developing tissue, such as belowground parts, emerging culms, and expanding leaves which possess high photosynthetic rates. The planthopper progeny benefit if translocatory processes are stimulated, since this insures a rich food source at a critical stage in their life history when the mobility necessary to locate a suitable resource is minimal. Increased leaching of organic compounds from leaves pierced by planthopper ovipositors is another likely result of egg laying. It can be a significant drain, since a single leaf may harbor more than 10,000 eggs.

Prokelisia survivorship curves are moderately to steeply concave, essentially the Type III and Type IV survivorship curves described by Slobodkin (1962). Mortality of early nymphal stages is particularly heavy, at times, for late spring and summer populations of *Prokelisia* which occupy *Spartina* stands below the levees of creekbanks. Of approximately 200×10^3 eggs m^{-2} present in such a stand of grass on the southwestern corner of Sapelo Island in the last week of May 1979, 3×10^3 nymphs m^{-2}, primarily second and third instars, were still alive on June 6. Younger nymphs have greater difficulty than do older nymphs or adults in relocation on plants once they have been dislodged onto the surface of the water (Denno and Grissell, 1979). Killifish (*Fundulus*) are abundant in the tidal creeks during the summer months and actively prey on planthoppers that are slow to find another plant. Large schools of these fish often swarm vigorously among the *Spartina* culms, dislodging planthoppers onto the surface of the tidal water, then quickly consuming the fallen nymphs.

Adult *Prokelisia* exhibit wing polymorphism similar to that of other delphacid planthopper species occupying intertidal and freshwater habitats. However, the incidence of the large-winged macropterous forms is substantially higher in *Prokelisia* than in the other delphacid species (Denno, 1977, 1978; Denno and Grissell, 1979). In Sapelo Island marshes the brachypterous form, wherein mesothoracic wings are reduced and metathoracic wings nearly vestigal, is rare in relation, to the large-winged macropters, which represent at least 95% of the wing-morphs. The highest incidence of macropterous *Prokelisia marginata* adults among all *Spartina* marshes along the Atlantic coast and Gulf of Mexico in North America occurs in Georgia marshes. Denno and Grissell (1979) attributed this high incidence of macropterous morphs to the mosaic of stable (high marsh) and unstable (creekbank) habitats found in Atlantic coast marshes. Creekbank stands represent an unstable resource; in summer tall grass is abundant in the stands, providing an optimum nutritional resource, oviposition space, and a refuge from tidal inundation, but during the winter months there is little live grass tall enough to remain emergent during the many high tides. The high marsh, used as an overwintering site by northern populations of *Prokelisia*, is suitable as a refuge from high tides. The Gulf coast stands of *Spartina* do not exhibit the extreme differences in height that are found in the grasses of the Atlantic coast marshes, and the suitability of resource for *Prokelisia* is presumed to be less variable in the Gulf coast marshes. Therefore planthoppers have less need of the large wings that would enable them to disperse over long distances. Approximately 90% of the adults in cordgrass marshes from the Florida Gulf coast are brachypterous. The large-winged morph requires more energy for development, possibly reducing its fecundity. Tsai et al. (1964) and Nasu (1969) reported reductions of 15% and 42%, respectively, in the fecundity of macropters of the delphacid, *Laodelphax striatella*, in comparison to the fecundity of brachypter. A 13% reduction in fecundity of macropters in the delphacid, *Nilaparvata lugens*, was reported by Manjunath (1977).

The dynamics of *Prokelisia* in creekbank stands on Sapelo Island resemble the behavior of this species in the New Jersey marshes (reported by Denno, 1976; Denno and Grissell, 1979). Planthoppers are rare in the low marsh from autumn through early spring. In late spring macropterous adults migrate from the high marsh to streamside vegetation growing below the creekbank levee. During the warm months only a low density of *Prokelisia* is supported by high marsh grass, with populations seldom exceeding 1000 individuals per square meter. This decline in planthopper densities at a time when the live standing stock of cordgrass is at a peak in the high marsh could be caused by changes in the quality of the grass, perhaps the result of changes in amino acids or vitamins. Such changes in the material translocated by the host plant could be caused by the higher temperatures and drier conditions of summer. There are also fewer young shoots and senescing plants in the high marsh during summer. Since such culms tend to be active translocators, the decline in planthopper populations in summer may simply be a response to a decrease in the amount of translocate available. This topic poses some fascinating questions and could be an area of fruitful research.

In addition to the suggested tracking of an optimum resource, or an optimum microclimate, the seasonal movements of *Prokelisia* from high marsh to streamside vegetation also lessen the predation pressure from the relatively immobile arachnid predators found mainly in the high marsh. The population densities of the spiders and pseudoscorpions increase in the presence of *Prokelisia* during the late autumn and winter months and subsequently plummet to low densities in the summer (Figure 5.3). At the time of the mass dispersal of *Prokelisia* in the spring, the density of macroarachnids–spiders and pseudoscorpions, but not mites–in high marsh grass is much greater than it is in streamside vegetation below the levee.

Whether accumulation of photosynthate within a leaf causes a reduction in the net photosynthetic rate of that leaf has been long debated (Neales and Incoll, 1968) and remains unresolved. Way and Cammell (1970) showed such inhibition of photosynthesis in cabbage and suggested a positive benefit accruing from aphids feeding on such plants and removing the excess photosynthate, thus stimulating production. Roots and

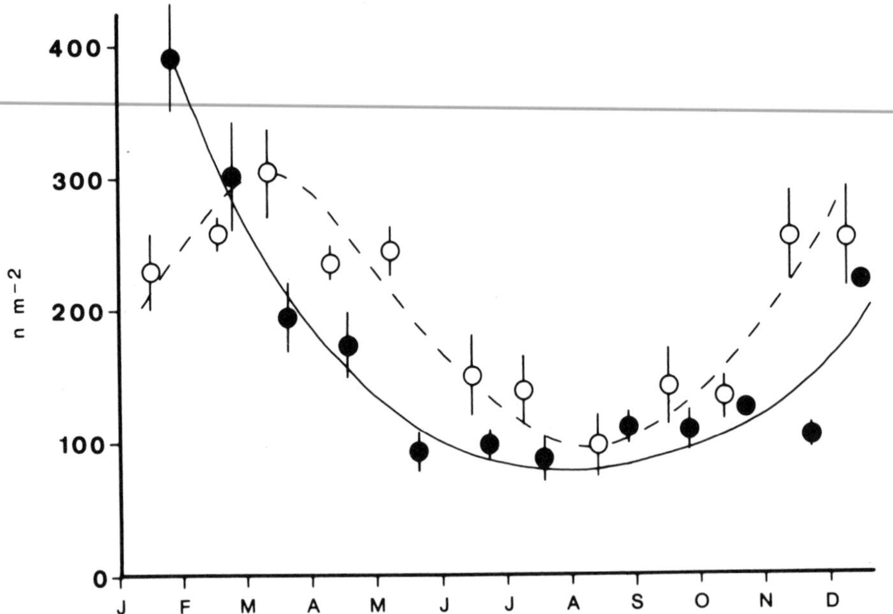

Figure 5.3. Changes in the numbers of macroarachnids with season in both short (•) and levee (○) *Spartina* marsh at Sapelo Island.

rhizomes serve as major sinks for the removal of photosynthate from *Spartina* leaves, but the growth and activity of these structures is depressed in the winter (Gallagher et al., 1977). A field experiment to investigate the effect of *Prokelisia* on photosynthetic rates in *Spartina* during the winter was inconclusive.

The annual consumption and production of *Prokelisia* is far larger than that of all other herbivores associated with *Spartina*. Thus, the total annual aboveground herbivore consumption and production estimates for the Georgia cordgrass marshes mainly reflect the estimate for planthoppers. The only previous energy budget for this species alone was Smalley's (1959). Unfortunately, since Smalley did not present the sampling methods and population densities on which his energy-flow estimates were based, a critical assessment of his values is impossible. To estimate annual assimilation, Smalley assumed a respiration/assimilation ratio of 0.75. The annual secondary production of this planthopper was then obtained by summing the annual estimates for assimilation and respiratory energy loss. Wiegert and Evans (1967), using Smalley's data in a comparison of the energy flow in several different grasslands, assumed an assimilation efficiency (assimilation/ingestion × 100) of 66%. This efficiency had been measured for a population of xylem-feeding cercopid Homoptera (Wiegert, 1964). Estimates of energy assimilation efficiencies for phloem-feeding aphids vary widely, ranging from 9 to 83% (Randolph et al., 1975). In view of these uncertainties, the cordgrass herbivore consumption and production estimates in Table 5.1 must be regarded as provisional.

The delphacid planthopper, *Delphacodes detecta*, is the most numerous sap-sucking insect inhabiting *Spartina patens* stands in North Carolina and New Jersey marshes, as well as being an extremely abundant parasite of *Distichlis spicata* marshes in North Carolina (Davis and Gray, 1966; Denno, 1977). But Davis and Gray, as well as Denno, reported much lower densities of this species on *S. alterniflora*. In New Jersey, macropterous adults, the rare alary morph of this species, colonize *S. alterniflora* in May and July, but only very sparse nymphal populations are present until the winter months. Denno (1977) believed *S. alterniflora* did not provide a suitable resource for the development of *Delphacodes* nymphs. In Sapelo Island marshes, this species is most common in the high marsh during the winter months, with populations generally fluctuating between 100 to 500 individuals m^{-2}. During this period *Spartina* appears to support active growth of the nymphs.

Trigonotylus uhleri utilizes both *S. patens* and *S. alterniflora* in North Carolina (Davis and Gray, 1966). This green mirid achieves the highest prorated densities of any species of Hemiptera in Sapelo Island marshes and, with the possible exception of *Prokelisia*, causes the most readily discernible feeding damage to cordgrass. This species has at least four generations per year in Georgia and overwinters in the egg stage. Two generations of this bug occur annually in New Jersey (Denno, 1977). The large brown mirid, *Trigonotylus americanus*, forms a very dense population in *Distichlis spicata* stands in North Carolina (Davis and Gray, 1966) but rarely appears in Sapelo cordgrass marshes. When present, it is nearly always represented by the adult stage.

Ischnodemus badius, a lygaeid bug, is almost exclusively restricted to intermediate and tall *Spartina*, generally residing between the stem and live leaf sheaths, and feeding on the maturing seedheads in autumn. This species has wings reduced in size to minute pads, a condition designated as microptery (Slater, 1977). No other wing morphs were found in a sample of more than 200 adults collected near Sapelo Island. Marked wing reduction in terrestrial insects is thought to reflect a resource base and habitat of relatively uniform quality in space and time as perceived by the consumer species (Southwood, 1961, 1962; Brinkhurst, 1963; Sweet, 1964; Denno, 1977, 1978; Denno and Grissell, 1979). Exactly how the spatial and material resources of this species differ so from those of *Prokelisia* and *Trigonotylus* is not known. Despite the cryptic habits of this lygaeid, the aposematic coloration of the nymphs, and the strong, offensive odor emitted by disturbed nymphs and adults, it comprised 7.2% of the annual prey volume in the stomach contents of the long-billed marsh wren (Kale, 1965).

The armored scale, *Haliaspis spartinae*, feeds on the adaxial surface of tall *Spartina*. Tippins and Beshear (1971) found colonies of this armored scale residing low enough on creekbank plants to be inundated for one hour during every high tide. This species was reported to overwinter in the egg stage on dead vegetation.

Other sap-sucking insects, such as the stink bug, *Rhytidolomia senilis*, and the leafhoppers, *Draeculacephala portola* and *Sanctanus aestuarium*, appear only sporadically in Sapelo Island marshes. A species of aphid develops sparse populations on *Spartina* from spring through autumn.

5.4.4 Miscellaneous Herbivorous Guilds

Davis and Gray (1966) remarked that adult picture-winged flies of the genus *Chaetopsis* associated with cordgrass marshes in North Carolina have sponging mouthparts and take up plant exudates–and probably components of the *Aufwuchs* community–from both live and senescing leaves. The occurrence of adult *Chaetopsis* on leaves with *Prokelisia* suggests that the flies may also be lapping up the sugar that must be abundant in the exudate of the phloem-feeder. Marples (1966) observed this leaf-licking habit in both species of *Chaetopsis* found in Sapelo marshes, but found different ^{32}P-uptake dynamics between the two species when the radionuclide was injected into grass stems. Despite Marples's disclaimer, larval uptake of labeled phosphorous may have occurred and may have been reflected in the body burden of adults collected five weeks after the initial pulse labeling. Marples's observation of disparate ^{32}P-uptake dynamics is noteworthy in light of the different δ^{13}C values exhibited by the adult forms. For a sample of pooled individuals collected from the high marsh in September of 1975, *Chaetopsis apicalis* showed a δ^{13}C value of -13.8, whereas *Chaetopsis aenea*–very possibly *Chaetopsis fulvifrons* (see Davis, 1978)–yielded a value of -15.2. The adults of these latter two species appear to coexist spatially and temporally and to feed on similar resources. Marples's data, together with these δ^{13}C values, suggest that coexistence of the two congeneric flies may result from the fact that the larvae occupy different feeding niches.

The cryptic stem-boring guild exploiting *S. alterniflora* is seldom mentioned in the literature. Kale (1964) reported two species of lizard beetles found in stomachs of long-billed marsh wrens. The pyralid moth, *Chilo sp.*, probably represents the most common stem-borer in Sapelo Island marshes.

A diverse array of insects, including numerous flower beetles, flies, and wasps exploits flowers in the marshes. The large golden-brown tumbling flower beetle, *Mordellistena splendens*, is a particularly conspicuous and abundant visitor to *Spartina* floral spikes. The marsh ant, *Crematogaster clara*, feeds on sap exuding from wounded plant tissue and may exploit flowers and seeds of *Spartina* to some extent. This ant, despite utilizing a somewhat unusual habitat in the marsh, is very much the generalist in distribution, having been reported in freshwater marshes, pine forests, and oak galls (Wheeler, 1910; Van Pelt, 1956).

Belowground herbivory of *Spartina* has not been investigated, but the probable consumption of roots by *Lissorhoptrus* may constitute the most significant belowground herbivory by insects. The larvae of some ephydrid flies may pierce the roots in order to obtain oxygen from the air spaces within the tissue; Simpson (1976) related this behavior for the larvae of *Dimecoenia spinosa* and for members of the genus to which *Notiphila bispinosa* belongs. Kale (1964) found both of these species in the stomachs of marsh wrens.

5.5. Vertebrate Predators

5.5.1 Birds

The long-billed marsh wren, *Telmatodytes palustris griseus*, is one of the most common of salt-marsh birds. During the breeding season the song of this wren is heard everywhere in the tall grass along the creek banks. The bird itself is seen less often, but patient observation is soon rewarded with a glimpse of a wren moving about within or just over the grass, searching for insects. Its habitat is shared with the seaside sparrow, *Ammospiza maritima*. Distinct differences in preferred foraging territory seem to exist. Kale (1965), in his monographic study of the ecology of the marsh wren, observed that, whereas the seaside sparrow foraged primarily on the marsh surface, the marsh wren occupied the grass canopy. He testified, "I have never collected a marsh wren with muddy feet, or a seaside sparrow with clean feet."

More long-billed marsh wrens, seasonal immigrants from the North, join the resident population during the winter, foraging over wide expanses of marsh, even into the stands of *Juncus*. The prey items reported by Kale (1965) from wren stomachs are summarized in Table 5.4. During the breeding season from April to August, prey items are divided evenly among a variety of arthropod orders. Following the decline of many insects from September to March, two species, *Prokelisia* and *Ischnodemus*, constitute 50% of the volume of the marsh wren diet. According to Kale (1965), the annual energy ingested by wrens comprises herbivorous insects (73 kcal m^{-2}), predaceous insects and spiders (37 kcal m^{-2}), and detritus feeders (16 kcal m^{-2}). Of the total 126 kcal m^{-2} ingested, 88 kcal m^{-2} are lost by the wren population as respiration and 38 kcal m^{-2} as feces. Energy used for new marsh wren tissue amounts to only 0.5 kcal m^{-2}.

Kale postulated a major role for marsh wrens as regulators of arthropod predators, especially of spiders and wasps. For example, the mean daily ingestion of the marsh wren population amounts of 345 cal m^{-2} or 63 mg m^{-2}, assuming 5500 cal g dry weight^{-1} of prey. Applying this value during the breeding season of the wren, when foraging activity is concentrated in creekbank grass, and employing the percentage of spiders–4.5%–found in wren stomachs over the corresponding interval, the wrens daily ingestion of spiders, 9.5 mg dry weight m^{-2}, equals 5.1% of the mean standing

Table 5.4. Percentages of the Total Prey Volume from the Stomach of the Long-Billed Marsh Wren Represented by Various Taxonomic Categories[a]

Taxonomic category	April-August	September-March
Mollusca	3.5	4.0
Araneae	15.1	6.2
Orthoptera	5.6	0.8
Hemiptera	5.4	10.0
Homoptera	13.0	40.1
Coleoptera	11.6	12.6
Lepidoptera	14.6	2.9
Diptera	8.9	7.7
Hymenoptera	17.3	12.4
Undetermined Insecta	4.5	3.3
Other Arthropoda	1.8	0.9

[a] From Kale (1965).

stock of spiders present on levee vegetation from April through August.

No studies of the seaside sparrow in Georgia marshes match the breadth of Kale's research on the marsh wren. However, an ecological analog of the seaside sparrow, the song sparrow, (*Melospiza melodia*), resides in California *Spartina foliosa* salt marshes and has been studied extensively (Marshall, 1948, 1948a; Johnston 1956, 1956a). In those marshes the song sparrow also displays a preference for the marsh surface in grass swards bordering creeks (Marshall, 1948a). The song sparrow consumes snails, small nereid polychaetes, insects, and other small invertebrates, as well as *S. foliosa* flowers (Marshall, 1948). This diet appears to reflect what would be expected from the seaside sparrow in Georgia marshes, since invertebrate populations, other than insects and spiders, are abundant on the surface of the vegetated creekbank. Sprunt says the seaside sparrow consumes various worms, shrimp, crabs, grasshoppers, moths, flies, spiders and cordgrass seeds (Bent, 1968). However, of the food items brought to seaside nestlings in a New York *Spartina* marsh, 99.5% were insects, chiefly Hemiptera, Diptera, Trichoptera and Lepidoptera (Post, 1974). Marshall (1948) thought the density of *S. foliosa* stems limited the surficial foraging activities of the song sparrow in the high marsh. This explanation could also account for the preference of the seaside sparrow for the tall vegetation in Georgia marshes. The seaside sparrow apparently does not nest exclusively in *Spartina*, but selects nesting sites in *Juncus roemerianus* and groundsel bush (*Baccharis halimifolia*) stands (Tomkins, 1941).

The clapper rail, *Rallus longirostris*, also a permanent resident of *Spartina* marshes, consumes only small quantities of insects and spiders. Cutworm moths (Noctuidae) constitute the predominant insect prey (Oney, 1951). In coastal Georgia, *Sesarma reticulatum* and *S. cinerum* form 54% of the stomach contents of this rail. The herbivorous grapsid crab serves as a food resource for an array of herons, ibises and egrets, including the abundant cattle egret.

Various nonpermanent insectivorous birds frequent Georgia marshes. Sharp-tailed sparrows (*Ammospiza caudacuta*) overwinter, generally from late September through late April, in the high marsh along the southeastern coast. Tree swallows (*Iridoprocne bicolor*) and, to a lesser extent, rough-winged swallows (*Stelgidopteryx ruficollis*) occasionally fly just above the upper canopy of grass during the autumn and winter. During spring high tides, large flocks of tree swallows repeatedly swoop over patches of grass still emergent above the flooding waters. Such patches of emergent grass harbor dense concentrations of planthoppers which have accumulated in these islands as a result of their passive dispersal by the incoming tide. During the late spring and summer, red-winged blackbirds (*Agelaius phoenicius*) are active in tall grass, and barn swallows (*Hirundo rustica*) occasionally glide over the meadows. According to Teal and Teal (1969), laughing gulls capture insects from *Spartina* marshes during high tides. Wiegert has observed gulls feeding on salt-marsh grasshoppers driven to the tips of grass by extreme spring tides.

5.5.2 Mammals and Fishes

The rice rat, *Oryzomys palustris*, is the only permanent resident rodent in Georgia *Spartina* marshes, and it prefers the regions of tall grass (Sharp, 1967). This species is nocturnal, dwelling in abandoned nests of long-billed marsh wrens or retreats of its own design during the daylight hours. Stem-borer (Pyralidae) larvae and crabs of the genera *Sesarma* and *Uca* comprise the major portion of rice rat stomach contents during the summer months along with small amounts of flies and beetles. Sharp believed that this rodent was too localized and scarce to have an appreciable impact on the arthropod populations.

The major source of egg and nestling mortality of the long-billed marsh wren stems from predation by rice rats, raccoons (*Procyon lotor*) and minks (*Mustela vison*), surpassing the contribution of high tides and storms to the mortality of the wren's early life stages (Kale, 1965). However, most of the destruction of seaside sparrow nests in New York marshes was the result of flooding, destruction by predators being of only minor importance (Post, 1974). Johnston (1956a) studying song sparrows in California marshes, attributed 80% of egg and nestling mortality to predation, flooding of nests, and desertion by the parents.

Cyprinodont fishes, in particular the schooling killifish, *Fundulus heteroclitus* and *Fundulus luciae*, capture insects and spiders from the surface of intertidal water. Salt-marsh insects have been found in the stomachs of *F. heteroclitus* from North Carolina (Kneib and Stiven, 1978) and Maryland (Baker-Dittus, 1978) salt marshes, of *F. confluentus* from Florida intertidal zones (Harrington and Harrington, 1961, 1972), of *F. luciae* from Virginia tidal marshes (Byrne, 1978), and of *F. majalis* and *F. diaphanus* collected from Maryland marshes (Baker-Dittus, 1978). In addition to collecting various species of *Fundulus* from marsh habitats adjacent to Sapelo Island, Dahlberg (1972) collected *Cyprinodon variegatus* from tidal pools and tidal ditches adjoining the high marsh.

5.6. Arthropod Predators

Spartina marshes support a richer diversity of predaceous arthropods than casual observation might suggest (Table 5.5). Predaceous insects associated with southeastern marshes were compiled from Davis's (1978) and Kale's (1964) lists using the same

Table 5.5. Number of Predaceous Species from Arthropod Orders Collected in the *Spartina* Marshes in North Carolina, South Carolina, and Georgia

Class	Order	Family	Common name	Species
Arachnida	Pseudoscorpionidae	Cheliferae	Pseudoscorpions	1
	Araneae	Dictynidae	Dictynid spiders	1
		Gnaphosidae	Running spiders	2
		Clubionidae	Sac spiders	1
		Thomisidae	Crab spiders	3
		Salticidae	Jumping spiders	5
		Pisauridae	Nursery-web spiders	3
		Lycosidae	Wolf spiders	3
		Theridiidae	Comb-footed spiders	1
		Araneidae	Orb-weaving spiders	8
		Tetragnathidae	Long-jaw orb weavers	3
		Micryphantidae	Dwarf sheet-web spiders	4
	Acarina	Trombidiidae	Trombidiid mites	1
Insecta	Odonata	Aeschnidae	Darner dragonflies	1
	Orthoptera	Mantidae	mantids	1
	Hemiptera	Miridae	Leaf (plant) bugs	2
		Reduviidae	Assassin bugs	1
		Nabidae	Damsel bugs	1
	Colepotera	Staphylinidae	Rove beetles	1
		Melyridae	Soft-winged flower beetles	3
		Cleridae	Checkered beetles	1
		Coccinellidae	Ladybird beetles	3
	Diptera	Pipunculidae	Big-headed flies	2
		Conopidae	Thick-headed flies	1
		Chamaemyiidae	Chamaemyiid flies	1
	Hymenoptera	Braconidae	Braconid wasps	4
		Ichneumonidae	Ichneumon wasps	6
		Eulophidae	Eulophid wasps	1
		Encyrtidae	Encyrtid wasps	1
		Eupelmidae	Eupelmid wasps	1
		Pteromalidae	Pteromalid wasps	3
		Eurytomidae	Seed chalcid wasps	2
		Scelionidae	Scelionid wasps	2
		Formicidae	Ants	1
		Vespidae	Paper (potter) wasps	3
		Pompilidae	Spider wasps	1
		Sphecidae	Sphecid wasps	2
	9 orders	37 families		81 species

criteria employed in Section 5.4.1. Enumeration of predaceous arachnid species was based on Barnes's (1953) study of spiders in North Carolina marshes, Kale's (1964) annotated compilation of marsh wren prey, and Pfeiffer's personal records. Table 5.5

lists 45 predaceous species of insects and 34 species of spiders. Cameron (1972) collected 13 species of predaceous insects from his study area of *Spartina foliosa* marshes in California.

5.6.1 Insects

Little information exists on the predaceous insects from *Spartina* marshes. What does exist is limited mainly to distributional records and anecdotal remarks. There have been no detailed life history studies of these predaceous insects. The present level of knowledge does not even allow unequivocal statements to be made concerning the particular resources preyed upon by the various species. However, based upon the tendencies of related species or genera, we suggest the following rudimentary sketch of the trophic habits of these marsh species.

Prokelisia eggs are an abundant resource, particularly during the late fall and winter months. Mirid bugs of the genus *Tytthus* are known to prey on Homopteran eggs (Denno, 1977). *Tytthus parviceps* and *T. vagus* reside in North Carolina marshes (Davis, 1978), and one unidentified species of *Tytthus* was recovered by Kale (1964) from the stomach of a marsh wren. Kale also retrieved *Nabis capsiformis* from two marsh wren stomachs. In Japanese rice fields, species of this damsel bug genus prey on the eggs of an abundant delphacid planthopper (Kuno, 1973). Hemipterans pierce the eggs with their stylets and suck out the contents. This is a particularly efficient method of feeding on eggs embedded in plant tissue, since the stylets can be readily inserted through the aperture originally created in the plant tissue by ovipositing *Prokelisia.*

Flies and wasps are potential parasites of *Prokelisia*. Big-headed flies, many species of which as larvae are parasitic on leafhoppers and mirid bugs (Borror and DeLong, 1964; Kuno, 1973), are represented in southeastern tidal marshes by two species. Potential parasitic wasps include one species of encyrtid wasp and a sphecid species belonging to a tribe, Alyssonini, some species of which provision their subterranean nests with leafhoppers (Borror and DeLong, 1964). Scelionid wasps are parasites of insect and spider eggs. Two of the braconid genera, *Apanteles* and *Chelonus,* which elsewhere parasitize the larvae of pyralid moths, have representative species in the *Spartina* marshes (Borror and DeLong, 1964). The eulophid genus *Tetrastichus* found in marshes has at least one member which is a hyperparasite of a seed chalcid wasp (Borror and DeLong, 1964). Some species of seed chalcid wasps are arthropod parasites; others feed in the stems of grasses. Pompilid wasps generally provision their nesting cells with spider prey. The remaining families of wasps listed in Table 5.5 represent a wide spectrum of host preferences and further conjecture is unwarranted.

Adult ladybird beetles appear only sporadically in *Spartina* marshes, presumably preying on the relatively sparse aphid populations. Larvae of the chamaemyiid fly genus *Leucopsis* (Kale, 1964) are also generally predaceous on aphids (Borror and DeLong, 1964). Thick-headed flies usually employ wasps and bees as hosts for their larvae. The remaining predaceous insect families are comprised of species which typically prey on a broad range of insects.

Table 5.6. Mean Annual Density (N) and Mean Annual Standing Stock (B) for the Spider and Pseudoscorpion Species Inhabiting a *Spartina* Stand Near Sapelo Island, Based upon 12 Monthly Sampling Dates. Density Units are Individuals m^{-2}, and Biomass Units are mg dry weight m^{-2}.

	High Marsh		Levee	
	N	B	N	B
Grammonota trivittata (Micryphantidae)	75.3	37.4	86.4	46.8
Clubiona littoralis (Clubionidae)	15.4	31.2	34.3	91.9
Lycosidae	1.8	16.8	0.7	7.0
Paisochelifer nov. *sp.* (Cheliferidae)	43.6	14.1	32.6	11.0
Poecilochroa sp. (Gnaphosidae)	4.3	5.5	4.3	5.6
Eustala anastera (Araneidae)	2.5	5.5	0.2	0.4
Hyctia pikei (Salticidae)	3.0	4.9	1.0	1.0
Hyctia bina (Salticidae)	0.4	1.7	1.0	7.2
Tetragnatha caudata (Tetragnathidae)	0.8	1.2	0.3	1.0
Dictyna savanna (Dictynidae)	2.6	0.7	3.5	1.0
Philodromus peninsulanis (Thomisidae)	0.3	0.4	1.9	1.5
Hentzia palmarum (Salticidae)	0.2	0.1	9.3	11.0
Drexelia (*Larinia*) *directa* (Araneidae)	0.1	0.0	0.6	1.4
Singa keyserlingi (Araneidae)	0.0	0.0	3.9	9.3
Gertschia (?) *sp.* (Salticidae)	0.0	0.0	2.2	1.1
Ceraticelus paschalis (Micryphantidae)	0.0	0.0	0.8	0.4
Miscellaneous	0.5	0.2	0.4	0.1
	150.8	119.7	183.4	197.7

5.6.2 Predaceous Arachnids

Spiders are by far the most numerous predators of the marsh insects, while mites are the dominant predators of the microarthropod community that flourishes on dead vegetation during the autumn and winter. These predaceous Acarina include a variety of mesostigmated mites and the larvae of trombidiid (chigger) mites, ectoparasites of spiders and, less frequently, of insects.

Different plant communities in the maritime zone of North Carolina possess distinct spider assemblages, but Barnes (1953) discovered that the species composition of the spider assemblages varied little within 10 widely separated grass swards sampled over a two year period in the vicinity of Beaufort. Indeed, spider assemblages inhabiting marshes near Sapelo Island (Table 5.5 and 5.6) closely resemble those described by Barnes.

The marsh surface is occupied mainly by cursorial species, in particular wolf spiders and nursery-web, or fishing, spiders. With the exception of the accumulated drifts of *Spartina* wrack, litter accumulation on the surface of the marsh is prevented by the tides. This absence of litter, along with the absence of the arthropod community typically associated with grassland litter, may account for the low densities and reduced species richness of spiders in the marsh-surface stratum.

The majority of the predatory arachnid species live in or on the leaves and sheaths of live and standing dead plants. Duffey (1962) remarked on the importance of the standing dead community of a limestone grassland in providing dwellings for a large segment of the spider community. The furled dead leaves, sheaths, and hollow stems of *Spartina* provide refuge from avian predators, parasitic wasps, and other predaceous arthropods. The recesses trap air during tidal inundation, and dead vegetation serves as the major site for spider egg nests. Barnes (1953) reported that the species richness of the orb-weaving guilds increased along an elevational gradient extending from intertidal macrophytes to the maritime climax forests. This tendency is explained by the increased potential for space and the increased support for the web construction in communities dominated by shrubs and trees. In contrast, in the intertidal zone the exposure to wind and tidal action decreases the suitability of *Spartina* for web construction.

Populations of macroarachnids, excluding Acarina, were investigated by Pfeiffer during 1974 in levee and high marsh grass adjacent to Sapelo Island. Triplicate samples were taken monthly from each of the two regions. A plywood frame ($0.98 \times 0.86 \times 1.70$ m), open on one side, was placed around a point selected at random. Enclosed grass was harvested and bagged, along with litter and any arachnids active within the confines of the frame. The arachnids were later hand-sorted from the thoroughly stripped vegetation. In both short and tall grass, macroarachnid densities, excluding eggs and nestlings, and standing stock biomass, including eggs and nestlings, peaked during the winter; whereas minimum densities and standing stocks were found to occur during the summer and autumn (Figures 5.3 and 5.4). Tall grass on the levees had a significantly greater biomass of spiders during the year than did short grass; Wilcoxon matched-pairs signed-ranks test gave results of $P < 0.0025$. Differences in macroarachnid densities between two regions of the marsh, however, were not significantly different over the course of the year ($0.05 < P < 0.10$). Cameron (1972), employing a clear-cut harvest and extraction in Berlese-Tullgren funnels of spiders inhabiting *Spartina foliosa* salt marshes in California, reported a peak density of only 20 m^{-2} in July. In both the high marsh and levee regions of the Sapelo Island marshes three genera–*Grammonota, Paisochelifer*, and *Clubiona*–accounted for over 80% of the mean annual macroarachnid densities (Table 5.6). In the high marsh and levee stands examined, *Grammonota* and *Clubiona* together comprised 57% and 70%, respectively, of the mean annual macroarachnid standing stocks. Most of the disparity in biomass between high marsh and levee vegetation is attributable to the concentration of *Clubiona* in the levee area.

Juvenile mortality is characteristically high in spiders and seems particularly severe after the population densities of the dwarf sheet-web spider, *Grammonota*, peak in the high marsh during the winter. In January 1974, the density of actively feeding *Grammonota* equaled 241 individuals m^{-2}, while egg and nestling densities within egg sacks amounted to 771 individuals m^{-2}. Four weeks later, after the spiderlings had left the egg sacks, the trophically-active population had declined to 176 individuals m^{-2}. During this period many of the early instars may have been cannibalized by larger members of their species. Such intraspecific predation has been observed frequently in this species and has been reported for many other species of spiders. In addition to intraspecific and interspecific predation, ballooning, that is, the net dispersal of the early instars of *Grammonota* from the study area by wind, could be another factor contributing to the abrupt population decline observed.

The two major regulatory factors influencing densities of spiders in Georgia marshes appear to be predation and periods of food scarcity. Cannibalism may be of particular

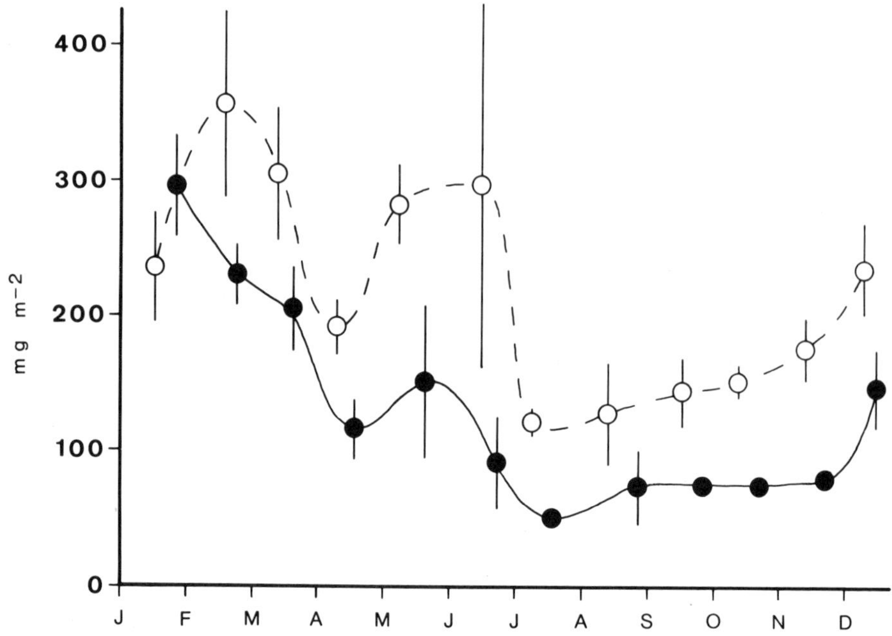

Figure 5.4. Variation in biomass of macroarachnids with season in both short (●) and levee (○) *Spartina* marsh at Sapelo Island.

importance when spider populations attain high densities, since at that time the probability of intraspecific encounters increases. The decline in suitable refuges that accompanies the degradation and removal of dead vegetation enhances such encounters and also increases the likelihood of exposure to avian predation and parasitism by wasps. A decrease in the abundance of dead leaves during the late spring and summer months also reduces the microarthropod populations, which may constitute a substantial portion of the prey consumed by the pseudoscorpion, *Paisochelifer*, and the immature stages of some spiders, especially *Grammonota*. In the high marsh in summer, the scarcity of prey, particularly of planthoppers, coincides with warm temperatures that impose higher maintenance costs on the spiders and restricts the allocation of energy to growth and reproduction. The consequent decline in spider densities is reversed when planthoppers again become abundant. But, since the spider growth potentials are not large enough to control the growth of the prey, the planthopper quickly achieves densities satiating the spiders. Then predation again supercedes food scarcity as the major regulator of spider densities.

5.7. Tidal Influences

Tides are the major characteristic distinguishing *Spartina* marshes from most other grasslands. The vast majority of species belonging to the grass-grazing food web of the marshes is fundamentally terrestrial in nature and, for the most part, lack aquatic adaptations (Davis and Gray, 1966). Arthropod species, which are not adapted for remaining submerged and cryptic during high tides, are eaten by aquatic predators when they fall to the surface of the water or are forced there by rising water. Even when the flooding waters do not completely submerge the grass, the concentration of the ascending arthropods in the upper canopy of cordgrass creates favorable foraging opportunities for predaceous arthropods and birds. *Orchelimum* will dive into the water to escape immediate danger, swimming to a submerged leaf or stem and remaining there under water. Sibley (1955) observed gulls and marsh hawks (*Circus cyaneus*) preying upon small mammals during spring high tides in *S. foliosa* marshes of California. In some years the flooding of nests can also be responsible for a substantial egg and nestling mortality (Kale, 1965; Post, 1974). The breeding cycle of the song sparrow in *S. foliosa* marshes may be synchronized to periods of lower tidal amplitudes (Johnston, 1956a). Terrestrially-adapted arthropods may be prone to excessive mortality if exposed to inundating tides during stages of egg emergence and ecdysis. Foster and Treherne (1976) suggested that such stages of the life history of the salt-marsh arthropod may be confined to interludes of neap tides.

Contrary to the observations of Arndt (1914), Davis and Gray (1966) found that *Prokelisia* is not submerged during high tide. Instead, this planthopper moves up the stalks and leaves until no space remains for further advancement, whereupon the insect abandons the vegetation and floats upon the surface film until emergent leaves are again encountered. During very high spring tides, which completely submerge large areas of the high marsh meadows, *Prokelisia* populations accumulate on the few remaining patches of emergent vegetation. With the ebbing tide, some of the accumulation of *Prokelisia* is withdrawn from these islands of refuge and redeposited over the marsh. The overall outcome of this sequence of passive transport is the redistribution, without the expenditure of energy involved in active locomotion, of a portion of the planthopper population to the areas of the high marsh where cordgrass growth has been more robust and where greater food resources are presumably available. Passive transport has been reported in other marsh-inhabiting Homoptera. For example, the root aphid (*Pemphigus trehernei*), which inhabits English tidal marshes, apparently employs tidal transport to achieve population dispersal in the first nymphal stage (Foster and Treherne, 1976, 1978). However, tidal activity does restrict planthopper exploitation of the young, nitrogen-rich shoots that develop below the creekbank levees. Such young culms maintain a spatial refuge from the relatively immobile nymphal stages of *Prokelisia*. Indeed, none but the upper leaves of infra-levee plants exceeding one meter in height are normally colonized by the nymph. Adult planthoppers can parasitize such net importers of stored reserves, but this feeding necessitates an expenditure of energy and an increased exposure to predation concomitant with active transport to a higher level in the intertidal zone during flooding tides.

The reactions to tidal flooding exhibited by *Prokelisia* are characteristic of many insects associated with *Spartina*. When conditions are not conducive to web construction or active search of prey, the two most common species of spiders, *Grammonota* and *Clubiona*, as well as the pseudoscorpion, *Paisochelifer*, remain submerged under tidal waters. But when the tides are moderately high and conditions are favorable, *Grammonota* and *Clubiona* move to the upper canopy to construct webs or to forage. In cases of extremely high tides they eventually cease all such activity as emergent vegetation becomes scarce and seek refuge on vegetation below the surface of the water.

Although these spiders may utilize air pockets trapped within leaf sheaths and furled dead leaves, Arndt (1914) found that, even without access to air pockets, *Grammonota* and *Clubiona* could withstand submergence in glass vessels up to 24 and 30 hours, respectively, with little apparent ill-effect.

5.8. Spatial and Seasonal Influences

The marshes adjoining Sapelo Island support a highly productive community of arthropod herbivores, which in turn support a diverse array of predaceous arthropods and vertebrates. The ecology of these consumers, particularly their influence on the dynamics of *Spartina*, is not understood well. *S. alterniflora* and its associated herbivores extend along a broad latitudinal range on the Atlantic coast from Newfoundland to northern Florida, a range which spans disparate climatic conditions and regions of dissimilar marsh structure and expanse. Along this coastal zone there should exist a varying interplay of biotic and abiotic regulatory agents which influence the distributions and abundances of the primary and secondary consumer populations. Whittaker (1971), for instance, provided evidence that population regulation of a cercopid Homoptera species was accomplished by control by abiotic agents in the peripheral areas of the species distribution and by feedback control by biological agents within areas typifying its optimum habitat.

The major habitat distinction perceived by this predominantly terrestrial assemblage is that between the tall *Spartina* growing on the creek banks and levees and the short *Spartina* in the high marsh. Major seasonal shifts in density occur differently in each of these zones and migrations of herbivores between the two are common. Changes in prey availability affect the seasonal density patterns of predators and parasites.

The total complement of primary consumers in the salt marsh ingests less than 10% of the annual net primary production. Determining whether or not this rather small percentage adequately reflects the impact of these consumers and their attendant predators will require much additional observation and experimentation.

The major herbivorous species associated with *Spartina* specialize heavily, if not exclusively, on this macrophyte. This observation concurs with the Rhoades and Cates's (1976) hypothesis that spatial and temporal predictability of a host plant should select for specialization in its attendant herbivores. The spatial and temporal predictability of *Spartina* is manifested by the broad monospecific expanses and the continuous growing season available for this macrophyte in Georgia.

6. Aerobic Microbes and Meiofauna

R.R. Christian, R.B. Hanson, J.R. Hall, and W.J. Wiebe

Heterotrophic microorganisms in a detritus-based salt-marsh system are a major link in the mineralization and transformation of organic matter. The biomass and activity of heterotrophic microorganisms in the salt-marsh estuary at Sapelo Island have been examined with regard to both habitat and metabolic roles. Habitat separation followed the divisions established in Chapter 1 by concentrating on microorganisms in tidal water, on living and dead emergent macrophytes, and in the soils and sediments. Metabolic differentiation was made on the basis of aerobes versus anaerobes. In this chapter we consider some of the organisms responsible for the aerobic transformations of organic matter and the factors influencing their regulation and interaction.

Aerobic microorganisms include bacteria, fungi, protozoans, and microscopic metazoans. The aerobic-anaerobic interfaces within the salt-marsh soils are spatially complex. Because of tidal flooding, the soil is generally saturated with water and only the top few millimeters of soil are oxidized. However, even here anaerobic microzones are found. Aerobic zones also occur in the soil along the surface of animal burrows and within the rhizosphere of living plant roots. Oxygen diffuses from the roots of *Spartina*, creating an aerobic zone at the soil-root interface. Although most of the salt-marsh soil is anaerobic, all materials entering or leaving the anaerobic soil must pass through the aerobic zone, where much of the nutrient transformation and the most rapid degradation of organic matter occur.

The studies at Sapelo Island were designed to answer three general questions:

1) What are the temporal and spatial distributions of microbial standing stocks?
2) What are the rates of utilization of organic matter by the aerobic microbial communities?
3) What factors regulate the metabolism of aerobic microbial communities?

6.1. Microbial Standing Stocks

Adenosine triphosphate (ATP) was first used as a measure of microbial biomass in seawater by Holm-Hansen and Booth (1966). Since then, because of the ease, sensitivity, and rapidity of this technique, ATP determinations have been made in a variety of aquatic and terrestrial environments. There are a number of problems with the method, and it is constantly undergoing refinement. Nonetheless, the technique remains useful, if interpreted with care.

A small pool of dissolved ATP that is not associated with living organisms exists in aquatic environments (Azam and Hodson, 1977), but its presence in soils or sediments has not been adequately examined. The ratio of ATP to carbon or biomass is sometimes assumed to be constant, but the stability of the ratio is suspect. Holm-Hansen (1973) proposed a C:ATP (w:w) of 250 based on laboratory and field measurements of algae, bacteria, and micro- and macrozooplankton. However, others have found different ratios, and there are large variances around mean ratios (Ausmus, 1973; Witkamp, 1973; Bancroft et al., 1976; Kaczmarek et al., 1976; Sikora et al., 1977; Karl et al., 1978; Oades and Jenkinson, 1979). As a result, comparisons of "microbial" carbon, as determined by ATP analyses, to the carbon associated with other portions of the community or with "nonliving" carbon should be viewed critically (Christian et al., 1975; Jassby, 1975; Sikora et al., 1977). Wiebe and Bancroft (1975) have suggested that the variation associated with ATP as an adenine nucleotide index of biomass may be reduced by the use of the total adenylate pool (ATP + ADP + AMP). This will not separate microbial adenylates from those of algae or metazoans where nonmicrobial biomass is dominant, as in aerobic sediments and most euphotic natural waters.

Spatial and temporal distributions of microorganisms have been estimated from ATP concentrations in the salt-marsh soils and estuarine waters at Sapelo Island. Christian et al. (1975) monitored ATP concentrations in soils of both tall (TS) and short (SS) *Spartina* marshes (Figure 6.1.), in a tidal creek, the Duplin River, and Doboy Sound. ATP was extracted from soils with boiling sodium bicarbonate by the method of Bancroft et al. (1976) and assayed by the firefly lucifern-luciferase system. In all salt-marsh soils ATP concentrations decreased with depth (Christian, 1976), but the ATP concentrations in the TS region decreased less with increasing depth than those in the SS region. Within the uppermost 10 cm, the concentrations in the SS region always exceeded those in the TS region, while below 15 cm, the reverse was true. This pattern also occurred in a North Carolina salt marsh (Rublee et al., 1978) and estuarine sediments (Ferguson and Murdoch, 1975; White et al., 1979). A similar pattern emerged for nematode and diatom numbers and biomass at Sapelo Island (Williams, 1962, Teal and Wieser, 1966), and for viable and direct bacterial counts in other salt marshes (Sivanesen and Manners, 1972, Rublee et al., 1978). ATP fluctuated seasonally, particularly in the SS soils. In both TS and SS soils the maximum concentrations of ATP in the top 10 cm occurred during July, although below that depth the maximum concentration was less marked. In a North Carolina salt marsh the maxima of ATP concentration, bacterial number, and bacterial biomass occurred in the fall (Rublee et al., 1978).

The microbial biomass was estimated in surface waters from three estuarine locations: Doboy Sound, 2 km from its mouth, the Duplin River, approximately 0.5 km

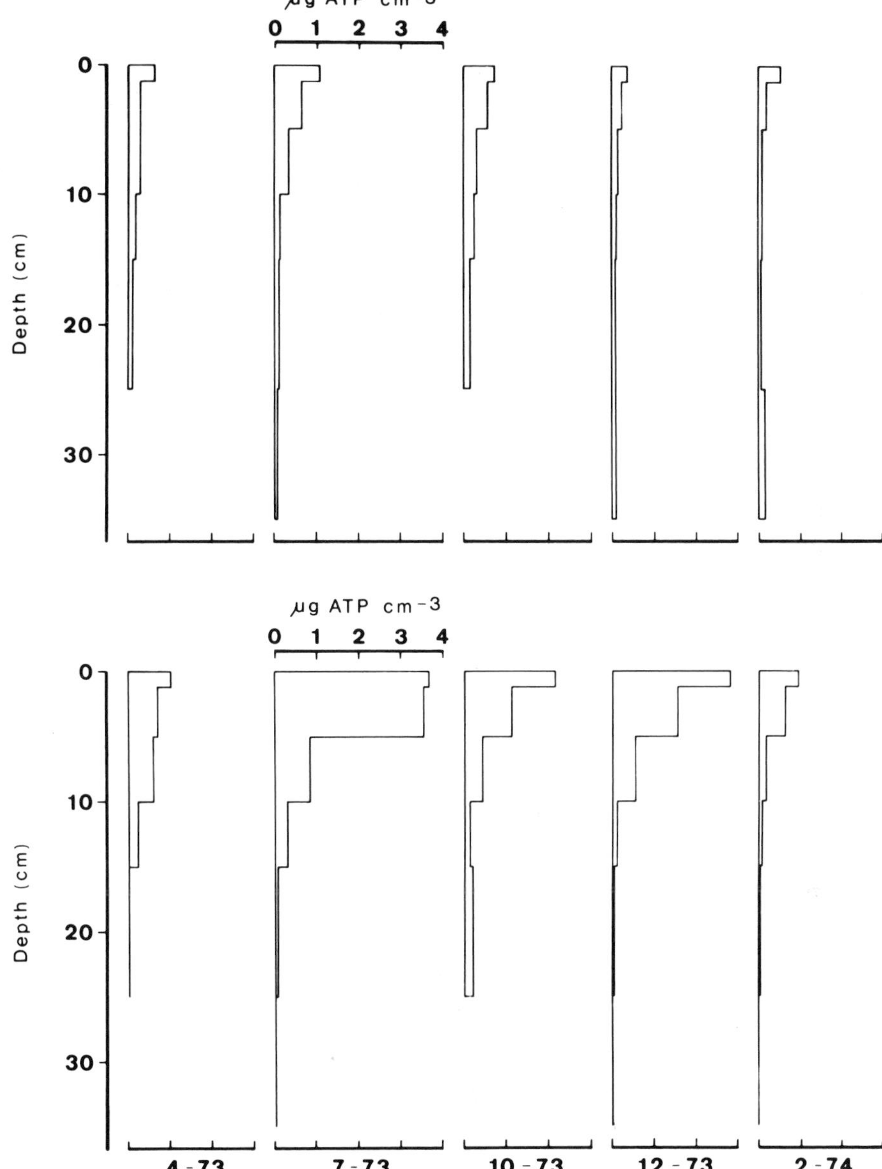

Figure 6.1. Vertical and seasonal distribution of ATP in salt marsh soils from tall *Spartina* (upper) and short *Spartina* (lower) regions. Adapted from Christian et al. (1975) with permission of Williams and Wilkins Company. Copyright (1975) by the Williams and Wilkins Company.

from its mouth, and Study Creek, a first-order marsh creek 2 km from the mouth of the Duplin River (Hanson and Snyder, 1979). ATP concentrations at all locations varied seasonally, and concentrations peaked during the warmer months (Figure 6.2). Although POC concentrations and viable bacterial densities decreased with distance from the marsh, ATP concentrations were similar at all locations. In Study Creek and Duplin River water, approximately 85% of the total ATP was associated with particles < 64 μm and 67% with particles < 10 μm. In Doboy Sound 66% of total ATP was associated with particles < 64 μm and 33% with particles < 10 μm.

The vertical distribution of ATP concentrations within marsh soils depended on marsh type, differing between TS and SS soils. ATP concentrations in the water did not differ consistently in a horizontal transect from a tidal creek in the marsh to Doboy Sound. Soil concentrations of ATP were as much as three orders of magnitude greater than the concentrations in the tidal water. ATP concentrations in both the soils and in the water differed with season, with the largest concentrations occurring during warmer months. After correction for area and volume of both tidal water and soil in the Duplin River estuary, 79% of the standing stock of the microbial community is associated with the soil in the marsh.

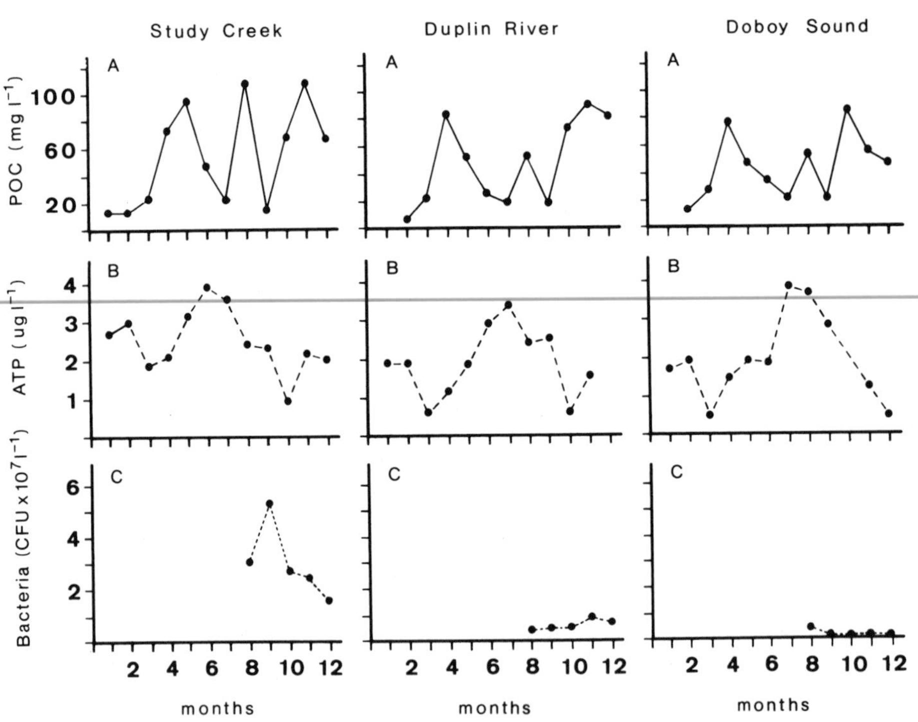

Figure 6.2. Seasonal and spatial relations of particulate organic carbon (POC), biomass (ATP), and bacteria (CFU) concentrations in unfractionated water samples in Study Creek (A), Duplin River (B), and Doboy Sound (C) during 1976. Adapted from Hanson and Snyder (1979) with permission of Ecology. Copyright (1979) by the Ecological Society of America.

6.2. Aerobic Utilization of Organic Matter

Metabolic measurements in the marsh are hampered by methodological difficulties, because the organic matter that feeds an aerobic microbial community is a complex of compounds, in different physical states and from different sources. This complex may consist of common biochemical monomers, their polymeric forms, and condensates. These may be particulate, colloidal, or dissolved; they may have plant, animal, or microbial origins. It is impossible, with current techniques, to trace the utilization of this complex completely. Metabolism of the standing dead *Spartina* community and subtidal sediment surfaces was measured by 0_2 uptake. Aerobic microbial community metabolism in the rhizosphere was not studied, but is a significant area for future research.

6.2.1 Soils and Sediments

The rates of oxygen consumption by the microbial communities in TS and SS soils and subtidal sediments were sampled using sets of eight cores, three of which were poisoned with formalin in order to correct for chemical oxygen demand. Determination of the rate of change of oxygen lasted three hours. To measure Q_{10} values, cores were acclimated for four hours at the desired temperature. The rates (Table 6.1) are equal to or lower than those reported previously (Pomeroy, 1959; Gallagher and Daiber, 1974; Hopkinson et al., 1978). Either variations in methodology or differences in site selection could be the cause of the disparities. Respiration rates of subtidal sediments and of soils of the tall *Spartina* region were high and similar, while rates in the short *Spartina* region were slightly lower, and the rates of sand were the lowest. All sites varied seasonally, probably in response to temperature differences.

The Q_{10} values for biological oxygen uptake in the TS and SS soils were 1.20 ± 0.08 (1 standard error) and 1.27 ± 0.09. These values are similar to those reported by Gallagher and Daiber (1974) for marsh soils in Delaware. The utilization of specific organic compounds by the microheterotrophs within SS soil surfaces was determined by radioisotopic labeling (Christian and Hall, 1977). Organic compounds labeled with ^{14}C were made available to the microheterotrophs through gently aerated water overlying the surface of intact soil cores. The time course of decrease in radioactivity in the water and the evolution of $^{14}CO_2$ were observed. Uptake followed a first-order kinetic model, as it did with anaerobic uptake of ^{14}C-glucose in slurries (Chapter 7). As a result, turnover time (standing stock of substrate divided by uptake rate, or the inverse of the specific uptake rate) was a measure of the potential heterotrophic activity of the soil community (Christian and Hall, 1977; Christian and Wiebe, 1978). No seasonal patterns emerged with either turnover time of substrate or percent mineralization (uptake and subsequent respiratory loss expressed as percent respired per hour). Turnover times of glucose ranged from 1.80 to 9.15 hours, and percent mineralization varied from 9.2 to 36.4.

To measure carbon turnover in an SS marsh soil, ^{14}C-glycine was introduced to the soil surface *in situ*. The intial uptake of glycine by the soil microheterotrophs was rapid and similar to that found in cores in the laboratory, but the turnover was slow.

Table 6.1. Soil and Sediment Oxygen Consumption, mg O_2 m^{-2} h^{-1}, in the Marshes of Sapelo Island and the Duplin River[a,b]

	February (12°C)	April (18°C)	July (28°C)	October (25°C)	December (15°C)
Subtidal sand (Duplin)	10.63± 4.80	14.22± 8.20	20.08± 6.73	14.66± 9.08	9.67± 3.47
	n = 31	n = 16	n = 26	n = 23	n = 32
Subtidal mud (Duplin)	23.88±10.17	32.97± 9.84	40.08±12.34	38.50±10.41	26.58± 8.62
	n = 18	n = 25	n = 30	n = 26	n = 24
Tall *Spartina*	18.89± 7.44	38.44±12.71	42.74±10.13	40.33± 8.23	28.28± 9.17
	n = 30	n = 28	n = 26	n = 20	n = 27
Short *Spartina*	16.05± 6.88	21.74± 9.11	34.21± 9.42	31.94± 9.42	21.28± 4.36
	n = 20	n = 28	n = 26	n = 30	n = 27

[a] Data expressed as mean ± one standard deviation, corrected for chemical oxygen demand. Measurements were made on 5 cores for periods of 0.5 – 3.5 h, and n = the number of rate measurements.

[b] Previously unpublished data of J. R. Hall.

After 31 days, 50% of the initially incorporated glycine remained in the soil. Thus, the uptake of organic substrate is rapid, but once that substrate has been incorporated as microbial biomass, the turnover of carbon is much slower.

6.2.2 Macrophytes

The microbial community on the leaves and stems of the plants in the marsh initiates detritus formation. On decaying *Spartina*, fungi dominate initially, with colonization by bacteria and small animals occurring later (Gessner et al., 1979; Gessner and Goos, 1973). Gallagher and Pfeiffer (1977) measured underwater oxygen consumption of standing dead tall *Spartina*, short *Spartina*, and *Juncus*. Respiratory rates of both growth forms of *Spartina* were similar (Figure 6.3), while respiratory rates on the standing dead *Juncus* were lower than those on *Spartina*. Respiratory rates were lower in winter, increased in the spring, decreased again in late June and July, and increased in late summer and early fall. The reason for the summer decrease is not known. Gallagher and Pfeiffer (1977) ruled out temperature and water stress but could not suggest any alternative reasons for the decrease. Similar results have been found for living *Spartina* biomass and DOC concentrations in the Duplin River (Reimold et al., 1975) No seasonal trends were observed when respiration rates were measured for each community throughout the year, and Q_{10} values ranged from 1.54 to 2.27.

6.2.3 Plankton

The contribution of microheterotrophic plankton to secondary productivity was studied in relation to tide, season, and particle size. The microheterotrophic plankton may be free in the water column or associated with detritus particles (Wiebe and Pomeroy, 1972; Christian and Wiebe, 1978). As Figure 6.4 indicates, detritus in the range of 10 to 180 μm suspended in the estuary took up labeled glucose more rapidly than did detritus fractions < 10 or > 180 μm (Hanson and Wiebe, 1977), but 90% of the biomass (ATP) was in the < 10 μm fractions. Thus, far more bacteria were unattached than were attached. The attached and unattached microbial plankton maintained different degrees of cellular activity; of the bacterial colony-forming-units (CFU), 81% were in the < 10 μm fraction, but relatively more heterotrophic activity was found with attached bacteria. In order to be active, bacteria in the estuary may require a particle for attachment, and particle origin (plant matter or fecal matter), as well as particle size, may influence both biomass and metabolic activities of the associated microorganisms. Some of the particles responsible for heterotrophic activity probably are fecal (Pomeroy and Deibel, 1980).

As Figure 6.5 shows, the uptake of glucose varied during the tidal cycle (Hanson and Wiebe, 1977). Always greatest on an ebbing tide and lowest at slack tide, activity was apparently related to the resuspension of the detritus. There appears to be a close coupling of the salt-marsh sediments and the sediment bed load in the tidal creeks and rivers with the suspended sediments in the water (Chapter 9.) proportionately more large particles than small ones were suspended and deposited during a tidal cycle.

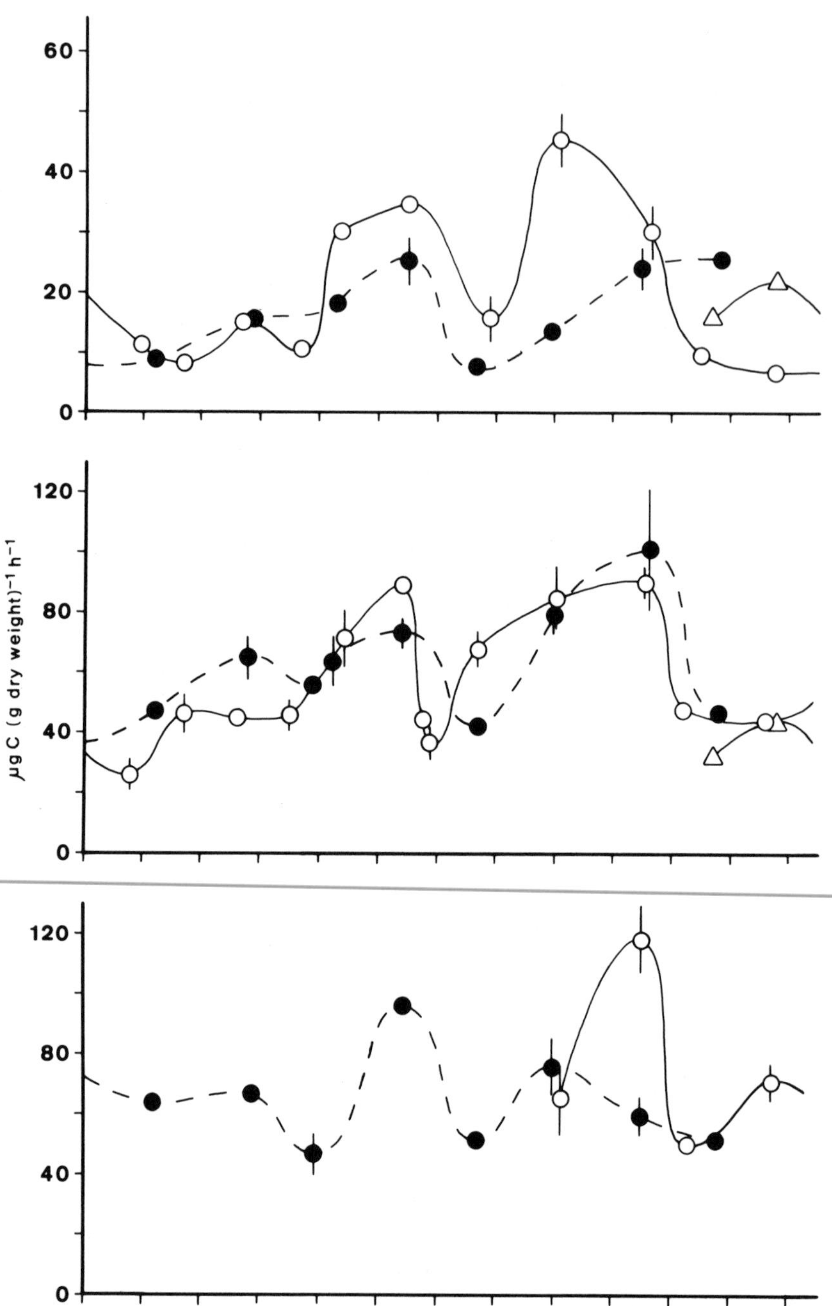

Figure 6.3. Aquatic respiration of the standing dead plant community in *J. roemerianus* (upper), tall *S. alterniflora* (middle), and short *S. alterniflora* in the Sapelo Island marsh. Bars = ± 1 standard error of the mean (n = 3), 1972 (△). 1973 (○), 1974 (●). Adapted from Gallagher and Pfeiffer (1977) with permission of Limnology and Oceanography. Copyright (1977) by the American Society of Limnology and Oceanography.

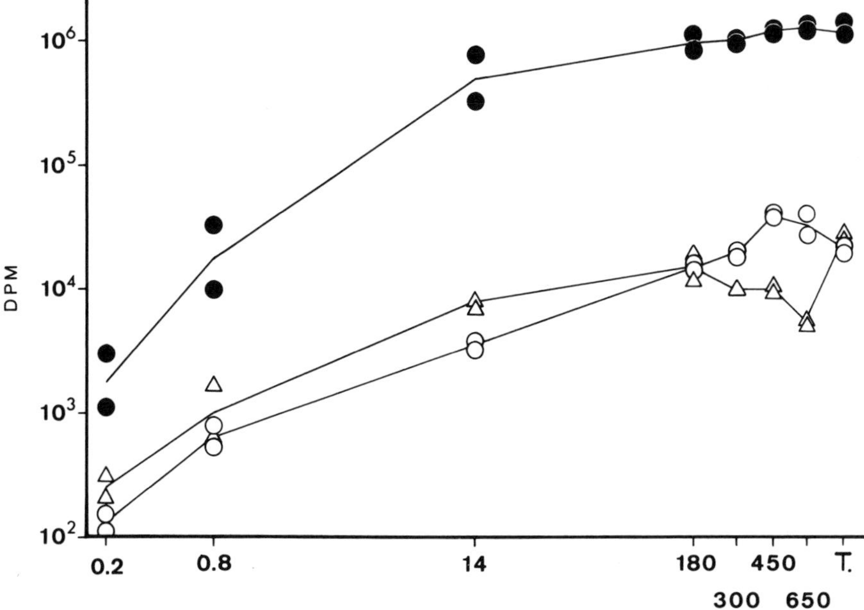

Figure 6.4. Heterotrophic activity of various suspended particle sizes (μm) in the water of the Duplin River at mid-ebb tide (●) mid-flood tide (○), and high tide (△). Adapted from Hanson and Wiebe (1977).

The studies just described did not specify activity within each tidal segment of the Duplin River (Chapter 2). A study of glucose concentration and uptake by the microheterotrophs can serve as a model of labile DOC dispersion in the Duplin River (Hanson and Snyder, 1980). Over several tidal cycles the glucose fluxes, but not the glucose concentrations, were relatively constant from day to day in the river (Figure 6.6). Concentrations were largest near the mouth and head of the Duplin River and lowest in the center. Some component of the marsh, probably *Spartina*, is a glucose source, since there was a longitudinal gradient in glucose concentration from the marsh to river. However, the fact that the gradient was not strong suggests rapid consumption of labile DOC by the heterotrophic community. Another source of glucose in the estuary may be the transformation of plant and algal detritus by benthic microorganisms. However, since glucose concentrations in a small stream running off the marsh at low tide were lower than those in the Duplin River, microorganisms had apparently consumed the glucose from the water draining over the marsh surface.

The major source of glucose in the river is probably phytoplankton or bacterioplankton. This is supported by the detection of high glucose concentrations during an algal bloom, and by the higher glucose concentrations found in Doboy Sound and Duplin River entrance waters than in the middle of the Duplin River (cf. Chapter 2 on hydrology). As Figure 6.7 indicates, microheterotrophs took up labeled substrates more rapidly in the marsh creek than in the river or sound (Hanson and Snyder, 1979, 1980). Maximum uptake velocity, V_{max}, was used as an index of potential heterotrophic activity. The apparent increase in potential activity in the creeks was attri-

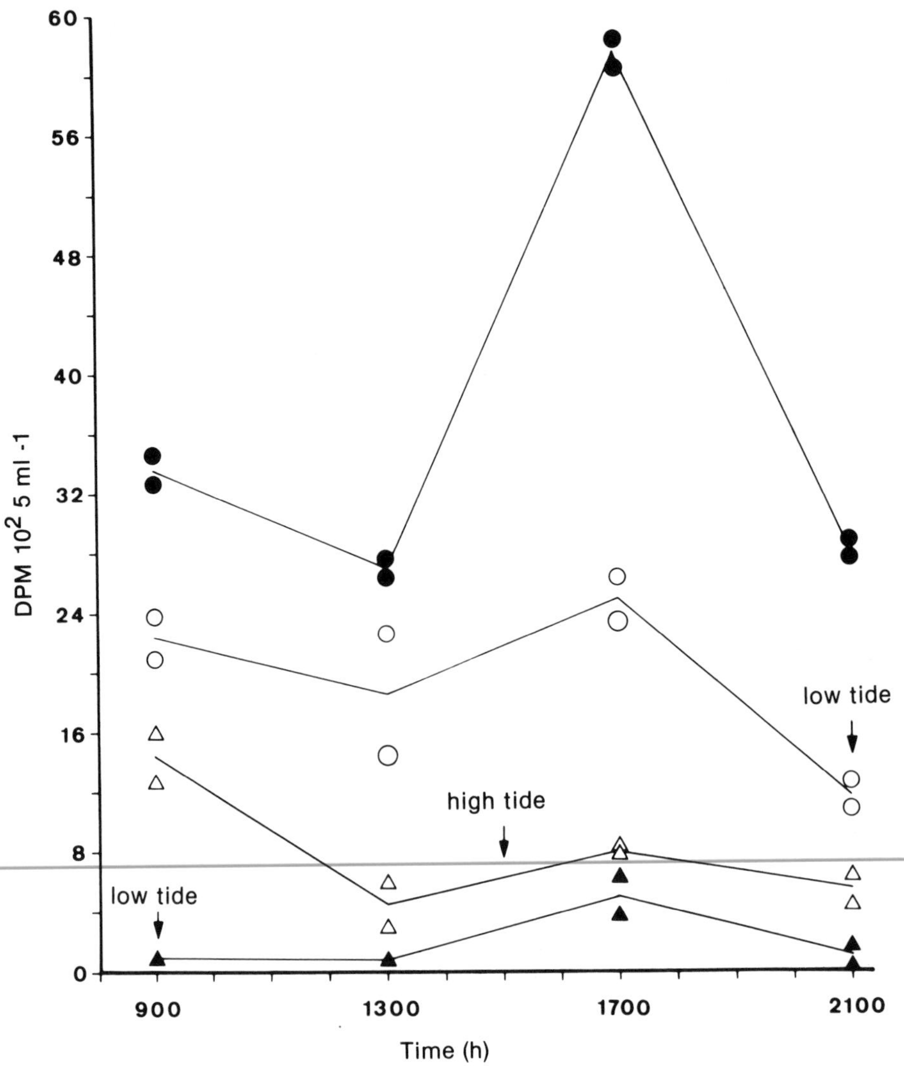

Figure 6.5. Heterotrophic activity of particles of various sizes in Duplin River water during one tidal cycle. The fractions are total activity (●), < 180 μm (○), < 14 μm (△), and < 0.8 μm (▲). Adapted from Hanson and Wiebe (1977). Copyright Springer-Verlag, Berlin.

buted to large populations of bacteria. Bacterial counts (CFU) were positively correlated with glucose uptake (Hanson and Snyder, 1979), whereas particulate organic carbon (POC) and ATP concentrations showed little or no change with location (Figure 6.2). Preliminarly evidence (Hanson and Snyder, 1979) and further unpublished data suggest that grazing pressure might cause some of the variation. For example, removal of organisms > 64 μm prior to incubation resulted in an increase in glucose V_{max} in Study Creek. Little is known about competing substrates which

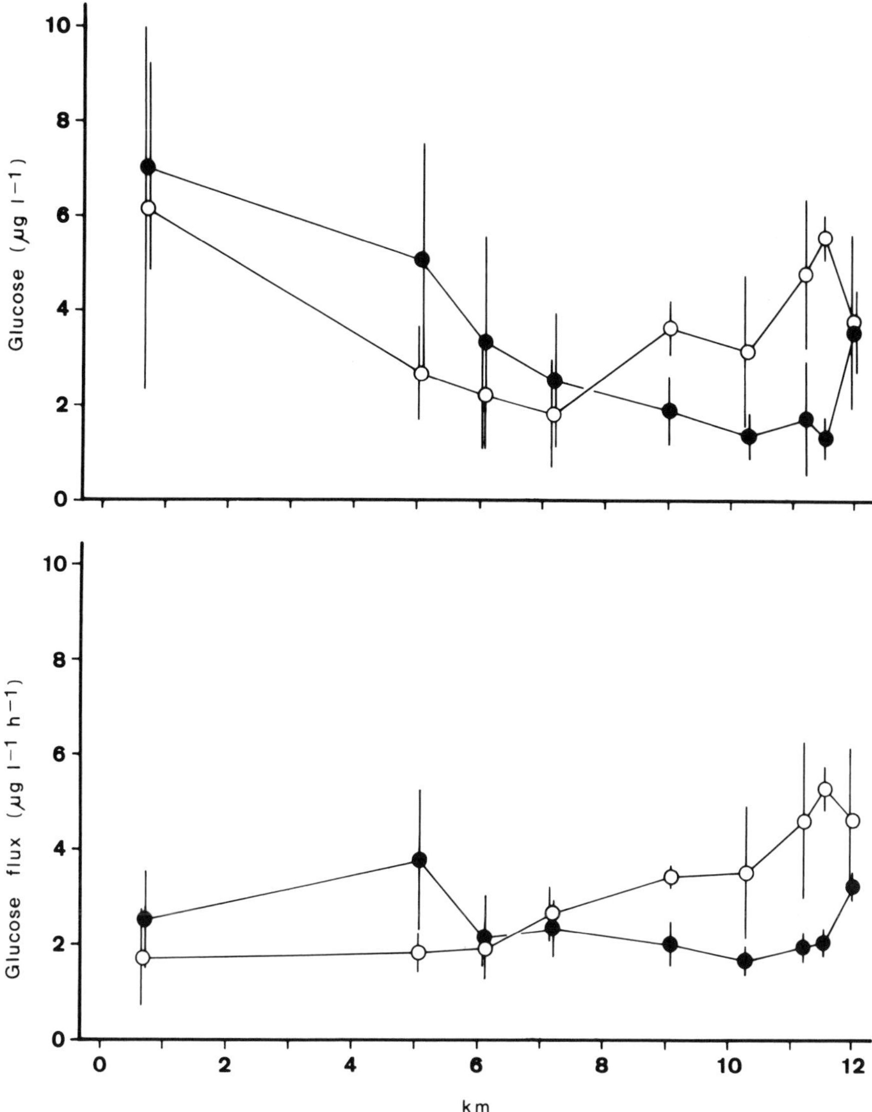

Figure 6.6. Glucose concentration, $\mu g\ l^{-1}$, and glucose flux, $\mu g\ l^{-1}\ h^{-1}$, in the Duplin River, August 1979, at low tide (○) and high tide (●). Adapted from Hanson and Snyder (1980) with permission of Limnology and Oceanography. Copyright (1980) by the American Society of Limnology and Oceanography.

might have influenced these results. The microheterotrophs in the marsh creeks were 100 times more active in the summer than in the winter (Figure 6.7). Water temperature was 8 to 10°C from November through March and 25 to 30°C from June through August. In the river and sound there were no significant seasonal changes in heterotrophic activity.

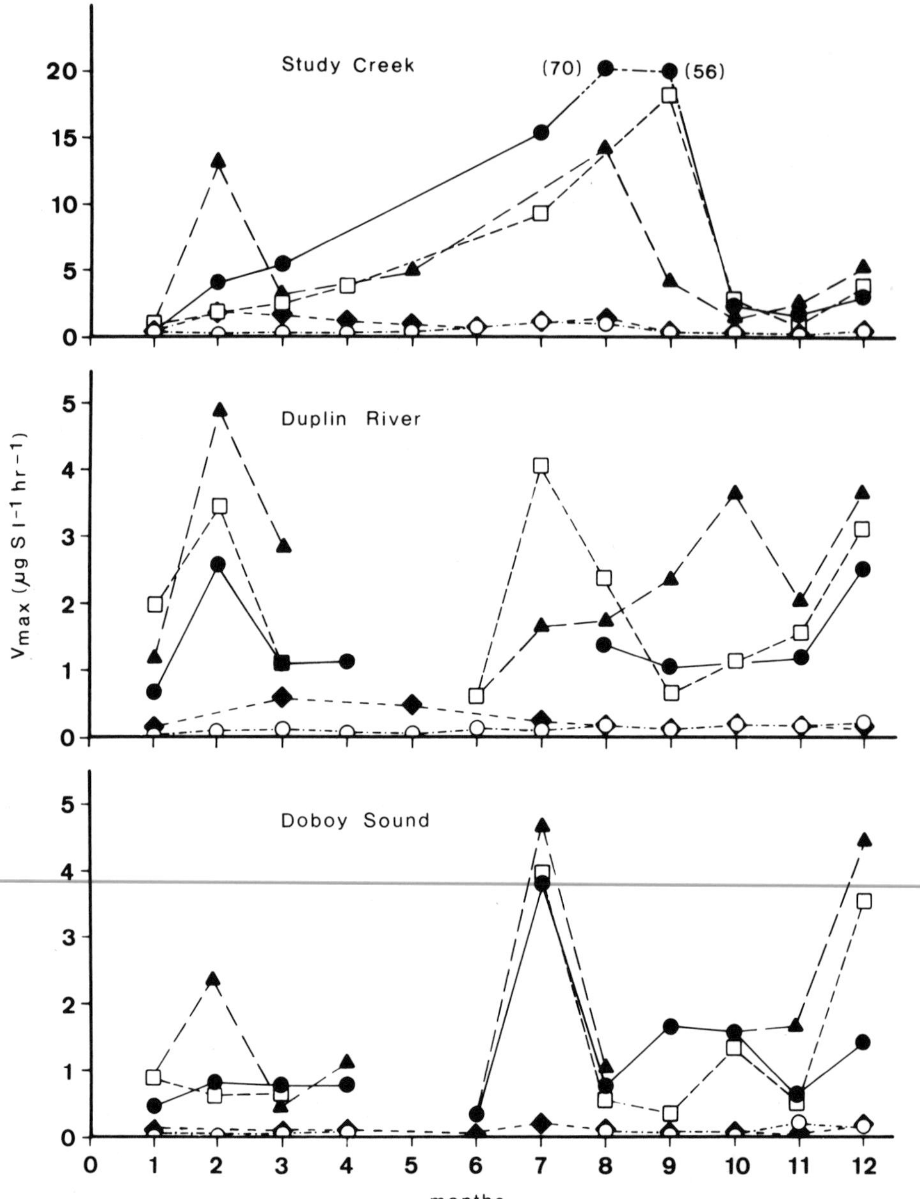

Figure 6.7. V_{max} values for glucose (■), aspartic acid (♦), glycolic acid (▲), lactic acid (●), and alanine (○) in unfiltered water from Study Creek, the Duplin River, and Doboy Sound during 1976. Lactic acid concentration in Study Creek goes off scale to the values shown in parentheses.

Microheterotrophs in the Duplin River and marsh creeks, the major consumers of labile DOC, utilized glucose, glycolic acid, and lactic acid more readily than aspartic acid and alanine (Figure 6.7). The maximum rate of utilization was similar to rates in other estuaries. For example, glucose V_{max} values ranged from 0.04 to 25 mg C m^{-3} h^{-1} in waters near Sapelo Island, whereas in a North Carolina estuary the range was 0.06 to 9.6 mg C m^{-3} h^{-1} (Crawford et al., 1974). However, in the Sapelo Island area, aspartic acid and alanine (0.005 - 1.4 mg C m^{-3} h^{-1}) were not utilized as quickly by the microheterotrophs as in the North Carolina estuary (0.15 to 69 mg C m^{-3} h^{-1}). In spite of the differences in the rate of uptake between these two estuaries, the organic compounds were respired at similar rates. The turnover times of labile organic compounds were on the order of a few hours. For example, turnover times of glucose, alanine, aspartic, glycolic and lactic acids were generally less than five hours during the entire year, showing that labile DOC is turned over rapidly in the estuary.

Michaelis-Menten kinetics were used to characterize glucose turnover. The calculated $(K + S_n)$* value provided an upper limit of both K and S_n. The $(K + S_n)$ values for the various substrates used in the waters around Sapelo Island ranged from 1 to 50 $\mu g\ l^{-1}$ (Hanson and Snyder, 1979). For glucose, $K + S_n$ was 10 $\mu g\ l^{-1}$. Glucose concentration in these waters was found to be 10 $\mu g\ l^{1}$ (Hanson and Snyder, 1980), suggesting that K is small compared to S_n. A similar result of $(K + S_n) \cong S_n$ was reported for aspartic acid in a North Carolina estuary (Crawford et al., 1974).

The bacteria are probably the major glucose sink. Bacteria, with their relatively efficient transport system, are able to consume compounds at low concentrations (10^{-8} M). Algae, on the other hand, are generally not considered to be significant consumers of labile DOC because of their higher K (10^{-7} M) and the low concentrations of labile compounds in natural waters.

The glucose dynamics in the estuary suggest a temporal and spatial coupling of production of soluble organic compounds by *Spartina* and phytoplankton with consumption by bacteria. Labile organic compounds are rapidly consumed. Consequently, little if any of the labile DOC produced by the marsh leaves the Duplin River as DOC. The principal population involved in this heterotrophic consumption is bacterial. Although a majority of the bacteria are free floating within the estuarine waters, most of the organic matter is consumed by organisms attached to particles. The rate of uptake varies with the tide. Because of settling and resuspension of particles, seasonal variation occurs, a variation probably related to temperature. Thus, the planktonic microheterotrophs regulate and control the dynamics of the labile component of DOC in the Duplin River.

6.2.4 Dissolved Oxygen

A different way to understand the heterotrophic activity in the Duplin River is to measure the concentration, distribution, and rate of change of dissolved oxygen in the water. The three tidal segments of the Duplin River provide sufficient isolation from

---

*The symbol K is the half-saturation concentration of the organic substrate, i.e., the concentration of substrate at which the rate of uptake is one-half of V_{max}. S_n is the observed substrate concentration.

Doboy Sound and the sea to allow the estuary to reflect the influence of the salt marsh. The Duplin River usually is undersaturated with oxygen, especially in the summer (Frankenberg, 1975). Moreover, oxygen consumption increases in a linear manner from mouth to headwaters (Frankenberg, 1975). This suggests that there is indeed a significant input of labile organic matter from the marsh, and that this organic matter, in addition to organic matter produced by indigenous phytoplankton, is consumed in the river. Quantitatively, most of the respiration probably is bacterial, and a substantial part of bacterial respiration, but not of microbial biomass, is associated with particles. This is reflected in the large variance found in measurement of respiration (Frankenberg, 1975), as well as in the heterotrophic utilization of glucose observed by Hanson and Wiebe (1977). Therefore, there is evidence of a net export of organic matter, and of energy, from the marsh. However, most of it is consumed within the Duplin River and its tributary creeks, except for refractory lignocellulosic POC and humic and fulvic DOC (see Chapters 2 and 9).

In addition to respiratory oxygen uptake in the water there is a substantial oxygen demand by both subtidal and intertidal sediments (Frankenberg and Westerfield, 1968; Pomeroy et al., 1972). Sediment oxygen demand is included in the very high total respiration values reported by Frankenberg (1975). This gives the impression that the water is more heterotrophic than it really is. Sediment respiration appears to be 30 to 90% of total respiratory oxygen uptake, and respiration in the water ranged from 0.4 g C m^{-2} day^{-1} in winter to 4 g C m^{-2} day^{-1} in summer (Pomeroy et al., 1972). Ragotzkie (1959) reported a mean planktonic respiration rate of 0.64 g C m^{-2} day^{-1}. Respiration in submerged sediments calculated by Pomeroy et al. (1972) averaged 1.5 g C m^{-2} day^{-1}, while the results presented in this chapter averaged 0.2 g C m^{-2} day^{-1}.

Aerobic respiration in the Duplin River watershed is summarized in Table 6.2. Oxygen consumption was prorated on the basis of relative areas of tall, mid-height, and short *Spartina* areas given by Reimold (1972). The values were corrected further for total marsh, using the values of 79% marsh and 21% tidal creeks and river. The respiratory rate of the microbial community in the standing dead *Spartina* was described earlier in this chapter. This decomposer group accounted for one-half of the total oxygen demand, a demand equivalent to 30% of *Spartina* shoot production, assuming an RQ of 1.0. The marsh-soil microorganisms accounted for only 30% of the measured oxygen demand or 6% of total production, a much lower percentage than that of Teal and Kanwisher (1961). (See discussion in Wiegert, 1979.) To obtain the value of 90 g C m^{-2} yr^{-1} in Table 6.2, the respiration rates in the soils of the tall and short *Spartina* region were increased by 22% to allow for increased surface area caused by crab burrows (Montague, 1980). Values were converted to carbon, using an RQ of 1.0, and prorated for relative areas of various marsh types. We believe that the major cause of the discrepancy between our values and those obtained by Teal and Kanwisher was the correction for chemical oxygen demand used to obtain the data of Table 6.1. This correction was not used by Teal and Kanwisher. Aerobic respiration in the soils and sediments is a demand equivalent to only 10% of the net primary production, as shown in Table 6.2. Anaerobic fermentation, methanogenesis, and the physical transport of plant detritus from depth to the surface are other processes which account for disappearance of the carbon produced by the roots and rhizomes of *Spartina*. The magnitude of these alternative routes is discussed in Chapters 7 to 10.

Table 6.2. Aerobic Microbial Respiration of Carbon in the Duplin River Watershed

Microbial Community	$g\ C\ m^{-2}\ yr^{-1}$	% Total Respiration	% Aerobic Primary Production[a]	% Total Primary Production[b]
Standing dead *Spartina*	180	53	22	12
Marsh soil surface	90	26	11	6
Subtidal sediment surface	20	6	2	1
Water	50	15	6	3
Total	340	100	41	22

[a] 780 $g\ C\ m^{-2}\ yr^{-1}$ used as value base.
[b] 1380 $g\ C\ m^{-2}\ yr^{-1}$ used as value base.

6.3. Regulation of Microbial Communities

Measurement of microbial biomass and metabolic rates provides basic information on standing stock and flow of nutrients and energy through the salt marsh. But to understand why these occur as they do, it is necessary to know which factors regulate or control microbial distribution and activity. Much of our research has been directed toward evaluating the sources and nature of regulation, and in this section we describe some of the research as it relates to the microbial communities in marsh soil and water. Other discussions of regulatory processes are found in Chapters 3, 5, 7, 9, and 10.

6.3.1 Regulation in Marsh Soil

When ATP concentration was compared to the concentrations of carbon and nitrogen in soil in the Sapelo Island salt marsh (Christian et al., 1975), microbial carbon and nitrogen represented less than 1% of the total organic carbon or nitrogen pools (Table 6.3). Rublee et al. (1978) found the same ratio in a North Carolina salt marsh. Plant material can account for a large fraction of the total carbon and nitrogen in the soil system (Gallagher, 1974). But soil organic carbon and nitrogen are often refractory and not available to the microbial community. Since an undoubtedly large component of carbon, and perhaps nitrogen, is associated with refractory soil humus, it should not be assumed that the large pool of organic carbon or nitrogen indicates a lack of limitation to microbial growth by either element.

Sources of fixed organic carbon on the marsh surface are algae, chemosynthetic bacteria, and *Spartina* litter. Below the marsh surface the major ultimate source of organic carbon is *Spartina*, either in the form of DOC or of dead roots and rhizomes. The DOC is readily available for microbial consumption (Grineva, 1963; Hale et al., 1971). The POC available from roots and rhizomes comprises a number of compounds, from intracellular sugars and amino acids to structural lignocellulose. Decomposition converts much of the structural material to humus (Oglesby et al., 1967), a refractory material.

To ascertain the determinants of control of microbial biomass in marsh soil, we designed experiments in which small plots of natural marsh were perturbed (Christian, 1976; Christian et al., 1978). In one such experiment, the degree of linkage between short *Spartina* primary production and the microbial community was tested by removing the capacity for production. Aboveground plant parts were clipped from plots, roots and rhizomes were pruned from around the edges of the plots, and plastic barriers were inserted into the soil around the plot edges to prevent regrowth from outside. If root exudates were the major source of organic matter to the microbial community, microbial ATP concentrations would have decreased shortly after the cessation of primary production. If the major source of organic matter were decomposing roots and rhizomes, a longer time period would have been needed to observe an effect. Finally, if soil humus were the major source, a decrease in ATP concentration would not have been seen at all. Although a lack of response by the microbial community would be suggestive of this last hypothesis, it would not strongly support it, and any response to such seemingly drastic perturbations should be viewed with caution. Many other factors might limit the soil microbial community. Because we screened out roots and rhizomes, discussion here is limited to the microbial community not associated with the rhizoplane.

Microbial ATP concentrations in the soil did not decrease for 12 months after initial clipping and pruning (Christian et al., 1978). This was also true for total adenylate concentrations (TA), microbial community adenylate energy charge (CAEC), and certain anaerobic metabolic processes (Chapter 7). After 18 months ATP concentrations in the clipped, pruned, and enclosed plots did decrease significantly as compared to concentrations in control plots. We do not validate the hypothesis of root exudates as a major limit on the microbial community. Moreover, Gallagher (Chapter 3) was unable to measure exudation of dissolved organic carbon from *Spartina* roots. Decreases in root and rhizome biomass were found in clipped and pruned plots, indicating active decomposition, and the pool of decomposing plant matter may have remained large enough to support normal maintenance of microbial biomass up to one year after clipping. While this experiment neither confirmed nor invalidated the release of DOC by *Spartina* roots, it demonstrated that decomposition of POC from roots and rhizomes can support microbial community metabolism throughout the year.

Enrichment of a marsh for one year with sewage sludge failed to increase ATP concentrations (Christian, 1976), and glucose additions to intact cores incubated in the laboratory failed to stimulate ATP production (Christian and Wiebe, 1979). However, addition of glucose to soil slurries did stimulate ATP production. It was concluded that the dilution of soil in a slurry may have relieved inhibitory regulation factors found in intact soils. Regulation through inhibition would have also explained the results of the previously described experiments.

Another *in situ* perturbation experiment was initiated in February 1976 (Christian et al., 1978). Forty 0.1 m^2 plots were arranged in four blocks, each block containing two rows of five plots. Five randomly selected plots within each block were clipped, pruned, and enclosed as previously described, and five were not. One perturbed and one control plot from each block were injected monthly with NH_4NO_3, glucose, both NH_4NO_3 and glucose, or distilled water. To one set of plots, rhodamine WT was added as a tracer of water movement. The plots were sampled during a neap tide period in July 1976. Two 3.2 cm cores were taken from each plot and subdivided into 0 to 5, 5 to 10, and 10 to 15 cm sections. Each section was assayed for salinity, then screened and assayed for ATP concentration. The control plots were clipped, and wet weight of shoots was determined. *Spartina* production in unclipped plots was stimulated by the addition of nitrogen and retarded by the addition of glucose. The microbial community resisted most of the perturbations (Figure 6.8), although ATP concentration did increase in the first 5 cm when nitrogen was added to the unclipped plots. In this one instance, plant aerial biomass was three times the control biomass, but the increase in ATP concentrations was considerably less. Pooling all the ATP concentrations from clipped treatments and comparing them with those from unclipped treatments showed slight, but statistically significant, decreases in the clipped plots in two of the three depth ranges. All differences found were within 50% of the control concentrations. Thus, addition of labile carbon and nitrogen compounds did not increase ATP concentration significantly in the long term. At this point, the hypothesis of nutrient limitation of the microbial community was rejected, but regulation by inhibition remained a viable hypothesis.

Table 6.3. Contribution of Microbial Carbon to Organic Carbon and Nitrogen in Salt-marsh Soil during February 1974[a]

Depth (cm)	Microbial C ($mg\ C\ cc^{-1}$) Carbon:ATP		Sediment Organic C ($mg\ C\ cc^{-1}$)	Percent Microbial C of Sediment Organic C	
	100:1	250:1		100:1	250:1
Tall *Spartina* Area					
0-1	0.052	0.13	19.4	0.27	0.67
1-5	0.023	0.058	19.3	0.12	0.30
5-10	0.013	0.033	18.2	0.07	0.18
10-15	0.013	0.033	19.7	0.07	0.17
15-25	0.11	0.028	19.0	0.06	0.15
Short *Spartina* Area					
0-1	0.096	0.24	23.5	0.41	1.02
1-5	0.064	0.16	22.3	0.19	0.72
5-10	0.026	0.065	22.2	0.12	0.29
10-15	0.010	0.025	19.8	0.05	0.13
15-20	0.008	0.020	18.6	0.04	0.11

Tall *Spartina* Area

Depth (cm)	Microbial N ($mg\ N\ cc^{-1}$) Nitrogen:ATP		Sediment N ($mg\ N\ cc^{-1}$)	Percent Microbial N of Sediment N	
	16.8:1	42:1		16.8:1	42:1
0-1	0.0087	0.022	1.52	0.57	1.45
1-5	0.0039	0.0097	1.60	0.24	0.61
5-10	0.0022	0.0055	1.44	0.15	0.38
10-15	0.0022	0.0055	1.51	0.15	0.36
15-25	0.0018	0.0047	1.46	0.12	0.32

		Short *Spartina* Area			
0-1	0.016	0.040	1.56	1.03	2.56
1-5	0.011	0.017	1.54	0.71	1.75
5-10	0.0044	0.011	1.52	0.29	0.72
10-15	0.0017	0.0042	1.08	0.16	0.39
15-25	0.0013	0.0034	1.57	0.08	0.22

[a] From Christian et al. (1975).

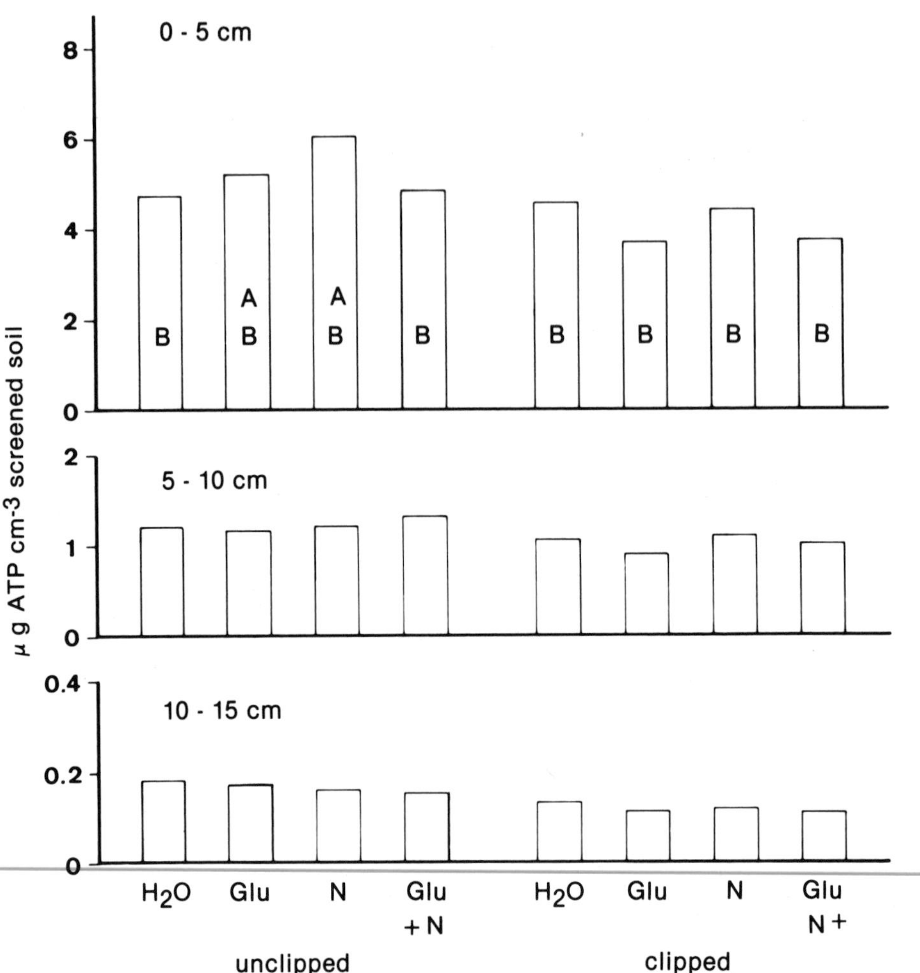

Figure 6.8. Effects of three perturbations on ATP concentrations in three depths in salt-marsh soils, 0-5 cm, 5-10 cm, and 10-15 cm. Plots were injected with distilled water (H_2O), glucose (G), ammonium nitrate (N), and glucose plus ammonium nitrate (G + N). A and B refer to SNK groupings at $P \leqslant 0.05$. Adapted from Christian et al. (1978) with permission of Duke University Press. Copyright (1978) by the Ecological Society of America.

Sampling over a neap tide period, we found that ATP concentrations were higher at the beginning than they were a week later. Because the marsh was not flooded, interstitial salinity increased and the soil dried by evapotranspiration (Christian et al., 1978). With drying and salinity increase, ATP concentrations decreased. The greatest drying and salinity increase were in the top 5 cm, as was the largest decrease in ATP concentrations. Based on these results and others, Christian and Hansen (1980) have postulated that a decrease in water availability in salt-marsh soils acts as a regulator of microbial activity. Other inhibitory regulating factors may include end products of fermentation and dissimilatory SO_4^{2-} reduction (Cappenberg and Prins, 1974).

Although most of the *in situ* perturbation studies were conducted in short *Spartina* marshes, Christian and Wiebe (1978) investigated the restriction of interstitial water movement in a tall *Spartina* marsh where potential microbial activity was higher (Chapter 7), due to the greater water flow in TS soils as compared to SS soils. Water movement was restricted for two months within coring tubes inserted into the soil with the bottom of the cores capped. Water exchange, traced through dye dilution, was reduced within the cores by as much as 80% compared to water exchange in non-restricted controls. ATP concentrations in a 1 to 5 cm depth range increased over the controls in one set of restricted cores but not in another. In no instance was ATP concentration decreased by the restriction of water flow.

The microbial community in salt-marsh soils was largely unaffected by perturbations of nutrient dynamics. When there were effects, they were small—less than 50% alterations in ATP content. The microbial community also responded slowly to alterations in *Spartina* production. Our experiments suggest that inhibitory processes, rather than nutrient limitation, may be the dominant control mechanisms on aerobic microorganisms in salt-marsh soil (Chapter 7).

6.3.2 Regulation in Standing Dead Plants

The availability of organic substrates appears to regulate respiratory oxygen uptake by the community of microheterotrophs on standing dead plants in the salt marsh. The organic substrates for these microbial communities may be plant structural material, such as lignocellulose, or the dissolved components of the plant cell sap made available during senescence and death. When Gallagher and Pfeiffer (1977) measured respiratory oxygen uptake and the annual cycle of release of DOC from standing dead *Spartina*, they found that rates of release were inversely proportional to oxygen uptake. The microbial community acts as a filter for DOC, preventing labile portions from entering the estuarine waters. Consumption of leached DOC was also inferred for the phyllosphere microbial community of living *Spartina* (Gallagher et al., 1976).

Oxygen uptake was more rapid in dead *Spartina* than in dead *Juncus*. The reasons for this are not known; however, Haines and Hanson (1979) had similar findings. In laboratory microcosms, the decomposition rate of *Juncus* detritus was slower than that of *Spartina* detritus, and this difference may account for the lower rates of metabolism in the standing dead *Juncus* community. Haines and Hanson (1979) suggested that slow decomposition is a property of C_3 pathway plants.

6.3.3 Regulation in Estuarine Water

Four factors regulating the activities of microheterotrophs in the Duplin River are temperature, nutritional substrate, substratum, and predation. Microbial activity and biomass are temperature dependent, and biomass is maximal during warmer months and minimal during colder months. Microheterotrophic activity follows this trend in Study Creek, but not in the Duplin River or Doboy Sound. Although a direct temperature effect on the microbes may have been responsible for a major portion of the observed seasonality, the possibility exists for indirect effects. For example, temperature-related seasonal responses in the producers of organic matter may increase the availability of substrate, or seasonal dynamics in grazer populations may alter bacterial density.

Qualitative and quantitative differences in nutritional substrate affect microheterotrophic activity. Hanson and Snyder (1979) found differing turnover times and maximum velocities of uptake for several organic compounds. Not included in this study was the suite of dissolved organic compounds, including humates and fulvates, known to be less available for rapid consumption. Thus, the composition of the DOC pool exerts a strong pressure on the total activity of the microheterotrophs. Also, the rate of uptake has been shown to depend on the quantity of any specific organic compound, showing a hyperbolic response of uptake rate to substrate concentration. The concentration of glucose in the Duplin River, for example, is almost always greater than the half-saturation constant; thus for glucose, at least, the rates of uptake approach the maximum uptake velocities, and its concentration is generally not limiting. It is not known whether this is also the case for other individual substrates. Even if this were true for most substrates, it cannot be inferred that growth is not limited by the availability of organic matter. Growth may still be limited by selected organic growth factors.

Planktonic microheterotrophs also use particulate organic carbon, and POC is probably converted to DOC prior to ingestion by most microorganisms. Polymeric compounds associated with detritus must be hydrolyzed extracellularly, and the organic matter sorbed to particles must be desorbed. These activities may occur so close to the cell surface that the turnover time of the released DOC is extremely short. Haines and Hanson (1979) compared the decomposition rates of three salt-marsh plant species–*S. alterniflora, Salicornia virginica*, and *J. roemerianus*–and the dynamics of the associated microbial community in laboratory microcosms. They measured the efficiency of conversion to microbial biomass, finding that *Spartina* detritus showed the highest efficiency (64.3%), although, *Salicornia* was degraded most rapidly. Thus, the quality of *Spartina* is the most conducive to microbial production, although it is not the most rapidly decomposed.

Particulate matter is a substratum as well as a substrate for microheterotrophs. Particulate detritus was once considered the major pathway of microheterotrophic carbon and energy flux in the estuaries (Darnell, 1967; Odum and de la Cruz, 1967), but recent studies show the greatest number of bacteria and highest microbial ATP concentrations in the water are not associated with particles (Wiebe and Pomeroy, 1972; Haines, 1977; Christian and Wetzel, 1978; Chapter 5). Even so, greater heterotrophic activity, defined as uptake of glucose, is associated with particles than with bacteria in free suspension; attachment seems to promote heterotrophic activity. One consequence of attachment, however, is the effect of the velocities of tidal currents on particle suspension. Heterotrophic activity is greater during flooding and ebbing tides than during slack water. Microheterotrophs and their particulate substrata spend time in both the water column and the sediment, dropping to the bottom at slack water. Access to dissolved nutrients such as inorganic nitrogen could be enhanced during the time in the sediment, whereas flushing of toxic end products could occur more readily in the water column. The free-floating bacteria of colloidal size are in permanent suspension.

The scarcity of bacteria on particles could either be the result of lack of growth and attachment or of stripping of bacteria from particles by detritivores (Christian and

Wetzel, 1978). The results of size fractionation experiments suggest the possibility of grazing control (Hanson and Wiebe, 1977; Hanson and Snyder, 1979). Water samples filtered prior to incubation may have had fewer predators than unfiltered water samples. In numerous instances the filtered samples demonstrated higher V_{max} values than unfiltered samples. The higher V_{max} values are an index of the large number of physiologically active organisms resulting from uncropped growth in the filtered water. This would be compatible with the hypothesis of strong grazing pressure.

Experimental work with water from the Duplin River containing naturally occurring microbial populations and dissolved substrates suggests that there is a near equilibrium between microbial growth and substrate production. To the extent that grazing by protozoa or stripping of bacteria from particles by detritivores holds down the size of the microbial populations, the supply of labile DOC increases. While, as in the case of glucose, the supply may reach concentrations permitting the maximal rate of uptake by bacteria, thus assuring a steady supply to bacterial populations, it never seems to reach sufficient concentration to permit transport of labile DOC out of the Duplin River by advection (Chapter 2). Both experimental and microscopic studies indicate that most of the heterotrophic activities in the water are associated with particles > 10 μm in size which contain relatively large, actively growing bacteria. These particles, many of which probably are fecal material from the plankton and benthos, have a probable residence time in the water of 24-48 hours. During that time, either the entire particle and its microbial flora will be eaten or ciliates will graze over and through it, removing nearly all bacteria and limiting bacterial growth by overgrazing. Thus, continuing production of particles supports a food web involving both particulate and dissolved organic matter in the estuarine water.

7. Anaerobic Respiration and Fermentation

W.J. Wiebe, R.R. Christian, J.A. Hansen, G. King, B. Sherr, and G. Skyring

For decades, ecologists have considered bacteria to be responsible, directly or indirectly, for much of the recycling of inorganic nutrients in nature and have regarded them as an important source of food for small animals. Apart from such important roles, these unseen but ubiquitous organisms perform many other activities. They can alter the chemistry of water and sediments, influence the distribution of plants and animals, and provide conditions for massive mineral deposition.

In salt marsh soils and sediments anaerobic habitats predominate. Baas-Becking and Wood (1955) pioneered studies on the importance of anaerobes and their processes in estuarine environments. They considered the anaerobic conversion of photosynthetic products in salt marsh-estuarine sediments to be *the* unique feature of these environments and the mode whereby substrates were released during periods of low primary production, thus extending the growth period for the aerobic organisms. Now the anaerobic microflora are thought to be responsible not only for release of organic compounds but also for the establishment and maintenance of a variety of the physical and biological environments within estuaries and salt marshes. In this chapter we examine the major anaerobic pathways in the salt marshes of Sapelo Island and the consequences of anaerobic microbial activities.

7.1. Processes

The major anaerobic processes in estuarine sediments and soil are fermentation, dissimilatory nitrogenous oxide reduction (DNOR), dissimilatory sulfate reduction (sulfate reduction), and methanogenesis. Organic matter from primary producers and aerobic heterotrophs enters the anaerobic cycle through fermenters and nitrogenous oxide (NO) reducers (Figure 7.1). Sulfate reducers and methanogens utilize relatively few substrates, most of which are end products of fermentation. Because NO reducers can utilize a wide variety of substrates, they may compete directly with fermenters for

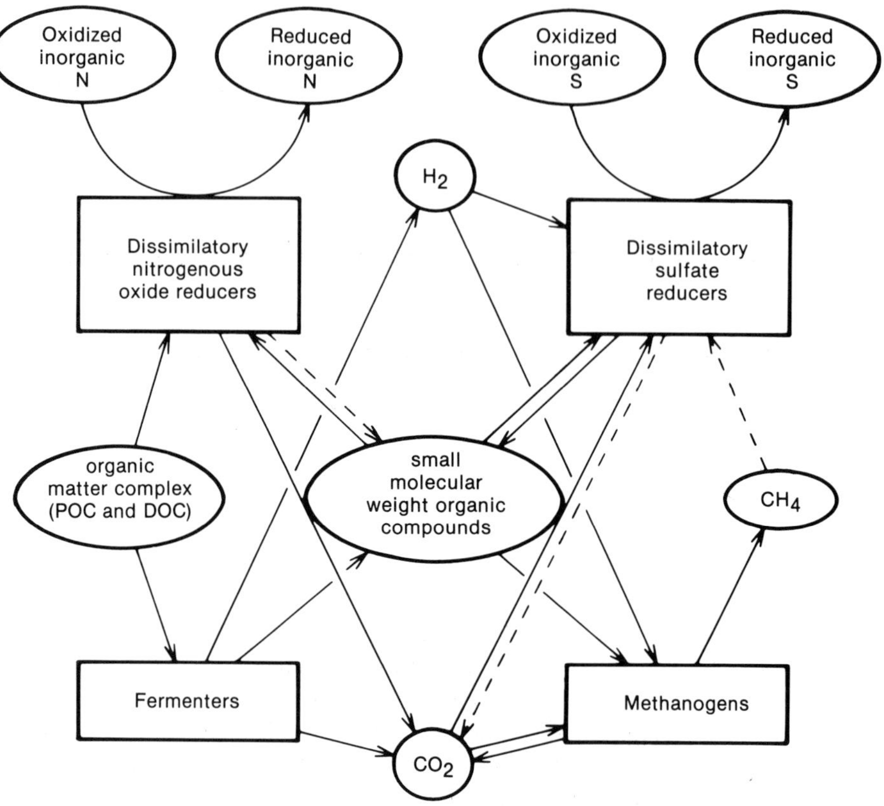

Figure 7.1. Conceptual model of the interactions of anaerobic microbial processes in salt-marsh sediments. Solid lines are confirmed fluxes; dashed lines are possible fluxes.

oxidizable energy sources. In addition, they may utilize fermentation end products and thus also compete with sulfate reducers and methanogens.

7.1.1 Fermentation

Fermentation is anaerobic dissimilation of organic matter in which the terminal electron acceptor is an organic compound. As a major consequence of fermentation, organic compounds of low molecular weight are produced; these can serve as substrates for the other anaerobic processes. Fermentation reactions are induced by a variety of both procaryotic and eucaryotic organisms living under a wide range of environmental conditions. The organisms may be facultative or obligately anaerobic; the former are capable of either aerobic respiration or fermentation, while the latter are inhibited or killed by oxygen.

Fermenters can utilize many substrates, including simple sugars, as well as cellulose, starch, pectin, alcohols, amino acids and purines. The end products include organic compounds, such as short-chain fatty acids and alcohols, as well as CO_2, H_2, and NH_3.

Glycolysis is the major pathway of carbohydrate catabolism. The products of glycolysis depend on both the compounds fermented and the organisms present. A major factor determining whether or not a substrate can be fermented is its oxidation state. Neither highly oxidized nor highly reduced compounds are easily fermented. Although a wide variety of organisms are capable of fermentation reactions, bacteria and yeasts are responsible for most of this activity in nature. Salt marshes are no exception, although microfaunal and meiofaunal fermenters do occur in anaerobic zones (Fenchel, 1967, 1969; Ott and Shiemer, 1973; Wieser and Kanwisher, 1961), and these animals are capable of anaerobic catabolism (Fenchel, 1969; Hochochka and Mustafa, 1972; Ott and Shiemer, 1973). Some meiofauna can sustain all life processes under these conditions, while others can tolerate anaerobiosis only transiently.

Measurement of fermentation in nature is difficult because of the diversity of organisms and organic compounds. Christian and Wiebe (1978) chose to add ^{14}C-glucose to samples of sediment and follow its incorporation into CO_2, particulate matter, and ether-soluble end products of fermentation. Glucose was selected since it is readily metabolized by most saprophytic microorganisms, and it is the only monomeric sugar released upon the hydrolysis of cellulose. The analytical procedures have been described elsewhere (Christian, 1976; Christian and Wiebe, 1978, 1979). Briefly, soil samples from cores were diluted 1:2 with seawater in an anaerobic glove bag. One ml of uniformly labeled ^{14}C-glucose (22 to 457 ng glucose) was added to each slurry. Incubation *in situ* for 5 to 30 minutes was terminated by injecting acid-formalin; samples were then analyzed for the amount of ^{14}C in CO_2, particulate matter, and ether-soluble extract. The turnover time for each fraction was calculated, and the kinetics of uptake were best described by a first-order model. Turnover times—or their inverse, the specific uptake rates—were independent of substrate concentration, as in a linear donor-controlled flux. Thus, $v = k[s]$, where v is the rate of uptake, k is the specific rate in units time ^{1}, and [s] is the substrate concentration. The uptake rate is dependent upon k, which is a measure of the biological potential of uptake, and [s], which is the pool size. Without knowledge of the natural pool size, the rate of uptake cannot be determined. However, the specific rate or the turnover time will give all of the information required to define the biological component of the rate. These data yield an estimate of *potential* rather than *actual* fermentation activity, since natural substrate concentrations were not determined.* The effect of temperature on the potential fermentation activity was studied by Hansen (1979) following the same protocol, except that several temperatures were used for the incubations. Turnover times of glucose uptake were always more rapid in tall *Spartina* (TS) than in short *Spartina* (SS) soils (Figure 7.2). At both sites activity was highest at the surface and decreased with depth. Turnover times in TS soils were generally minutes to approximately an hour, while in SS soils they were generally several hours.

The proportion of ^{14}C label in the three fractions—CO_2, particulate matter, and ether-soluble end products—also differed. In SS soil 30 to 70% of the radioactive carbon was found in the ether-soluble fraction, which contains the fermentation end products, while TS soils yielded virtually no activity in this fraction. Christian and

---

*For more complete interpretation of the kinetics, see Christian and Hall (1977) and Christian and Wiebe (1978).

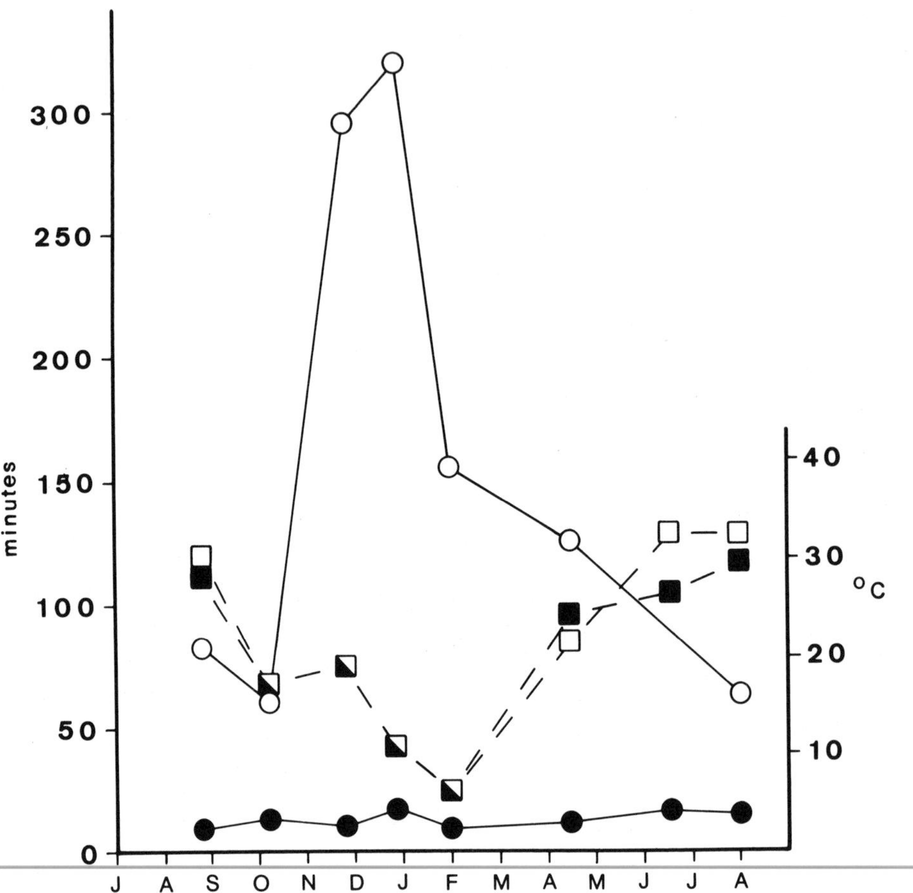

Figure 7.2. Seasonal variations of soil temperature and the turnover time of glucose in salt-marsh soil. Solid symbols = tall *Spartina;* open symbols = short *Spartina.* Circles = turnover times; squares = soil temperatures. Adapted from Hansen (1979).

Wiebe (1978) suggested rapid uptake of end products in TS soils as the major difference between the two zones. Subsequently, Skyring et al. (1979) measured sulfate reduction in both soil types and found much higher activity in TS soils. These results follow from the hypothesis of rapid end product uptake. Alternatively, rapid NO reduction in the TS zone could have accounted for the small fraction of ether-soluble label. However, since Sherr and Payne (1978) found that DNOR activity was low in TS soils; these soils are unlikely to be a sink for ether-soluble products.

Fermentation, as we have mentioned, is the initial process in anaerobic metabolism, so it was important to determine its source or sources of substrate. A series of plots were set up and maintained in SS soils to evaluate the potential sources of substrates: root exudate, dead roots and rhizomes, and the dissolved and particulate organic matter within the soil. The plots were modified as follows:*

*See Christian et al. (1978) and Section 6.3.1 for details of the experiment.

1) Clipped and pruned. *Spartina* was clipped at the soil surface and the roots and rhizomes were cut to prevent the entry of carbon from root and rhizome growth outside the plot.

2) Clipped, pruned, and enclosed. Plastic sheeting was inserted around the plot to prevent roots from entering and to reduce lateral water flow.

3) Control. No manipulation.

The results showed that glucose turnover times were unchanged after six months. Christian et al. (1978) concluded that the substrates for fermentation were mainly the particulate and dissolved carbon already in the soil and that the linkage of fermentation to plant production was indirect. Only water restriction experiments were performed in TS soils because the root mass constituted only a very small portion of a large total soil volume (see Chapter 3 on plant root distribution). After two months of restricted water flow, with cores inserted into the marsh and capped at the ends, turnover times of glucose were unaffected (Christian and Wiebe, 1978; Hansen, unpubl.) The substrates for fermentation in both zones appeared to comprise small-sized POC and the DOC. The potential activity of fermenters is so high in TS soils that the rate-limiting step may be diffusion-limited, rather than biologically controlled (Hansen, 1979). In such a case, substrates cannot diffuse far from their origin before they are metabolized, so those cells closest to sources would be the most active. In winter, the metabolic rate per cell would be much lower, but diffusion of substrate would extend over a greater area. The effect of low temperature on metabolism would be countered by an increasing number of metabolizing cells.

This hypothesis may also help to explain why there are differences in turnover times in TS and SS soils. Water flow through TS soils is much greater than through SS soils (Chapter 2). If diffusion is the limiting step, increased water flow would cause an increase in activity and probably an increase in the number of fermenters. Water restriction for two months, described above for TS soils, was probably not long enough to alter the population density of bacteria; thus the manipulation did not affect the rate.

Fermentation plays a central role in providing substrates for other anaerobes in sediments and in water-logged soils. Lack of adequate methods has prevented us from examining the *in situ* reactions, but we have been able to measure potential rates of activity spatially and seasonally. Techniques to study fermentation *in situ* are needed.

7.1.2 Dissimilatory Sulfate Reduction

Sulfur exists in seawater primarily as the sulfate ion, which is reduced by organisms. There are two kinds of sulfate reduction. In assimilatory reduction, sulfate-S is reduced to sulfhydryl-S, most of which is incorporated into S-amino acids. This pathway is ubiquitous among higher and lower forms of plant life. Dissimilatory sulfate reduction results from the anaerobic respiration of substrates, with sulfate being the terminal electron acceptor. This reaction is accomplished only by a small group of bacteria, called sulfate reducers, that release the sulfide produced into the environment. In some marshland soils large quantities of sulfide are formed, the reaction being dependent upon a plentiful supply of organic matter, mainly simple organic acids. Two to six moles of organic carbon are oxidized for every mole of sulfate reduced.

The pathway of dissimilatory sulfate reduction is shown in Figure 7.3. Many of the steps, those indicated by question marks, are postulated but not proven. Study is difficult because abiotic chemical conversions can take place at certain steps. Dissimilatory sulfate reduction, as carried out by *Desulfovibrio sp.* is:

$$2 \text{ lactate} + SO_4^{2-} \rightarrow 2 \text{ acetate} + 2CO_2 + S^{2-}$$

Sulfate reducers possess cytochromes for ATP generation. The reaction takes place only under anoxic conditions, but sulfate reduction can take place in oxidized sediments if anoxic microzones occur (Jφrgensen, 1977). A variety of substrates can support sulfate reduction, although until recently, lactate, pyruvate, dihydrogen and a few small-chain fatty acids were thought to be the sole substrates. Now acetate, a few long-chain fatty acids, and methane have been shown to provide energy for sulfate reducers (Widdel and Pfenning, 1977; Panganiban et al., 1979; Pfenning, pers. comm.). *Desulfovibrio, Desulfuromonas* and *Desulfomaculum*, are the only genera known to reduce sulfate, but because of the increased effort recently to isolate new types of sulfate reducers, other bacterial genera will probably be found.

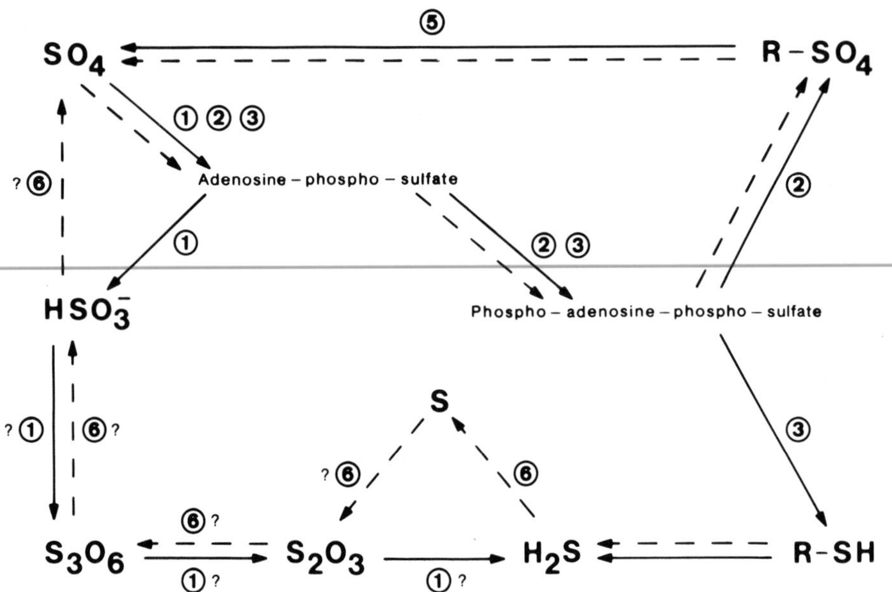

Figure 7.3. The sulfur cycle in estuarine sediments. Solid lines = reductive steps. Dashed lines = oxidative steps. Combined solid and dashed lines = organic-mediated, organic-organic or organic-inorganic transformations. ? = a postulated intermediate. Other pathways undoubtedly exist. 1. Dissimilatory sulfate reduction. 2. Assimilatory sulfate production. 3. Assimilatory sulfate reduction. 4. Sulfide mobilization. 5. Sulfate mobilization. 6. Sulfide oxidation. Adapted from Wiebe (1979) in *Ecological Processes in Coastal and Marine Systems*, edited by R. J. Livingston. Copyright (1979) by Plenum Press.

Sulfate was extracted from the soils at Sapelo Island by the method of Kowalenko and Lowe (1975) and determined by the turbidometric technique of Dodson (1961), as modified by Tabatabi (1974).* Concentrations of volatile (free H_2S, S^{2-}, HS^-) and acid-volatile (iron-bound) sulfide in the soils were measured in samples taken from the surface to 35 cm depth by the methods of Skyring and Chambers (1976), as modified by Oshrain (1977). Pyrite (FeS_2) is not released by this treatment.

The rate of sulfide formation was evaluated, using the direct injection technique of Skyring *et al.* (1979). Briefly, a known quantity of $^{35}SO_4^{2-}$ was spotted onto a glass rod, one end of which had been inserted into a rubber stopper. After the solution dried, the rod was inserted into a sediment core contained in a 10 cc syringe. The stopper sealed the core at one end, while the syringe plunger sealed the other end. After incubation, the sediment was extruded into an oxygen-free chamber for extraction of sulfide. Total sulfate in extracts from the core, distilled sulfide, and the specific activity of the sulfide and sulfate were always determined.

The patterns of sulfate distribution differed markedly between TS and SS soils (Figure 7.4). In TS soils sulfate concentrations were nearly constant with depth, while in SS soils they decreased with depth (Oshrain, 1977). Interstitial water salinity profiles also differed in the two zones; in TS soils salinity was constant with depth, but in SS soils it increased with depth. Thus the total sulfate depletion in SS soils was greater than that originally calculated (Oshrain, 1977).

The sulfide concentrations in the two soils, as shown in Figure 7.5, displayed no consistent pattern of change with depth. However, two differences between the zones were evident: total sulfide concentrations were much greater in TS than SS soils, and virtually no volatile sulfide was present in TS soil, while more than one-third of the sulfide was volatile in SS soil (Oshrain, 1977). In both soils, but particularly in SS soils, there was a great deal of variation in sulfide concentration at any given depth; the range of the ratio for all cores was 10^3. However, when the total sulfide concentration was integrated over all depths, the differences between individual cores were only fourfold. We believe that local events, such as animal burrowing and root growth, can redistribute material vertically, but the system exhibits a more conservative or constant amount of sulfate reduction, perhaps related to the overall rate of water exchange.

The distribution of sulfate and sulfide was examined in perturbation plots described in Section 6.3.1. Perturbations included clipping and pruning of *Spartina* and nutrient enrichment of the soils. There was no statistically significant effect of the perturbations on either sulfate or total sulfide distribution for six months (Oshrain, 1977).

Sulfate reduction rates, measured with $^{35}SO_4^{2-}$, provided further evidence of a distinction between TS and SS soils. Active sulfate reduction in TS soils occurred from the surface to the deepest layer measured—35 cm. However, activity in SS soils was barely detectable below 10 to 15 cm (Skyring et al., 1979). Sulfate reduction was associated with distribution of roots in the two soil types in a manner similar to that of the fermentation process and to ATP concentration (Christian et al., 1975). Although the similarity of the results for potential fermentation activity and sulfate reduction certainly does not prove a cause-and-effect relationship, it is consistent with the hypothesis that the two processes are linked.

---

*See Skyring et al. (1979) for details of the procedure.

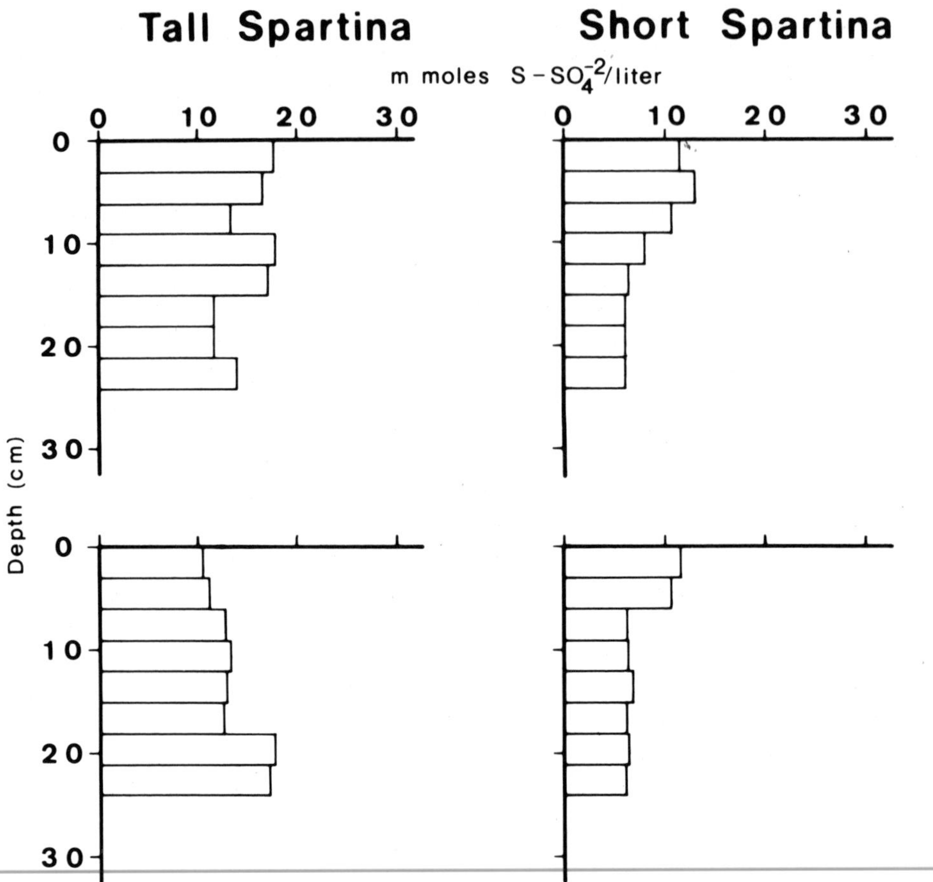

Figure 7.4. Distribution of sulfate in interstitial water of salt-marsh soils of tall and short *Spartina* regions. Each graph presents the result from one core taken in September 1976.

One technical aspect of this work requires comment. When samples were placed in containers and incubated in the laboratory, conditions could change and affect the measured rate of reaction. A single, long incubation time would not permit correction for such effects. To exclude this possibility, time-course experiments were conducted. The production of ^{35}S-sulfide for replicate samples incubated for up to 72 hours was linearly correlated ($r = 0.93$) with the period of incubation (Skyring et al., 1979). The regression line intercepted zero, so the manipulations did not appear to affect the rate of sulfide formation. Our standard practice was to run a time course for each set of incubations. This provided direct evidence that linear rates were occurring, and of course all points could be used to calculate the hourly rate.

The TS and SS soils differed consistently in the distribution of sulfate and sulfide and in the rates of sulfate reduction. Sulfate concentrations could be maintained at approximately those of seawater in TS soil, even if sulfate reduction was rapid, but sul-

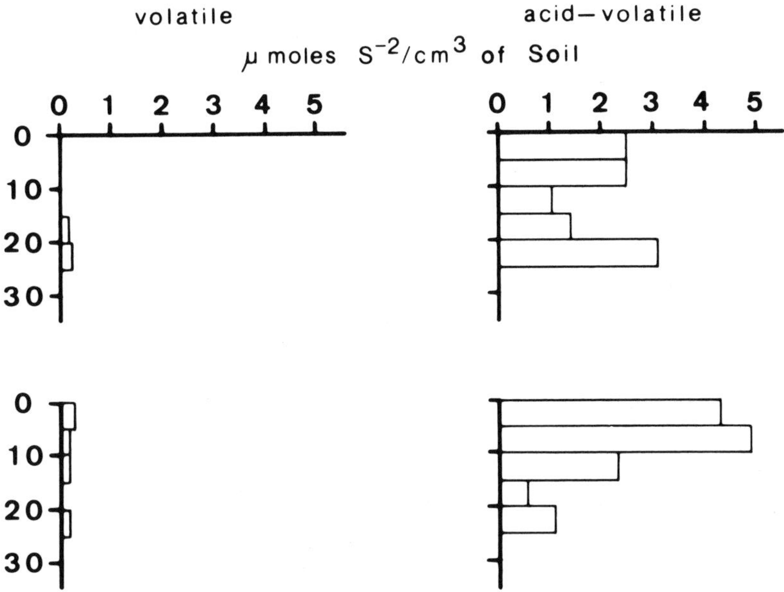

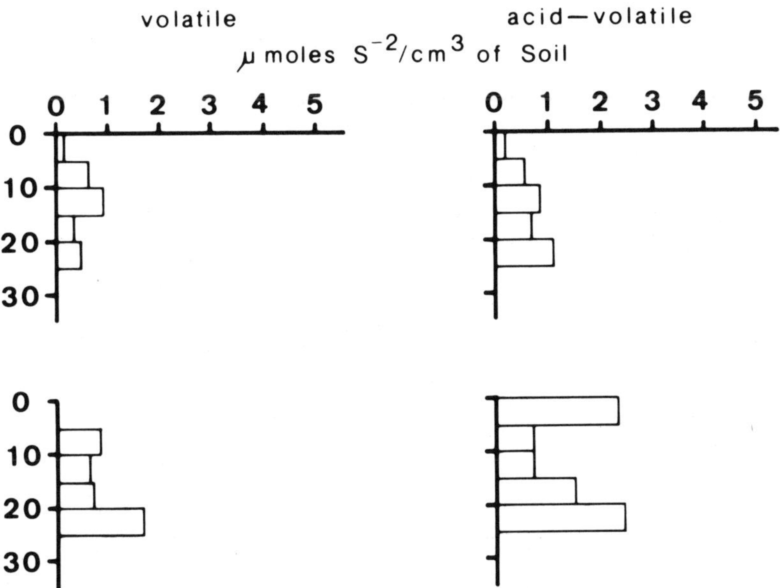

Figure 7.5. Distribution of volatile and acid-volatile sulfides in salt-marsh soils. Each pair of graphs presents data from one core taken in September 1976.

fate concentrations were depleted in SS soil (Oshrain, 1977). As discussed in Chapter 2, interstitial water exchanges in two to three days in TS soil, while exchange below the first 10 to 20 cm in SS soil requires weeks to months. King and Wiebe (in press) postulated that the differences in distribution and rates of production of sulfide were related to the contrasting patterns of interstitial water flow (Section 7.1.4).

The differences in water flow also could explain the contrasting sulfide distribution in the two soils. TS soils contained little volatile sulfide, whereas over one-third of the total sulfide was free, or volatile, in SS soils. Whether sulfide is volatile or not depends upon the amount of iron present, because sulfide actively combines with iron to form FeS. In TS soil, the rapid passage of seawater through the soil and rapid sedimentation could maintain adequate iron to react with sulfide. However, in SS soil, exchange of interstitial water and sedimentation are too slow to provide an adequate supply of dissolved free iron, and volatile sulfide accumulates. To test the effect of water movement on sulfate concentration, King and Wiebe (in press) placed cores in the TS soil for two months. Some were left open while some were capped (Section 7.1.4). After two months, sulfate concentration had decreased 30 to 40% in closed cores. Control sites and open cores showed no sulfate depletion. Thus, the rate at which water flows through the sediments may determine, to a large extent, the sulfate concentration.

The discrepancy between the rates of sulfate reduction in TS and SS soils is similar to that for rates of fermentation. One explanation for these differences could be that sulfate reducers were directly limited by substrate, or diffusion rate, because fermenters probably provided their substrates. However, in the absence of substrate pool size data, the *in situ* rates of transfer of organic matter from fermenters to sulfate reducers are not known. There is also the possibility that sulfate availability in SS soils limited activity. Although sulfate was depleted by 30 to 40%, this should not have been enough to limit sulfate reduction; however, within microhabitats sulfate could be depleted to a point where its concentration was limiting. Preliminary results of experiments in which lactate or sulfate was added to incubating soil slurries suggest that substrate is the more likely limiting factor, but of course sulfate could be limiting in microzones, even when it is apparently abundant in the soil as a whole.

7.1.3 Dissimilatory Nitrogenous Oxide Reduction

Dissimilatory nitrogenous oxide reduction (DNOR) is a general term which encompasses three specific pathways (Figure 7.6). Only bacteria have been shown unequivocally to possess this capability. It is an obligately anaerobic process, since oxygen concentrations as low as 0.2 ml per liter inhibit enzyme activity (Goering, 1968). The nitrogenous oxides serve as terminal electron acceptors during cytochrome-catalyzed oxidation of organic matter. When the end products are N_2O or N_2, the process is called denitrification. In agricultural soils, where the process has decided economic implications related to the loss of fixed nitrogen from the soil, there have been numerous studies of DNOR. There have been fewer studies in salt marshes and other marine habitats, yet salt-marsh soils may be considered ideal habitats for DNO reducing organisms. Large portions of the soil are anaerobic with limited oxygen diffusion capacity. Organic matter is abundant, the pH and Eh are optimal, and an extensive aerobic rhizosphere exists in which nitrate could be produced.

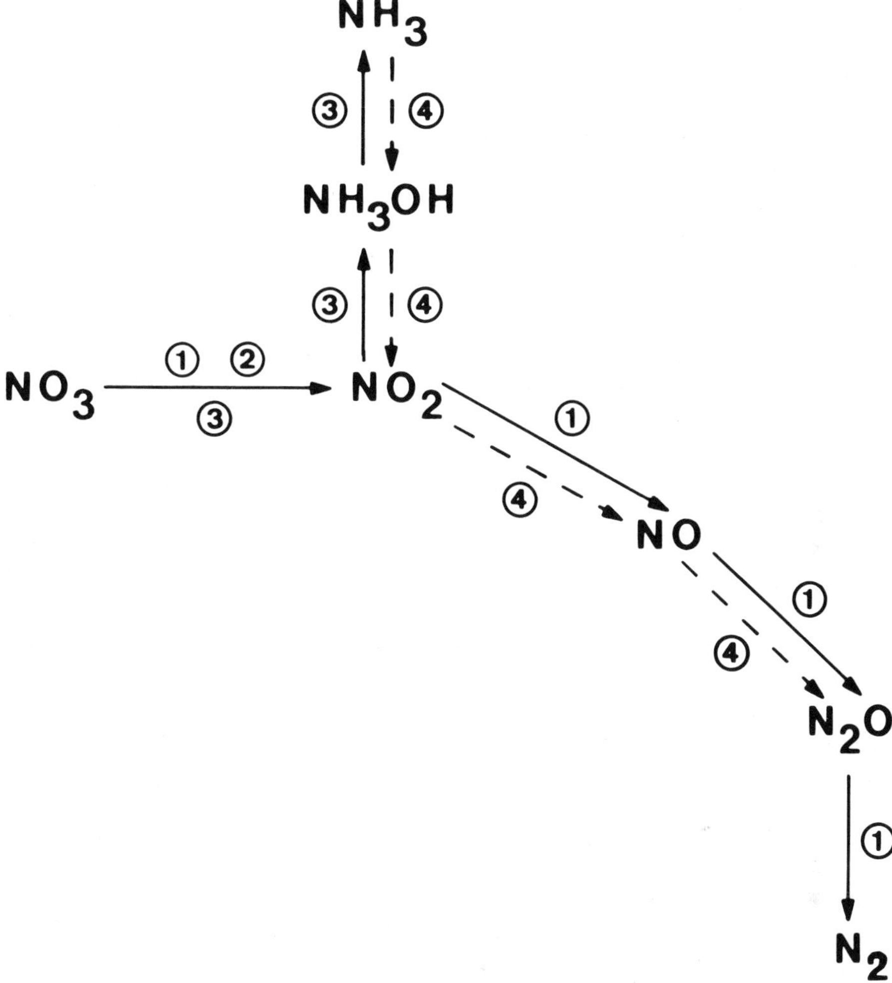

Figure 7.6. Pathways of dissimilatory nitrogenous oxide reduction. 1. Denitrification. 2. Dissimilatory reduction (terminates at NO_2). 3. Dissimilatory ammonia production. 4. "Nitrification" N_2O pathway: ammonia to nitrous oxide. Adapted from Sherr (1977).

DNO reducers can utilize a wide variety of substrates for growth. Theoretically they can utilize the compounds almost as efficiently as aerobic organisms, because the oxidation of substrates by DNO reducers is cytochrome catalyzed. The DNO-reducing enzymes are induced by the absence of oxygen, whether or not nitrogenous oxides are present. However, the *in situ* rate of reduction may be affected by the concentration of nitrogenous oxides (Cavari and Phelps, 1977). Payne (1973) discussed the possible controls of this enzymatic activity. Although all enzymes are induced under anaerobic conditions, their activity is regulated to some extent by the products formed at each step in the process, and in marine sediments accumulation of several intermediates has been demonstrated (Sφrensen, 1978a). There is evidence for enzymatic conversion of each intermediate in the pathway (Payne, 1973); in addition, Komatsu et al. (1978) reported the chemical conversion of nitrite to nitric oxide via reduced iron.

Dissimilatory reduction to ammonia (Figure 7.6) can be quantitatively important in nature (Buresh and Patrick, 1978), specifically in coastal sediments (Koike and Hattori, 1978; Sørensen, 1978). This process offers an alternative to the loss of fixed nitrogen during anaerobic respiration of nitrate.

Nitrous oxide, as well as dinitrogen, can accumulate *in situ* from denitrification. Many pure cultures of DNO reducers produce nitrous oxide rather than dinitrogen as the terminal product, and Sørensen (1978a) and Yoshinari and Knowles (1976) have found nitrous oxide accumulations in sediment and water. We do not know why denitrification will sometimes stop at nitrous oxide or at times proceed to the release of dinitrogen. In pure cultures, high concentrations of nitrate inhibit nitrous oxide reduction (Ingram and Payne, pers. comm.). Freney et al. (1978) suggested the further complication of nitrous oxide in soil derived from the oxidation of ammonia, and it is not known whether this pathway is anaerobic or not. Thus nitrous oxide can be produced from both nitrate and ammonia, and its presence in an environment may not be evidence for nitrate dissimilation.

Many different bacteria reduce DNO. Gamble et al. (1977) isolated and identified DNO reducers from a variety of environments. The dominant organisms isolated in all systems were gram-negative pseudomonad types, but the diversity within the group was high. Still, the isolation of DNO-reducing organisms from an environment is not proof that the isolates were reducing nitrogenous oxides *in situ* (Tiedje et al., 1978), and in virtually all environments the bacteria actually responsible for *in situ* DNOR remain unknown. Further taxonomic information about these organisms is provided by Payne (1973).

Denitrification was measured at Sapelo Island by the nitrous oxide reductase method of Sherr (1977) and Sherr and Payne (1978). Nitrous oxide was added to soil slurries and the subsequent decrease in nitrous oxide concentration was measured with time. The enzymes for DNOR could be formed in the absence of nitrogenous oxides, so this technique gave a potential rate for denitrification rather than an *in situ* rate. In the Sapelo Island marshes SS-zone, N_2O reductase activity was lower in winter than during other seasons (Figure 7.7, top). The activity corresponded roughly to the peak periods of *Spartina* growth. During the summer and fall, greatest activity was found between 0 to 10 cm. This peak disappeared in winter. In the TS sediments the nitrous oxide reductase activity showed no seasonal and little depth differences (Figure 7.7, bottom). Activity was never greater than that in the SS zone and was usually much lower.

Kaplan et al. (1977) found that temperature was the major controller of denitrification rates in a Massachusetts *Spartina* marsh and that seasonal variation in temperature caused selection for two or more different populations of denitrifiers. But at Sapelo Island, the TS and SS zones had virtually identical temperature regimes. Sherr and Payne (1978) attributed the difference in denitrification potential between TS and SS soils to the differential growth pattern of *Spartina* in the two region. In the top 25 cm of soil the macroorganic matter (MOM) per unit volume was much less in TS than in SS soils. The difference results from the growth pattern of the belowground plant structures. In TS soil, individual roots and rhizomes penetrate to more than one meter depth, forming a very loose network in the soil. In contrast, roots and rhizomes

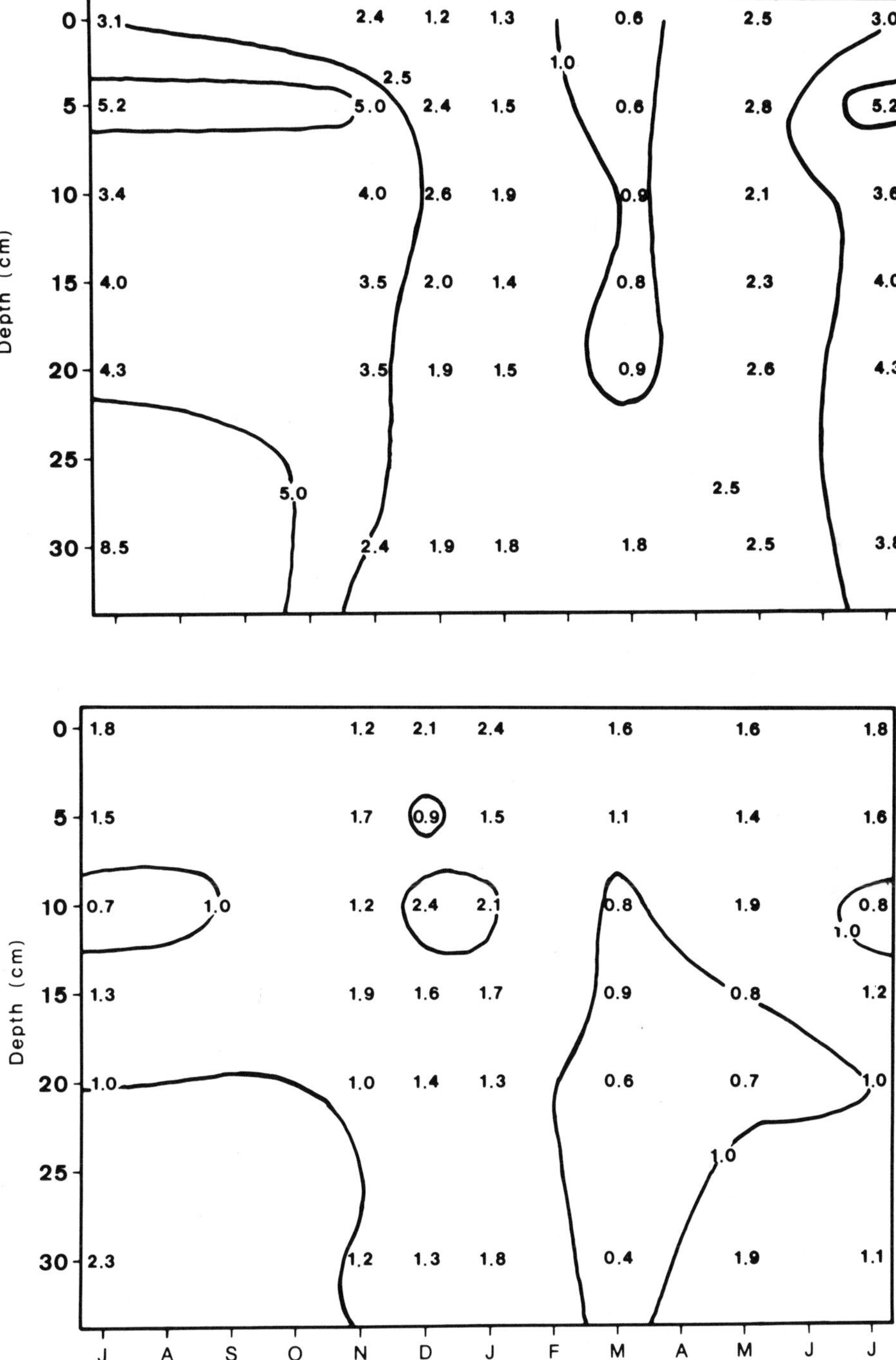

Figure 7.7. Seasonal depth distribution of potential N_2O reductase activity, μg N_2O reduced cm^{-3} soil h^{-1}, in short *Spartina* (a) and in tall *Spartina* (b) marsh. Adapted from Sherr (1977).

in SS soil grow in a narrow band to about 15 cm depth and form a tightly packed mat. Denitrifiers in the SS soil appeared to be coupled to the seasonal production of MOM.

Sherr and Payne (1978, 1979) examined in more detail the relationship between *Spartina* production and denitrification in the SS region using the same plots in which fermentation and microbial biomass were examined (see 6.3.1 and 7.1.1). Plots containing living *Spartina* or plots in which clipping and pruning occurred were maintained for 18 months. In the clipped plots, DNOR decreased significantly after 18 months but not after five months. Root exudation was not the major source of substrate for DNOR. Addition of ammonium nitrate or glucose (Sherr and Payne, 1979) did not affect DNOR rates in plots with intact plants but did affect rates in clipped and pruned plots. Sherr and Payne (1978) concluded that the plants serve to buffer the organisms against these forms of perturbations.

Availability of electron donors partially controlled the N_2O reductase rates in SS soil. On the other hand, the availability of nitrogenous oxides limited growth of the organisms (Sherr, 1977). In TS soil, electron donors did not limit the rate of denitrification, but both substrate and nitrogenous oxide additions affected growth rates. The differences between the responses of the TS and SS denitrifiers in laboratory experiments may help resolve the differences in the rates of nitrous oxide reductase activity measured in field samples. We believe that the differences in reductase rates between TS and SS soils are related to nitrate availability. Denitrifiers require nitrate for a terminal electron acceptor, and nitrate can only be produced in aerobic environments. Since root and rhizome mass in TS soil is smaller than in SS soil, nitrate production in TS soil should be much less than in SS soil. In the laboratory studies of TS soil, activity of N_2O reductase was not stimulated by the addition of organic substrates, but activity did respond to nitrate additions. In contrast, N_2O reductase activity in SS soil was stimulated by organic substrates, but not by nitrate. The simplest explanation of these data is that nitrate is much less available and is the major control of N_2O reductase activity.

7.1.4 Methanogenesis

Methane is produced under anaerobic conditions by bacteria using carbon dioxide or a methyl group as an electron acceptor. Although the microbial production of methane was recognized in the late 19th century, research on factors affecting the rate of production, the organisms involved, and the biochemistry and physiology of the process has increased only recently, probably for a number of disparate reasons. Methane significantly affects the chemistry of the upper atmosphere (Enhalt, 1974); methane generation for fuel could provide a partial alternative to the use of fossil fuel; methanogens and a few other bacteria form a distinct taxonomic group only distantly related to other procaryotes or eucaryotes (Balch et al., 1979); finally, methanogens are essential for the anaerobic oxidation of organic matter. Whatever its impetus, all of this recent work has altered our view of the biochemistry and organisms involved, the substrates used, the subsequent oxidation of methane, and the roles of methanogens in nature.

Unlike the fermenters, sulfate reducers, and DNO reducers, we know neither the mechanism nor the occurrence of oxidative or substrate-level phosphorylation in methanogens. Smith and Mah (1978) reported electron transport mechanisms in energy metabolism from dihydrogen, carbon dioxide, methanol, and acetate, but Zeikus (1977), in his review of the biology of methanogens, stated that b-type or c-type cytochromes have not been identified in any methanogen. Zehnder and Brock (1979) proposed a novel scheme for ATP production in methanogens that does not involve substrate-level or oxidative phosphorylation. The organisms in pure culture require an Eh of −350 mv and strict anaerobic conditions. However, Zehnder and Wuhrmann (1977) reported that, although methane production ceased in the presence of as little as 0.01 ppm dissolved oxygen, even high oxygen concentrations did not kill some methanogens.

The substrates that support methanogens are few, and they are most often the end products of the other anaerobic processes. All organisms isolated to date, except that isolated by Zinder and Mah (1979), use dihydrogen and carbon dioxide. Formate, methanol, methyl-amine, and acetate are the only other verified energy sources. Zehnder and Brock (1979) found that some methanogens not only produce methane, but also simultaneously oxidize a small amount of it (0.3 to 0.001%), some of which is incorporated into cell material, but most of which is converted to carbon dioxide.

A possible case of aerobic methane formation was reported by Scranton and Brewer (1977), who found a methane maximum in the western subtropical North Atlantic Ocean at about 100 m. They claimed physical transport from the atmosphere could not account for their results and suggested that algae might be capable of producing the methane or, alternatively, that anaerobic microzones in animal guts and zooplankton fecal pellets could serve as sites for production. They reported that axenic cultures of marine algae (*Thalassiosira pseudonana*) produced detectable amounts of methane in the light. Whether this is an important and general source of methane is not known. Recently Oremland (1979) reported methane production from previously aerobic samples of plankton that were incubated anaerobically. Thus methanogens can exist and produce methane even in what, at first inspection, appear to be aerobic environments. For the details of the biochemical pathway of methane formation see Balch et al. (1979) and Zeikus (1977).

Workers have isolated only a few species of methanogenic bacteria which vary greatly in morphology, in utilizable substrates, and in requirements for growth. Interest in the taxonomy of methanogenic bacteria was stimulated by the work of Fox et al. (1977). They believed that methanogens occupy a unique place among microorganisms, based on the oligonucleotide cataloging of the 16s RNA. According to Balch et al. (1979), methanogens antedated cyanobacteria. Their taxonomic revision of this group places the methanogens in a new group, Archaebacteria, which also contains halophiles and two thermoacidiophiles. This group is separated from all other bacteria and cyanobacteria.

Methane is assayed with a gas chromatogaraph, making methanogenesis the most easily measured of the anaerobic processes (King and Wiebe, 1978; King, 1978). At Sapelo Island, the emission of methane from the salt-marsh soil surfaces provided data on both the spatial distribution of methanogenesis and the carbon lost from the marsh. Methane was trapped using bell jars inserted into the various zones of the marsh, and its accumulation was measured over 12 to 24 hours by flame-ionization gas chromatography. Two manipulations permitted examination of aerobic methane oxidation: 1) 40 ml of 30% glutaraldehyde was added to eliminate the surface aerobic microflora, and 2) 5 to 10 mm of the surface was physically removed before insertion of the chamber. To examine the effects of water movement in TS soils on methanogenesis, aluminium core tubes, 7.5 cm in diameter and 25 to 30 cm long, were inserted into TS soil. Some tubes were left open, some were capped at the top, some at the bottom, and some on both ends. Cores were left in the field for approximately two months, from June 21 to August 23, 1978. Incubations of subsamples with sulfate, glucose, or acetate permitted evaluation of the effect on methane production.

Methane release was seasonal in SS and intermediate marsh (MM) zones, but not in the TS zone (Figure 7.8). The curve of average hourly release rates in SS soil correlated well with daily temperatures. Methane release over a year varied significantly in the different marsh zones: 53.1 gm^{-2} for SS soils, 14 gm^{-2} for MM soils, and 0.41 gm^{-2} for TS sediments. This represents a loss of 8.8% of the net carbon fixed in SS soil, 0.002% in TS soil, and about 5% for the entire marsh. The data illustrate the necessity for conducting seasonal studies, for Atkinson and Hall (1976) estimated that 1 $g\ m^{-2}\ yr^{-1}$ of methane was released from a SS marsh 60 miles north of Sapelo Island, using samples taken only in November. We have no explanation for the 50-fold difference in results, except for the lack of seasonal data. Our values are similar in magnitude to those for rice paddies, 25 g CH_4–C $m^{-1}\ yr^{-1}$ (Koyama, 1963); Maryland marshes, 105g CH_4–C $m^{-2}\ yr^{-1}$ (Conger, 1943); and a fresh-water swamp in Michigan, 40 g CH_4–C $m^{-2}\ yr^{-1}$

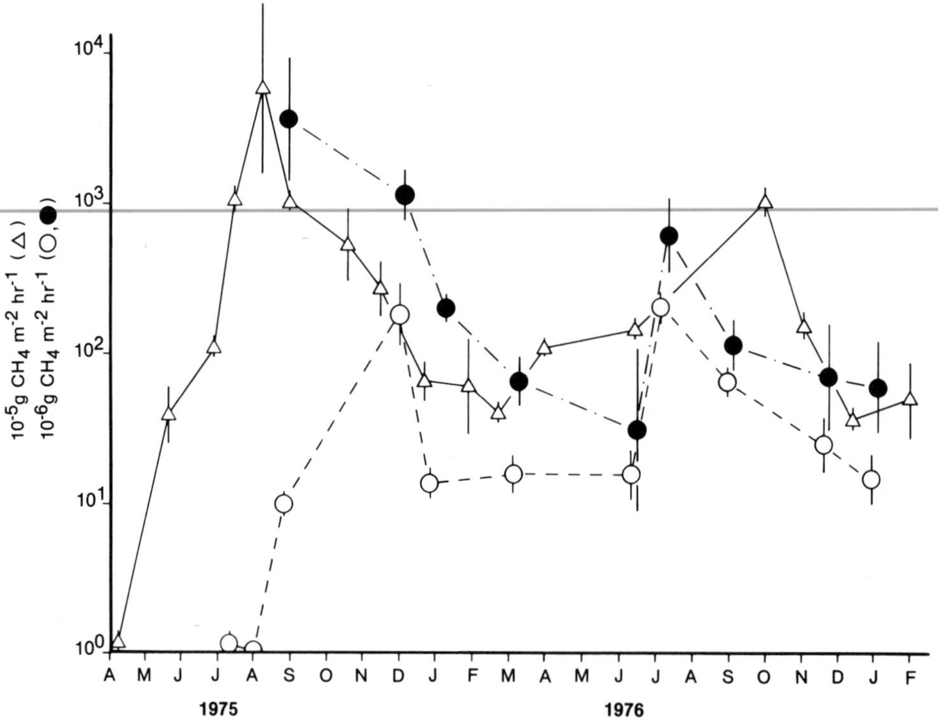

Figure 7.8. Seasonal release of methane from three zones of the salt marshes of Sapelo Island, Tall *Spartina* (○), *Spartina* of intermediate height (●), and short *Spartina* (△). Adapted from King and Wiebe (1978).

(Baker-Blocker et al., 1977). Rudd and Hamilton (1978) reported that 55% of the net carbon fixed in a Canadian lake was regenerated as methane and that, of this amount, 36% was oxidized in the water column. Similarly, Atkinson and Richards (1967) showed that much of the methane generated in anaerobic waters of a British Columbia fjord was utilized in the aerobic part of the water column. At Sapelo Island none of the treatments to eliminate potential aerobic methane oxidizers affected the rate of methane release.

The discrepancy between methane release rates in TS and SS soils perplexed us initially, for the rates of the other three anaerobic processes were generally equal, or else greater in TS soils. One major difference between the two zones is that water flows through TS soil at a much more rapid rate than through SS soil. Odum (1971) suggested that this flow was responsible for the large difference in *Spartina* growth in the two zones. To test whether water flow could influence methanogenesis as well, water movement was restricted by insertion of open and closed cores into TS soil. After two months, cores in which water flow was restricted showed enhanced methane production. Rates from open cores showed no differences when compared with adjacent, undisturbed sediment. Closed cores also exhibited sulfate depletion.

In vitro depth profile studies of methane production in SS soil gave no clear, reproducible indication of a single zone of maximal activity. Jones (1975) had suggested that the greatest production of methane was in the 5 to 7 cm depth zone, but he did not examine seasonal or site differences. Our finding seems to contradict other studies (Cappenberg, 1974; Claypool and Kaplan, 1974; Martens and Berner, 1977), which report spatial separation of methanogenesis and sulfate reduction. Roots in the soil and the concomitant assemblage of vertically mixed, oxygenated, and anoxic zones may account for this discrepancy. Microzonation could occur, but separation of the two processes would not necessarily be organized vertically.

7.2. Controls and Interactions

Knowledge of sites and rates of important processes constitutes only the beginning of an ecological investigation. Such information permits insight about the degree of heterogeneity in the system and establishes the importance of a process relative to flows of organic matter and nutrients. But questions of why the rate is as measured and how it can change demand additional information. The interactions of organisms with each other and with the physical environment must be sought and described. The occurrence and magnitude of a particular microbial process depends not only on the presence of appropriate organisms, but also on the environmental conditions. Three broad categories of factors affect anaerobic processes: 1) abiotic factors, including light, water movement, soil type, 2) aerobic biological processes, such as root exudation, animal burrowing, and oxygen production by benthic diatoms, 3) and interactions among the anaerobic microorganisms, for example substrate competition and the production of stimulatory and inhibitory compounds (Wiebe, 1979).

Production and maintenance of anaerobic conditions depend on the availability of sufficient reduced compounds not only to remove all oxygen, but also to prevent its diffusion in the soil. At Sapelo Island, soil-water flow appears to exert a major control on the type and magnitude of anaerobic processes. The texture of the soil (percent sand, clay, or silt), the slope of the soil surface, and the rate of sedimentation influence the water flow characteristics.

In the SS soil, all four anaerobic processes responded positively to increase in temperature. While some of this effect may have been indirect (e.g., plant growth yielding soluble substrates), a large component of the effect was the direct result of elevated temperature. Light plays a complex role as an indirect influence in anaerobic microbial activity, stimulating plant growth and even acting to increase soil oxygen through activities of benthic diatoms and roots. To some extent the effects of light can be separated from those of temperature. For example, in the spring of the year *Spartina* production begins to increase, as the result of increases in light intensity, before there is much change in temperature (Chapter 3).

Aerobic organisms and their activities affect both the function and distribution of the anaerobes. These effects may be unilateral, such as the burrowing of fiddler crabs, or the result of specific interactions between the aerobes and anaerobes, such as the aerobic utilization of anaerobic end products at the soil surface. Fermenters, sulfate reducers, and methanogens release organic products and sulfides into the soil which can diffuse to the aerobic zone (Wiebe, 1979; Howarth and Teal, 1980). The biological and abiotic oxidation of these materials can restrict oxygen diffusion into the subsurface soil layers. Fiddler crabs and other burrowing fauna, on the other hand, act to aerate and mix the soil (Chapter 4). *Spartina* releases oxygen through the roots and establishes an aerobic rhizosphere well below the soil surface (Teal and Kanwisher, 1966).

Anaerobes can affect the distribution and function of aerobes by altering the redox potential of the soil, thereby limiting the penetration of aerobes. Toxic products, such as hydrogen sulfide, can also restrict the intrusion of aerobes into the anaerobic zones (Theede et al., 1969; Powell et al., 1979; Reise and Ax, 1979). Anaerobes can restrict the distribution of aerobes, but paradoxically can also be responsible for their growth. For example, Patrick and Khalid (1974) reported that anaerobic soil buffers inorganic phosphate when soluble concentrations are low and sorbs it when soluble concentrations are high. Baas-Becking and Wood (1955) proposed the constant infusion of anaerobic end products into aerobic zones as a cause for the high, relatively constant aerobic heterotrophic activity in salt-marsh estuaries. They believed this infusion of material released the aerobic heterotrophs from direct dependence upon continuous primary production, because products continue to be released even during periods of low primary production. Rich and Wetzel (1978) found anaerobic release of DOC from lake sediments to be an important nutritient for aerobes in the water.

7.2.1 Fermentation

Fermentation requires organic substrates whose identity we must ascertain. Root exudates, POC, including decaying roots and rhizomes, and interstitial DOC are three likely sources of substrate. Because most of the DOC decomposed slowly, it was an unlikely source of carbon (Sottile, 1973). Root exudates appeared relatively unimportant; following the perturbation experiments discussed in 6.3.1 (Christian et al., 1978; Christian and Wiebe, 1978), fermentation showed little change for 12 months or more where plant roots were removed. In TS soil, experiments conducted for two months produced identical results. Thus POC emerged as the major source of substrate for fermenters, and fermentation was only indirectly coupled to primary production. Subsequently Hansen (1979) found that soil texture, clay or sand, affected the rate constants of fermentation in the TS zone. There was a large difference between POC in the TS and SS sites as well, so the specific controlling factor could not be identified.

Temperature often affects the rate of microbial processes either directly or indirectly through such factors as season, substrate release, or microbial activity. Fermentation rate constants in SS soils decreased in winter but did not decrease in TS soils (Hansen, 1979). When cores of SS soil were taken in winter ($< 10°C$) and preincubated at 25°C for 24 hours, summer rate constants were restored. Thus the resident microflora was present but inactive. We do not know why the temperature effect was missing in TS soil. One hypothesis has the bacteria in the TS soil limited by diffusion of glucose, such that the slower metabolic rates of individual cells at low temperature were still greater than the rate of diffusion. Even if this hypothesis were correct, it does not explain why there is a consistently higher rate of potential fermentation in TS than in SS soil, since the basic composition of both soils is similar and plant processes do not play a direct role in the supply of substrate. The hypothesis of removal of inhibitory end products of metabolism by the faster movement of interstitial water in the TS soil was not supported by the water restriction experiments. An alternative explanation invokes the more frequent drying of SS soils than TS soils. Fortnightly drying of soils in the SS zone may exert a water stress on the anaerobes and thus reduce their potential activity (Christian and Hansen, 1980).

We examined fermentation first, because the end products are likely substrates for sulfate reducers and methanogens (Figure 7.1), but the processes that produce the latter also can affect the rate of fermentation. Mah et al. (1977) found increased fermentation reaction rates in the presence of methanogens, caused by the removal of free dihydrogen from the medium. Sulfate reducers under some circumstances may also have the same effect, although interspecies hydrogen transfer in the marsh soils has not been investigated.

7.2.2 Denitrification

Several factors determining the potential for denitrification are similar to those controlling fermentation. Temperature does not affect N_2O reductase activity in TS soil, but in SS soil, rates fluctuated with seasonal changes in temperature. The source of substrates for both denitrification and fermentation appears to be POC and DOC. Restricting lateral interstitial water flow or clipping plant leaves in the SS zone produced no change in rates of denitrification for at least five months. The addition of ammonium nitrate or glucose had no effect on plots with intact plants, but it did increase the rates in clipped plots.

Substrate competition by denitrifiers probably does not greatly affect the other anaerobic processes in Sapelo Island marshes, although interactions with the other anaerobes are complex (Figure 7.1) and competition for substrates can occur. However, Sherr and Payne (1979) reported low activity in both TS and SS soils compared to the other anaerobic processes.

Denitrifiers could exert control on the other processes as well by removal of fixed nitrogen. Denitrification may exceed nitrogen fixation in salt-marsh soil at Sapelo

Island (Chapter 8). Recently Kaplan et al. (1977) reported that most of the denitrification in a Massachusetts salt-marsh system occurred in the creek bed, not in the marsh. Measurements in the creek beds have not been made at Sapelo Island.

7.2.3 Sulfate Reduction

Both biotic and abiotic factors influence the rate of sulfate reduction. Water flow through the soil plays a major role in controlling rates of reduction and distributions of end products. Acid-volatile, reduced sulfur end products, presumably iron sulfide, predominate in TS soil. A mixture of gaseous (HS^-, S^{2-}, H_2S) and acid-volatile products occurs in SS soil. Differences in distribution of sulfide in TS and SS soils probably result from differential inputs of iron from sedimentation and water flow. More dissolved and acid-soluble iron occurs in TS soils than in SS soils (King, unpubl.) Little work has been done on temperature effects, but sulfate reduction rates in SS soil cores were lower in summer than in winter. Nedwell and Floodgate (1972), Abdollahi and Nedwell (1976), and Howarth and Teal (1979) reported similar results. Seasonal changes in sulfide production are likely to occur in the SS soils of Sapelo Island marshes.

Sulfate reduction is of interest because several investigators regard this process as responsible for a large proportion of the organic mineralization in sediments and other marsh soils. Jorgensen (1977a) reported that sulfate reduction accounted for 53% of the total mineralization in a Danish bay. In a Massachusetts salt marsh, sulfate reduction accounted for consumption of 1800 g of organic C m^{-2} yr^{-1}, a value equal to the net primary production of the marsh (Howarth and Teal, 1979). Skyring et al. (1979) suggested that sulfate reducers could utilize 36% of the net Spartina production in the TS zone and 67% n the SS zone. Thus sulfate reduction is of major importance in carbon cycling in coastal marine sediments and soils. Sulfate reduction rates are higher in TS than SS soil because of differences in fermentation rates (Skyring et al., 1979). However, because the assay for the rate of fermentation was "potential" rather than actual (7.2.1.), we cannot produce a budget to examine this relationship directly for Sapelo Island marshes.

Certainly the data available suggest that a significant amount of the net primary production of salt marshes is degraded through sulfate reduction. Because of this possibility, we must consider some of the problems with the present methods for studying sulfate reduction and the interpretation of results. In all of these studies $^{35}SO_4^{2-}$ was added directly to undisturbed sediments by some type of direct injection procedure (Howarth, 1979; Howarth and Teal, 1979; Jϕrgensen, 1977; Skyring et al., 1979). Intuitively, this method seems more realistic than adding the label to a slurry, but a serious unresolved problem remains. Calculation of the sulfate reduction rate relies on using the bulk soil sulfate concentration for interpreting the specific activity. Localized micro- or macrozones of active sulfate reduction in a sample may, if they exist, have lower sulfate ion concentrations than inactive zones. Thus the sulfate reduction estimates produced so far may represent overestimates of the actual reduction rates, because the effect of estimating higher nonradioactive sulfate ion concentration would be to underestimate the specific activity. Further investigation is necessary before the published rates of sulfate reduction can be assumed correct.

Accounting for the fate of all of the sulfate was a problem in some studies. Howarth (1979) and Howarth and Teal (1979) interpreted their results, in which a large amount of the added ^{35}S label disappeared, as evidence that pyrite was formed. On this basis they claimed that rapid pyrite formation was an important result of salt-marsh sulfate reduction, but no direct measurements of pyrite were made, and such measurements would be difficult with present techniques. There are, however, other ways to "lose" $^{35}SO_4^{2-}$. Organic sulfates can form rapidly. Although Howarth (1979) and Howarth and Teal (1979) attempted to release and measure organic sulfate, their technique seems incomplete at best. At Sapelo Island, Skyring et al. (1979) accounted for all of the SO_4, except those samples of SS soil, 0 to 10 cm in depth, where the loss averaged 30%. Boiling in hydrochloric acid released up to 20% of the "lost" SO_4, but 10% of it was still unaccounted for. Skyring et al. believed that pyrite was formed, but also that gases, such as dimethyl sulfide and dimethyl disulfide, could account for the rest of the SO_4. These gases would not be trapped in the sodium hydroxide or 5,5'dithiobis-(2-nitrobenzoic)acid (DTNB) traps. Thus, whether rapid pyrite formation took place in Massachusetts salt marshes or gaseous end products were formed cannot be resolved with the existing data. What is clear is that at Sapelo Island the only possible site for pyrite formation would have been the 0 to 10 cm zone of SS soil, and that zone could account for no more than 10% of the activity. In all other sites, Skyring et al. (1979) could account for all of the ^{35}S. In this regard then, the products of sulfate reduction in the two marshes appear to be very different. Comparative studies on several sites and the investigation of the production of other possible compounds are needed to resolve this dilemma.

Volatile sulfide has been found to inhibit rice metabolism (Joshi et al., 1975). The presence of high concentrations of volatile sulfide in SS soils may be the reason that *Spartina* grows less well in SS than in TS soils. Sulfate reduction may thus directly affect the rate of primary production of *Spartina*.

7.2.4 Methanogenesis

Recently our understanding of the conditions that affect methanogenesis has improved greatly. For example, we know that nitrogenous oxides can inhibit methanogenesis (Balderston and Payne, 1976). Bollag and Czlenkowski (1973) reported the relative effectiveness of N compounds in soil as $NO_3^- > NO > N_2O$. But Winfrey and Zeikus (1977) listed the order of effectiveness in lake sediments as $NO > N_2O > NO_3^-$. In their study denitrification increased methanogenesis by limiting substrates for sulfate reducers.

Competition occurs between sulfate reducers and methanogens. Methane producers and sulfate reducers in a lake sediments could coexist only if substrates were available and were spatially separated *in situ* (Cappenberg, 1974). Martens and Berner (1977) did not find methane produced in coastal marine sediments until 90% of the sulfate was depleted. Sulfate also inhibited methanogenesis in salt-marsh soil (Abram and Nedwell, 1978, 1978a); however, this effect was relieved if dihydrogen, but not acetate, was added to the medium. As little as 0.2 mM sulfate inhibited methane production in lake sediments, but dihydrogen or acetate reversed this inhibition (Winfrey and Zeikus, 1977). The inhibition of methanogenesis by sulfate ion appears to result from competition for substrate between methanogens and sulfate reducers. If sufficient substrate is available, there is no inhibition.

Synergistic relationships between methanogens and sulfate reducers have been reported. A sulfate reducer, *Desulfovibrio desulfuricans* can oxidize $^{14}CH_4$ (Davis and Yarborough, 1966). Cultures of *D. vulgaris* grown in the absence of sulfate but mixed with *Methanobacterum formicium* grew more rapidly (Bryant et al., 1977). *D. vulgaris* supplied growth factors for *Methanosarcina sp.* (Zhilina and Zavarzin, 1973). Optimal growth of *Methanosarcina barkeri* was attained with the addition of 1.25 mM H_2S (Mountfort and Asher, 1979). Sulfate (0.85 mM) was required for the production of methane from cellulose by an inoculum of organisms from sewage sludge, but higher sulfate concentrations inhibited methane production (Khan and Trotter, 1978).

With regard to spatial separation of sulfate reducers and methanogens, Cappenberg (1974) found highest sulfate reduction in a lake sediment at 0 to 2 cm, but maximum methane evolution at 3 to 6 cm. However, the two processes do occur simultaneously in marine sediments (Oremland and Taylor, 1978). Methane production increased where sulfate reduction was active or sulfate concentration was low (Whelan, 1974). In the marsh soils at Sapelo Island, sulfate reduction and methanogenesis were not mutually exclusive processes in the spatial scales investigated, but small-scale heterogeneity of microzones remains possible.

In the marshes of Sapelo Island two major factors affect the rate of methaneogenesis—water flow and sulfate reduction. Methanogenesis was over 100 times greater in SS than in TS soil (King and Wiebe, 1978). Oshrain (1977) and Skyring et *al.* (1979) measured up to 40% decrease in sulfate concentration with depth in SS soil but only a slight decrease in TS soil. Stopping the interstitial water flow in TS soil for two months produced a dramatic reduction in sulfate concentration and an increase in methane evolution (King and Wiebe, in press). Addition of an 80% dihydrogen-20% carbon dioxide atmosphere, but not acetate, to TS soil slurries produced an immediate, large increase in methane production (King and Wiebe, 1980). These results are similar to those reported by Abram and Nedwell (1978, 1978a) for salt-panne sediments and by Oremland and Taylor (1978) for mangrove soil. All of these data support competition for substrates between sulfate reducers and methanogens as a major factor affecting methanogenesis.

Temperature changes played a role only in SS soil, where most of the methane production took place in the summer (King and Wiebe, 1978). King (1978) added an 80% dihydrogen-20% carbon dioxide atmosphere to SS soil slurries incubated at 10°C and produced a large increase in the rate of methane production. These results independently validate the work of Hansen (1979) who reported decreased potential fermentation-rate constants in SS soils in winter. Because fermentation end products are the most likely source of substrate for sulfate reducers and methanogens, the rates of production by these two anaerobic groups should follow fermentation, even if they are not affected directly by temperature changes.

The four anaerobic processes—fermentation, DNOR, sulfate reduction, and methanogenesis—are tightly coupled in the Sapelo Island salt marsh. The tall *Spartina* and short *Spartina* soil systems operate quite differently, and the major determinants of magnitude are interstitial water flow, together with substrate availability. Our studies demonstrate the advantage of studying all four processes simultaneously so as to explain changes in any one rate by the state of the other three. Certainly, such studies are necessary if one is to develop hypotheses concerning the nature and magnitude of *in situ* responses in natural ecosystems.

The Salt-Marsh Ecosystem

8. The Cycles of Nitrogen and Phosphorus

D.M. WHITNEY, A.G. CHALMERS, E.B. HAINES, R.B. HANSON, L.R. POMEROY, and B. SHERR

Both the salt marsh and the estuarine water at Sapelo Island are highly eutrophic (Figure 8.1). Nutrient limitations at some levels, or in certain microhabitats, do exist, but they are not a major concern in our efforts to understand how the elements, especially phosphorus and nitrogen, function within the system. The earliest ecological studies of essential elements focused on their role as limiting factors in the sense of Liebig. Nutrient-organism interaction was emphasized, but the fundamental unity or interconnectedness of nutrient pools in all forms of an element and in all parts of an ecosystem were not. We now recognize the value of analyzing nutrient flows as a descriptor of ecosystem function. In this way nutrient flows can be used in the same way energy flows are used to describe properties of the ecosystem, although flows of energy and nutrients will not be identical. Elements are of concern to ecologists not only as limits to system biomass and function, but also as indicators of formation, transformation, and degradation of organic matter, both living and nonliving. Our emphasis here is on a holistic perception of the flux of phosphorus and nitrogen through the salt-marsh ecosystem.

8.1. Phosphorus

Phosphorus is in some respects a more useful tracer than either carbon or nitrogen. There are virtually no atmospheric phases of the cycle; phosphorus stays largely in water and sediments. Also, the existence of short-lived radionuclides of phosphorus made it possible to do field studies of the flux of phosphorus in the salt marsh (Pomeroy et al., 1969). Other tracer studies with ^{32}P in the salt marsh and the estuary revealed the rapidity with which phosphate exchanges with sediments (Pomeroy et al., 1965) and the short residence time of phosphate in the water, a phenomenon attributed primarily to biological activity (Pomeroy, 1960).

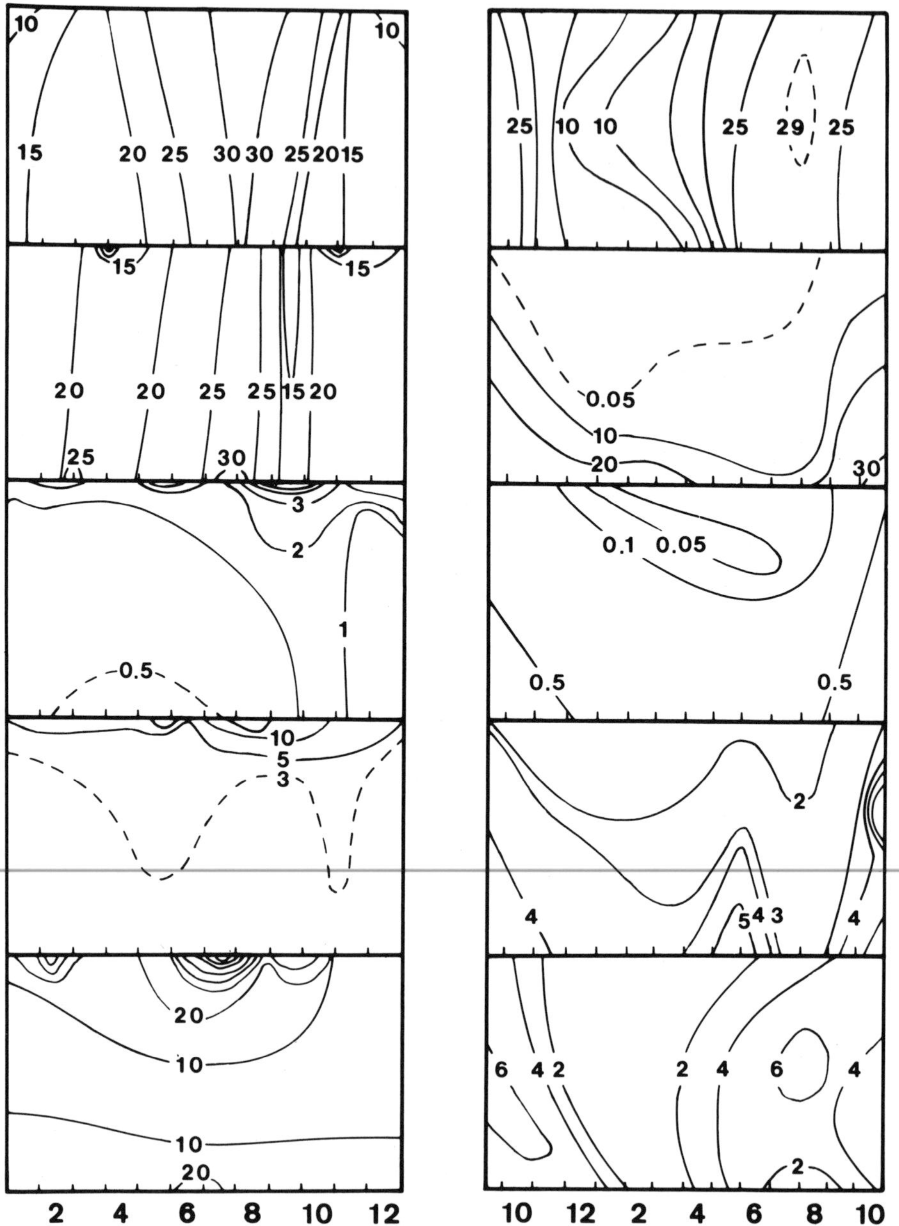

Figure 8.1. The regime of phosphorus and some related physicochemical factors in the Duplin River (left) and in Altamaha Sound (right). The latter is the principal egress for river water in the Altamaha delta. The vertical axis is distance along the estuary, from the inlet to the sea at the bottom of each panel to the upper end of the estuary at the top. In the case of Altamaha Sound this is defined as the point at which salinity approaches zero. From the top are shown: °C; salinity, $^0/_{00}$; phosphate, μM; total phosphorus, μM; Chlorophyll *a*, mg m^{-3}. Adapted from Pomeroy et al. (1972). Copyright (1972) by the American Society of Limnology and Oceanography.

Phosphorus is not limiting in salt marshes, and the turbid estuaries associated with them, because large reserves of phosphorus, much of it phosphate, are sorbed in clay sediments or peat. Clay is the dominant binding agent in the Georgia marshes. Although some of the phosphate is virtually permanently bound to the clay lattice structure, much of it can be removed by leaching. There is enough available phosphorus in the upper meter of the salt marsh to supply all that the marsh plants require for several hundred years (Pomeroy et al., 1972). Plants are probably not depleting the marsh of phosphorus, but such a large stock makes a steady state difficult to validate.

The amount of phosphorus which is tied up in the various components of the salt-marsh system is quite variable and, in some cases, quite large. An estimate of the average standing stocks of phosphorus in the Duplin River salt marshes is given in Table 8.1. The chief sink and the largest standing stock of phosphorus is that in the soil and sediment. The biota, primarily *Spartina* and the microorganisms, are the only other large standing stocks. However, smaller pools of phosphorus, along with the sediment bacteria, may contribute disproportionately to total flux within the system, because of the high turnover rates exhibited by these smaller compartments.

Phosphorus may be mobilized from the soils and sediments through several processes. Sediments suspended in water or water percolating through sediments will exchange phosphate. If the water is depleted in phosphate or if, as in the case of rain water, it has low ion strength, the net exchange will increase the phosphate content of the water (Carritt and Goodgal, 1954). If, on the other hand, the phosphate concentration of the water is high, net exchange will reduce the phosphate concentration in the water. These physicochemical exchanges with the sediments result in a buffer system in which the sediments are in equilibrium with water containing a little less than 1 μM P m^{-3} (Pomeroy et al., 1965).

Except in unusual circumstances such as storms, the exchange between sediments and the overlying water is probably slow and limited by the small amount of surface actually available for exchange. In Georgia marshes the percolation of water through the sediments is very slow (Nestler, 1977), and the amount of sediments actually suspended in the water on each tidal cycle is small. Therefore, the active uptake of phosphorus from the interstitial water by plants, especially by *Spartina*, may mobilize more phosphate than does direct sediment-water exchange.

Spartina obtains most of its phosphorus for growth of new tissue from the sediments. This phosphorus is recycled into the water and circulated through the system as the grass dies and decomposes or is eaten and digested (Pomeroy et al., 1969). A substantial amount of phosphate is lost from *Spartina* plants when they are covered by water during high tides (Reimold, 1972). However, this estimate was based on experimental leaching of *Spartina* leaves and on the assumption that *Spartina* is inundated daily, when, in fact, much of it is inundated only on fortnightly spring tides. Gardner (1975) disputed the role of *Spartina* as a source of phosphate for the estuary. Extrapolating from field measurements of the phosphate content of water running out of tidal creeks in South Carolina on the ebbing tide, he concluded that most of the excess phosphate in tidal creeks was derived from seepage. Neither Reimold's nor Gardner's calculations can be recapitulated successfully, and some question remains about the absolute magnitude of both *Spartina* leaching and sediment seepage. Both are probably primarily phenomena of spring and summer and vary fortnightly with the tidal regime

Table 8.1. Annual Mean Standing Stocks of Phosphorus in the Salt Marsh and Estuary near Sapelo Island[a]

Site	mg P m^{-2}	Study
Estuarine water	30	Pomeroy et al. (1969, 1972).
Sediment	5 times 10^5	Pomeroy et al. (1969).
Spartina shoots	523	Gallagher et al. (1980).
roots	744	Gallagher et al. (1980).
detritus	325	Reimold et al. (1975).
Phytoplankton	24	Whitney et al. (in ms.).
Benthic algae	12	Pomeroy (1959); Whitney and Darley (in ms.).
Bacteria in sediments	940	Wiegert and Wetzel (1974).
Altamaha River water	15	Windom et al. (1975).
Ocean water	15	Pomeroy et al. (1972).

[a] One meter of sediment is assumed to be in active interchange with the living components; mean water depth is taken as one meter.

as well. In any case, there does not appear to be a limiting rate of supply of phosphate to any part of the salt-marsh ecosystem.

The rate of mobilization of phosphate from the Duplin marshes can be estimated from the data in flux of water and phosphate in the Duplin River (Chapter 2), but the source of that phosphate is less certain. There are several potential sources of phosphate for the estuarine water. There is a large input of phosphate from the Altamaha River (Windom et al., 1975), but probably not more than 1% of that phosphate reaches the Duplin River, most of it going directly into coastal water. Because the water of the Duplin River is partly saline, on the order of 20 $^{o}/_{oo}$, there is a phosphate contribution from the ocean which exceeds that from the Altamaha River. The phosphate content of coastal water is at least as high as that of the river, and the flux of saltwater in the Duplin River exceeds that of fresh water (Pomeroy et al., 1972). Undoubtedly, some leaching of phosphate from the sediments occurs, as suggested by Gardner (1975) and by Pomeroy et al. (1965). Although there is some leaching of phosphate from *Spartina*, it is probably less than 10% of the amount suggested by Reimold (1972), considering the actual frequency of inundation of *Spartina*. Finally, microorganisms transform *Spartina*, with the eventual, but not necessarily direct, release of phosphate, most of which originated in the sediments. At least half the annual production of *Spartina* is degraded within the marsh, releasing on the order of 3 g P m^{-2}, exclusive of that released into the marsh soil. This is probably the largest single flux of phosphorus in the Georgia salt-marsh system.

Reimold and Daiber (1970), Reimold (1972), and Gardner (1975) consider the marsh to be a source of phosphorus for the estuary and the coastal waters. It is difficult to justify this conclusion in the absence of any identified primary source *within* the marsh. On balance, the marsh is receiving organic phosphorus and releasing inorganic phosphate, with the net flux of total phosphorus probably being *into* the marsh. Flax Pond salt marsh (Woodwell and Whitney, 1977) exhibits much the same pattern; phosphate is exported during the warm months and organic phosphorus is imported throughout the year. Because the phosphate concentration in the Altamaha River is generally low, relative to the general concentration in other river waters (Windom et al., 1975), phosphate will be flushed out of the estuary during periods of high river flow, as well as periods of heavy local rainfall (Chapter 2). Periods of low river flow and low local rainfall coincide with increased concentrations of phosphate in the Duplin River, indicating releases from the marsh which will ultimately be recycled. The concentration of total and organic phosphorus is higher in coastal water than in the Altamaha River (Pomeroy et al., 1972), so the largest single source of organic phosphorus for import into the Sapelo Island salt marshes is from the ocean. The Altamaha River is the second largest contributor of phosphorus to the system, while the contribution from rain is minor.

The net movement of phosphorus within the Duplin River and its marshes may be obscured by the rapid biological transformations within the system (Imberger et al., in press). Residence time of phosphate in the water is from hours to days, depending on rates of photosynthesis and respiration (Pomeroy, 1960; Pomeroy et al., 1972). Not only is internal recycling rapid, but exchanges with the sediments are rapid as well. Field experiments with ^{32}P indicate complete exchange between water and sediments daily. ^{32}P released on the rising tide entered the marsh but did not return with ebbing water. It was found on the surface of the sediments but not in *Spartina*, which has been shown to obtain most of its phosphorus from the subsurface sediments (Reimold, 1972). Labeled sediments were observed to move slowly out of the marsh along the creek bottom (Pomeroy et al., 1969). Because *Spartina* is a link between phosphorus in the subsurface sediments and phosphorus in the water, the growth and microbial transformations of *Spartina* are key biogeochemical processes. Other processes that complete the recycling include the deposition of sediments and organic detritus on the marsh surface and the seaward transport of particulate materials in storms. Improved quantification of the latter two processes would probably be helpful in understanding not only the flux of phosphorus but the biogeochemistry of the marsh in general.

From a purely ecological point of view, the large reservoir of phosphorus in the sediments, and its continued recycling through *Spartina*, appear to confer stability to the salt-marsh ecosystem (Pomeroy, 1975). The system is naturally eutrophic, and it has a high capacity to assimilate and store phosphorus. Although the Duplin River watershed is as pristine as an ecosystem can be in eastern North America, it contains an excess of phosphate. Still, because of its high storage capacity for phosphorus, the system would be able to assimilate substantial phosphate pollution without eutrophication. Other potential polluting elements, such as nitrogen or biodegradable organic matter, however, cannot be as readily assimilated. As it is, the system is probably very nearly as euthrophic as it can be without suffering adverse consequences of temporary depletion of oxygen in the tidal water (Chapter 6).

8.2. Nitrogen

The nitrogen cycle is more complex and more difficult to study than that of phosphorus or carbon. Part of the difficulty arises from the lack of a useful radioactive tracer and from the technological difficulties associated with the use of the stable isotope, ^{15}N. One of the most important features of the nitrogen cycle is that a few groups of microorganisms carry out some of the key transformations of nitrogen. Plants and bacteria are able to transform inorganic nitrogen into organic form. Aside from excretion of ammonia by invertebrates, most return of organic nitrogen to the inorganic form is a microbially mediated process. Some species of blue-green algae and bacteria are able to fix atmospheric nitrogen; nitrification and denitrification are strictly bacterial processes (see Chapter 7).

Much of the work on the Sapelo Island marshes has involved study of specific microbial transformations of nitrogen, such as nitrogen fixation and denitrification, and of changes in standing stock of nitrogen nutrients: ammonia, nitrate and nitrite. Observation of the concentration of nitrogen species in time and space under natural and experimental conditions has provided insights into processes controlling the distribution of nitrogen in estuarine ecosystems (Aurand and Daiber, 1973; Hale, 1975; Nixon et al., 1976). Research using ^{15}N labeled compounds to quantify specific pathways of nitrogen in coastal waters has increased in recent years and is providing much needed information about estuarine nitrogen cycling (Stanley and Hobbie, 1977; McCarthy et al., 1975; Harrison, 1978).

The quantity of nitrogen in salt-marsh soils is determined by three interrelated factors: tides, physical and chemical exchanges with water and air, and biological fluxes. In Great Sippewissett marsh in Massachusetts, marsh ground water has been found to be a major source of nitrogen (Valiela et al., 1978), but in Georgia this is negligible (Haines, 1979a). Although substantial sorptive exchanges occur, the net transformations within the soil are biologically controlled. Exchanges between soil and water are primarily abiotic, although they too may be controlled biologically. The tides affect movement of particulate organic nitrogen and dissolved organic and inorganic nitrogen both into and out of the marsh and from place to place within the marsh. Diffusion of dissolved forms of nitrogen or volatilization of gaseous forms from marsh soils are examples of other abiotic fluxes. We know the magnitude of some of these abiotic fluxes, but we do not know the rates of exchange at the soil-water interface or the factors controlling these exchanges.

Similar processes determine the quantity of nitrogen in estuarine water, which is an intermediary in the exchange of nitrogen between rivers, salt marshes, and continental shelf waters. Most of the salt-marsh estuaries along the Georgia coast are hydrologically dominated by tidal flushing, although river flow is significant at the mouth of the Altamaha and other rivers. In addition to tidally mediated exchanges, rainfall, biological transformations within the water column, and exchange with the sediments can affect the concentrations of various forms of nitrogen in the water.

8.2.1 Standing Stocks

Except for N_2, the largest pools of nitrogen in estuarine water are dissolved and particulate organic nitrogen. The concentration of dissolved organic nitrogen (DON) in Georgia estuarine and nearshore shelf water ranges from 2 to 20 μM and that of particulate nitrogen (PN) from 0.1 to 30 μM (Haines, 1979a). Concentrations in marsh creek water at low tide are higher, ranging from 4.4 to 38 μM for DON and from 13 to 239 μM for PN. In the open estuary and nearshore shelf waters, the concentration of ammonia and nitrate is generally low, below 5 μM (Haines, 1979a). In the mouths of major rivers, such as the Altamaha, the concentration of ammonia and nitrate may increase to 5 to 20 μM (Windom et al., 1975). These higher concentrations decrease rapidly in a nonlinear fashion on mixing with seawater, which suggests biological uptake rather than simple mixing processes (Haines, 1974). The relatively low concentrations of ammonia, nitrate, and nitrite measured in Georgia coastal waters at all times of the year (Haines, 1979a) result from a small input of river nitrogen and from high year-round rates of phytoplankton production (Thomas, 1966).

A substantial concentration gradient exists for NH_4^+, NO_2^-, and NO_3^- between salt-marsh soils and the estuarine waters which cover them at high tide. In Georgia, concentrations of NH_4^+ average 30 to 70 μM in the interstitial water of both high-marsh and low-marsh soils with little seasonal variation, although concentrations tend to be lowest during the early summer when *Spartina* growth is greatest (Chalmers, 1977). Nitrite and nitrate concentrations in marsh soils average 2 to 3 μM and 5 to 10 μM, respectively, although nitrate is sometimes found in concentrations as high as 100 μM. As with ammonia, there is little difference between the amount of NO_2^- and NO_3^- found in the high marsh and that found in the low marsh, although concentrations are highly variable throughout the marsh. The highest concentrations are usually found at or near the soil surface, although some NO_3^- can be found as deep as 30 cm in the soil. Concentrations tend to be higher inthe winter (Chalmers, 1977).

Inorganic nitrogen is less than 0.1% of the total soil nitrogen pool. The remaining nitrogen is organic, most of it in some refractory form. Total nitrogen concentrations in the soil fall in the range 0.2 to 0.5%. Percent nitrogen is slightly higher in the low-marsh soils than in the high-marsh, although the amount of nitrogen in a 30-cm soil column is approximately the same in both areas, 485 g N m^{-2} (Chalmers, 1977).

Sources of inorganic nitrogen–NH_4^+, NO_3^-, NO_2^-–in estuarine waters include diffusion from nitrogen-rich marsh soils and subtidal sediments (Haines, 1979a), influx via river flow (Windom et al., 1975), rainfall (Haines, 1976), and shelf waters (Dunstan and Atkinson, 1976), nitrogen fixation, and regeneration through ammonification and nitrification in the water. Sinks for inorganic nitrogen include uptake by algae and by heterotrophic microorganisms degrading nitrogen-poor plant detritus and denitrification in anixoc sediments. Sources of dissolved and particulate nitrogen in estuarine waters are leaching of sediment and resuspension of marsh soils and subtidal sediments (Oviatt and Nixon, 1975; Pomeroy et al., 1977; Roman and Tenore, 1978). Some fraction of what appears to be particulate organic nitrogen may simply be ammonium bound to clay particles. Phytoplankton production represents an *in situ* source of both DON and PN, quantitatively minor compared to the nonliving seston load, but qualitatively important in terms of nitrogen turnover in the water column. Plants supply most of the organic nitrogen to the soil, as detritus at the surface and as roots and rhizomes below the surface. Sedimentation accounts for some of the nitrogen input. Plant growth, detritus removal, and sediment resuspension in the water account for losses of organic nitrogen from the soil.

8.2.2 Abiotic Fluxes

Little is known about rates of abiotic fluxes of nitrogen in marsh soils or about the factors controlling them. Gardner (1975) found that silicate and phosphate were lost from South Carolina marsh soils when they diffused into a thin layer of water left on the marsh surface at low tide, a layer which was then picked up by the next flood tide. He indicated that a similar process occurred with ammonia. The high NH_4^+ concentrations in the water of tidal creeks at low tide, concentrations measured as high as 74 μM (Haines, 1979a). Dissolved organic nitrogen is probably lost in the same manner. Volatilization can also occur, depending on surface temperature, pH, and concentration of NH_4^+. Physical entrainment of sediment, detritus, and benthic algae by tidal flow can cause loss of substantial amounts of nitrogen from the marsh and consequent addition of nitrogen to estuarine water (Roman and Tenore, 1978; Oviatt and Nixon, 1975). Removal of dead *Spartina* in this manner takes 12 to 21 g N m^{-2} yr^{-1} from the marsh (Haines et al., 1977). The reverse process also takes place, with particulate matter being filtered out of the water by the grass. In fact, several studies have shown a net input of particulate matter to marshes and a net loss of dissolved forms (Haines, 1979a; Woodwell et al., 1977; Heinle and Flemer, 1976). Using a sedimentation rate of 2 mm yr^{-1} (Wiegert and Wetzel, 1979) and a content of 0.54 g N cc^{-1} in surface soil in the high marsh (Chalmers, 1977), it is estimated that sedimentation adds nitrogen to the soil of Sapelo Island marshes at the rate of 3.2 g N m^{-2} yr^{-1}. Rainfall adds 0.3 g N m^{-2} yr^{-1}

(Windom et al., 1975). However, most river water probably moves swiftly through the estuary. Exchanges between estuarine water and continental shelf waters have been poorly studied, but inner-continental shelf water can potentially contribute at least as much nitrate to the marsh as the Altamaha River.

8.2.3 Nitrogen Fixation

Nitrogen fixation provides a new source of nitrogen to salt-marsh soils of Sapelo Island at the rate of 15 g N m^{-2} each year (Hanson, 1977a). Bacterial fixation in the rhizosphere accounts for most of the input in this marsh as in others (Whitney et al., 1975; Teal et al., 1979). Fixation on the marsh surface and by epiphytes on *Spartina* contributes < 1% of the total amount fixed (Hanson, 1977a). However, at some times of year there are very high rates of fixation (0.7 to 7.6 mg N m^{-2} h^{-1}) in mats of *Calothrix* and *Anabaena* (Hanson, 1977a). Nitrogen fixation was greatest in the tall (40 g N m^{-2} yr^{-1}) and short (13 g N m^{-2} yr^{-1}) *Spartina* soils (Table 8.2). Fixation in other angiosperm habitats was low (<5 g N m^{-2} yr^{-1}). Nitrogen fixation is a significant source of nitrogen for *Spartina* and estuarine algae. The annual nitrogen flux through *Spartina* in on the order of 22 g N m^{-2} yr^{-1} (Table 8.2).

Nitrogen fixation is enhanced by organic compounds added *in vitro* in tall and short *Spartina* soils (Hanson, 1977). When first-order kinetic values (mM C_2H_4 mg carbohydrates^{-1} hr^{-1}) are compared, the activity of the nitrogen fixers in the tall *Spartina* soil is increased two to seven times over the activity in the short *Spartina* soil (Figure 8.2), suggesting nitrogen fixation is substrate limited. Raffinose, mannose, and arabinose are most effective as energy sources. Addition of organic substrates to short *Spartina* soil does not enhance nitrogen fixation, but this may be due to high *in situ* substrate concentrations in short *Spartina* soils (Gardner and Hanson, 1979; Hanson and Snyder, 1980). The long-term addition of moderate levels of nitrogen to the marsh did not suppress nitrogen fixation (Hanson, 1977a), but resulted in an increase in nitrogen fixation activity in the short *Spartina* soil zone of 5 to 7 cm (Figures 8.3, 8.4, 8.5). In these field experiments the added nitrogen was apparently consumed by the plants, algae, and microorganisms, relieving any temporary suppression of nitrogen fixation. However, nitrogen fixation is immediately suppressed by elevated concentrations (1 mM) of NH_4^+-N and NO_3^--N but not by amino acids (Hanson, 1977). The concentration of these nitrogen compounds in the field is less than 100 μM and is below the concentration required to suppress nitrogen fixation. Consequently, nitrogen fixation is not apparently influenced by nitrogen concentrations in the marsh. What may be controlling nitrogen fixation in Georgia salt marshes is the rate of water movement through the soil (Ubben and Hanson, 1980), *Spartina* production, and microorganic matter in the soil (Figure 8.6).

8.2.4 Denitrification

Nitrogen fixation is equaled or exceeded by denitrification, estimated to be 65 g N m^{-2} yr^{-1} (Sherr, 1977). There are significant differences in denitrifying activity between high and low marsh. In the high marsh, activity is substantially lower in winter, but rates in the low marsh do not vary seasonally. Denitrifier activity in the high marsh is

Table 8.2. Comparison of Annual Nitrogen Fixation, Net Aerial Primary Production, and Estimated Nitrogen Requirements of Major Salt-Marsh Angiosperms in the Marshes of the Duplin River Watershed

Salt-Marsh Angiosperms	Marsh Area[a] (10^4 m^2)	Annual Nitrogen Fixation ($g\ N\ m^{-2}$)		Annual Net Aerial Primary Production[b] ($g\ C\ m^{-2}$)		Annual Nitrogen Requirements[c] ($g\ N\ m^{-2}$)	
		($g\ N\ m^{-2}$)	(10^6 g N)	($g\ C\ m^{-2}$)	(10^6 g C)	($g\ N\ m^{-2}$)	(10^6 g N)
Short *Spartina*	978	13.1	128	650	6357	18.6	182
Tall *Spartina*	96	39.7	38.1	1650	1584	47.1	45.3
Juncus	68	4.0	2.7	1100	748	31.4	21.4
Total	1142		168.8		8689		248.7
Mean per unit area of marsh ($g\ m^{-2}$)		14.8		761		21.7	

[a]From Gallagher et al. (1972).
[b]From Gallagher et al. (1980), assuming carbon = 50% of dry weight.
[c]From Gallagher and Plumley (1979), based on C:N ration of 35:1.

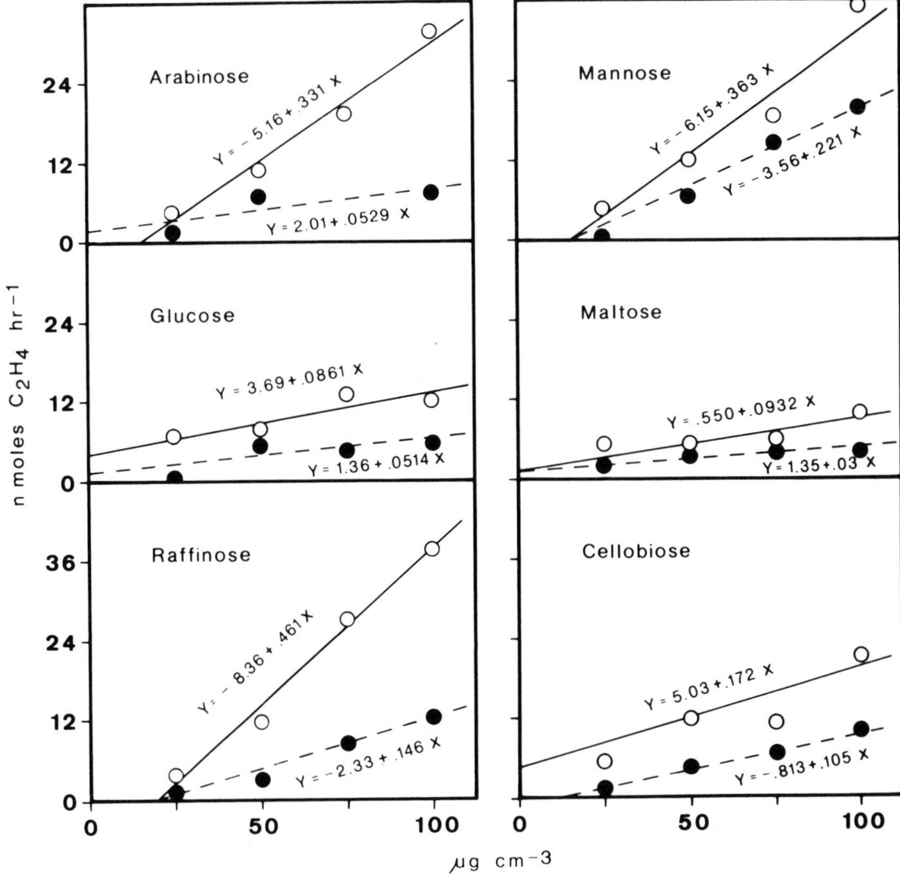

Figure 8.2. Effects of organic carbon substrates on nitrogen fixation in the tall (○) and short (●) *S. alterniflora* marsh-soil slurries. Linear regression lines were plotted. Acetylene concentration was held constant at 13% of gas phase in all flasks. Adapted from Hanson (1977). Copyright American Society for Microbiology.

tightly coupled to belowground production of *Spartina* (Sherr and Payne, 1978), and the seasonal variations in availability of belowground macroorganic matter in high and low marsh probably account for the seasonal differences in activity. In the high marsh, rates of denitrification are limited by the availability of carbon-energy substrates, but this is not true in the low marsh (Sherr, 1977). Growth of denitrifier populations in the high marsh appears to be limited by the availability of nitrogenous oxides. In the low marsh, population growth is limited both by the availability of energy substrates, the electron donors, and of nitrogenous oxides, the electron acceptors (Sherr, 1977).

Our present estimate of the rate of denitrification is very approximate. The number of samples is not large, considering the variability within the salt marsh over space and time, and the method itself tends to give a potential rather than an absolute *in situ* rate. It is reasonable to expect, however, that the anaerobic sediments of the salt

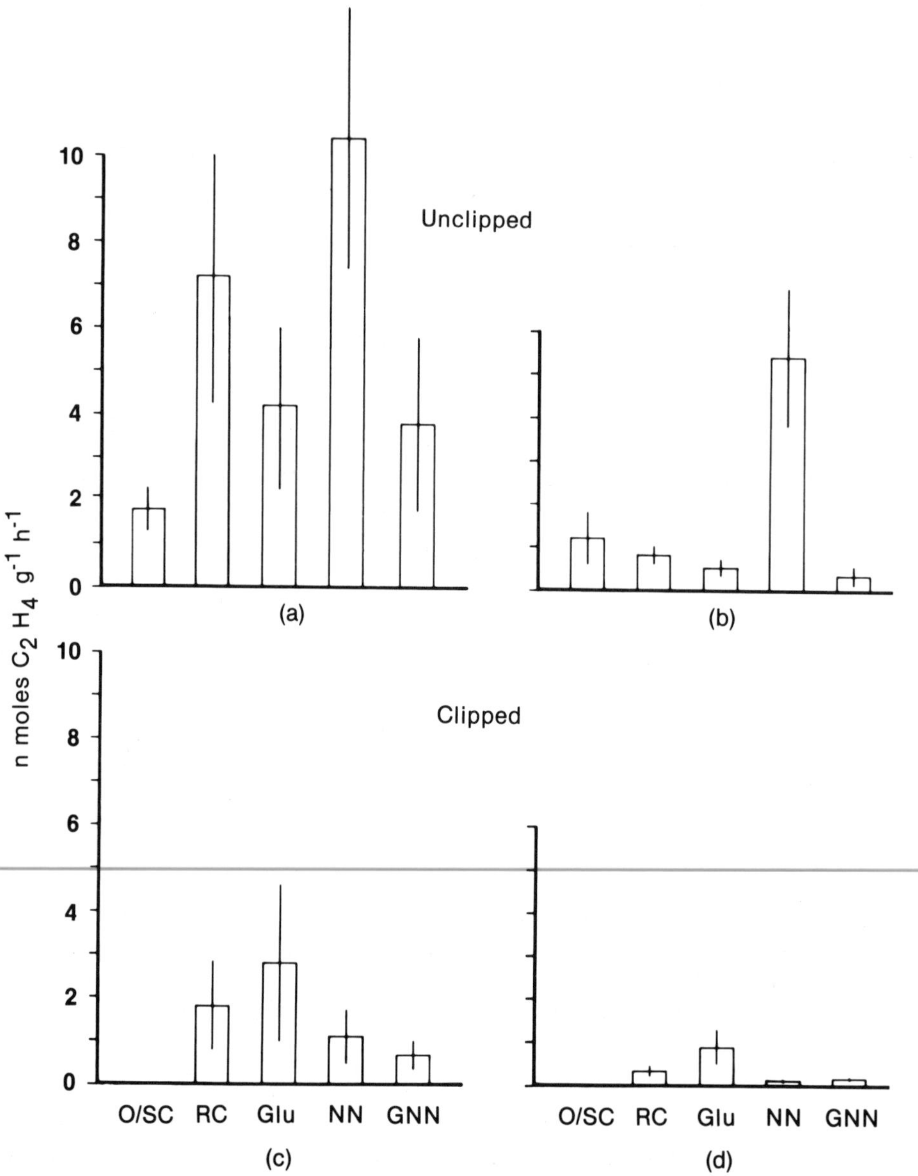

Figure 8.3. Nitrogen fixation rates in clipped (bottom) and unclipped (top) plots of short *S. alterniflora* in two depth zones, 0-5 cm (a and c) and 10-15 cm (b and d). Mean rates from four replicates. Vertical bars = 1 standard error of the mean. O/SC = outside control; RC = replicate control; Glu = glucose; NN = ammonium nitrate; GNN = glucose + ammonium nitrate. Adapted from Hanson (1977a). Copyright (1977) American Society for Microbiology.

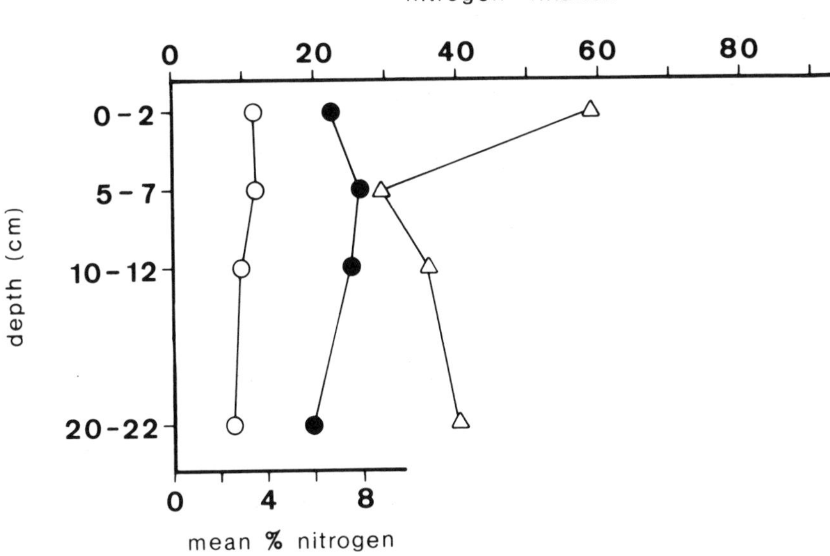

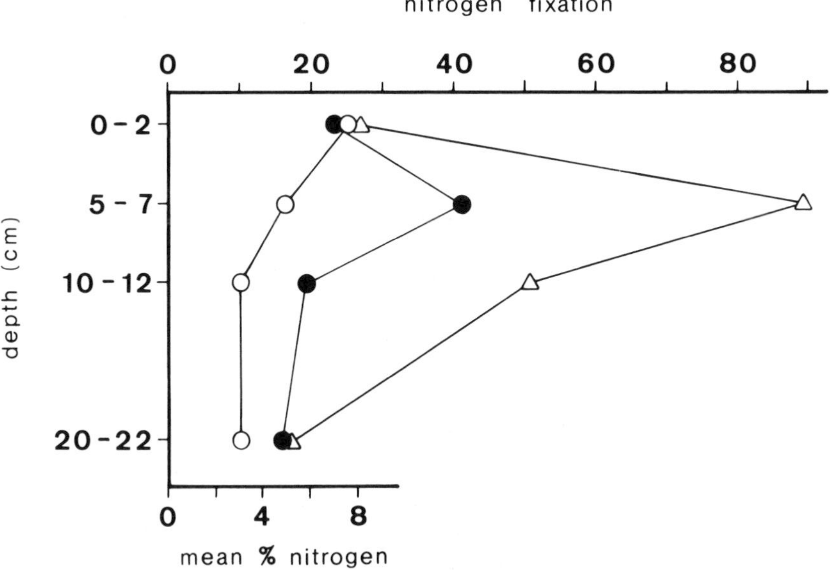

Figure 8.4. Mean monthly nitrogen fixation rate in short *S. alterniflora* marsh, both in a natural state (upper) and with the addition of sewage sludge (lower). The mean rates were normalized to nanomoles of C_2H_4 g^{-1} h^{-1} (●) and nanomoles of C_2H_4 mg N^{-1} h^{-1} (△) and plotted with the mean monthly percent nitrogen (○). Adapted from Hanson (1977). Copyright (1977) by the American Society for Microbiology.

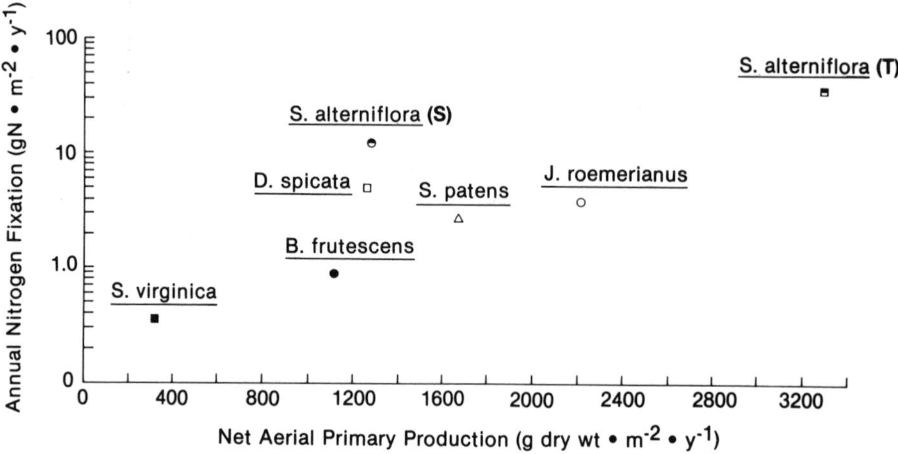

Figure 8.5. Relationship of annual nitrogen fixation to the net aerial primary production in habitats dominated by various salt-marsh angiosperms. The estimates of net aerial primary production are from Gallagher et al. (1980) and Linthurst and Reimold (1978, 1978a).

marsh, with a continuing input of organic substrates from *Spartina* production, are the site of significant denitrification. This appears to have a major impact on the coastal cycle of nitrogen.

8.2.5 Nitrification

The high concentrations of nitrate and nitrite in the soil relative to those in tidal water indicate nitrite and nitrate production in oxidized zones near the soil surface and around roots and rhizomes of *Spartina* (Chalmers, 1977). In other areas with concentrations in tidal water considerably higher than those found in Georgia, nitrate uptake by marsh soils has been observed (Aurand and Daiber, 1973; Axelrad et al., 1976; Engler and Patrick, 1974). In Georgia, however, *in situ* nitrification seems to be the primary source of NO_2^- and NO_3^-.

Other nitrogen transformations in marsh soils are less completely understood, in part because of the difficulty of conducting tracer studies. Consequently, most studies measure the concentrations of various forms of nitrogen through time or as a result of various perturbations. Studies of seasonal patterns of accumulations of nitrogen in plants are the most common. There is little difference found in the seasonal pattern of nitrogen accumulation between tall and short *Spartina*, but biomass and, consequently, g N m^{-2} are significantly higher in the low marsh throughout the year, as is the quantity of dead *Spartina*. In 1975 the amount of nitrogen in live *Spartina* shoots was 9.8 g N m^{-2} in the low marsh and 4.7 g N m^{-2} in the high marsh (Chalmers, 1977). The concentration of nitrogen in plant tissue was consistently higher in the tall *Spartina*. Using a net productivity of 1577 g C m^{-2} yr^{-1} (Chapter 3) and the mean C/N ratio for *Spartina* of 33.78 (Chalmers, 1977), the flux of nitrogen to *Spartina* is 46.7 g N m^{-2}

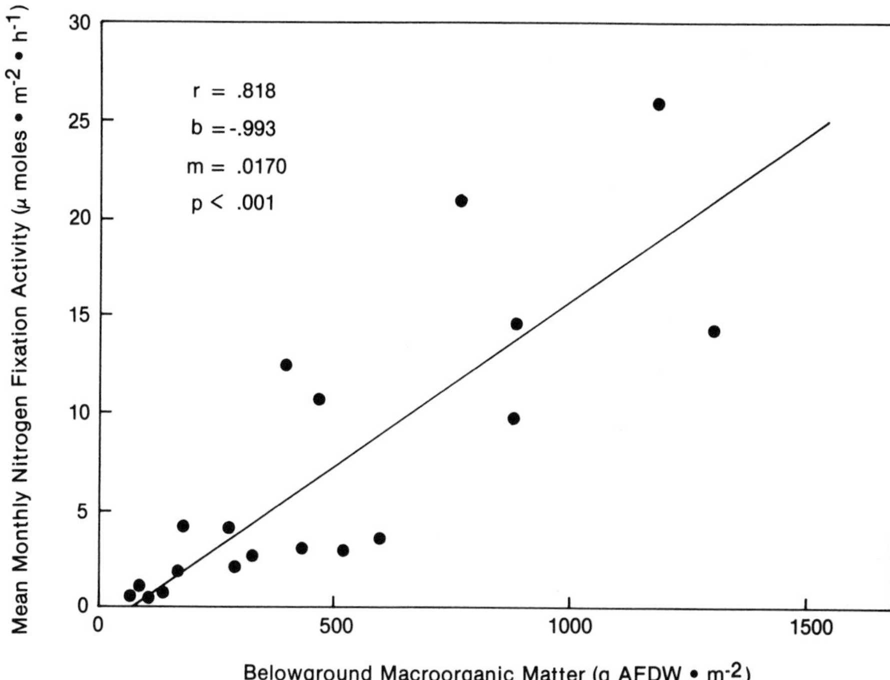

Figure 8.6. Correlation between the distribution of nitrogen fixation activity and belowground macroorganic matter in salt-marsh soils. Data on macroorganic matter distribution at each site are from Gallagher and Plumley (1979).

yr^{-1}. This includes belowground production, which was not part of the maximum accumulation of nitrogen cited above, and losses of dead *Spartina* as well. Since *Spartina* moves nitrogen from shoots to roots during senescence, this is an overestimate, but probably by 10% or less.

8.2.6 Nitrogen Limitation

Increased production follows nitrogen fertilization of short *Spartina* (Valiela and Teal, 1974; Sullivan and Daiber, 1974; Broome et al., 1975; Gallagher, 1975; Patrick and Delaune, 1976; Haines and Dunn, 1976), although similar fertilization of tall *Spartina* has no effect (Gallagher, 1975). The difference may be explained by the rate of mineralization, the rate at which organic matter is broken down by microbes and inorganic nitrogen is made available for use by plants. The rate of mineralization in the low marsh is four times that in the high marsh (Dunn, pers. comm.). Moreover, interstitial water of the low marsh exchanges with that in the tidal creeks, so a supply of nitrate from creek water is available.

Factors other than nitrogen availability play a role in limiting *Spartina* growth in the high marsh (Section 3.1.2). Soil salinity, which is higher in the high marsh than in the estuarine water, is one of these. High salinity can partially overcome the stimulating effect of ammonium (Haines and Dunn, 1976), since increased energy expenditures are required for ammonium uptake at high salinity. Fertilization with nitrogen indirectly increases soil salinity because of the higher rate of transpiration accompanying increased production (Christian et al., 1978; Chalmers, 1979). Thus nitrogen fertilization may increase plant growth, but also limit the amount of the increase. This feedback limits the amount of added nitrogen that can be utilized by *Spartina* in the high marsh (Chalmers, 1979).

The human population of coastal areas often enriches estuarine waters with nitrogen. In fact, the use of marshes as natural waste treatment systems was proposed (Gosselink et al., 1973), although little was known at the time about the capacity of the marsh to assimilate excess nitrogen. Effects on the marsh of nitrogen added in the form of dried sewage sludge were measured at Sapelo Island. Dried sewage sludge with a 2% nitrogen content was added biweekly at a rate of 100 g m^{-2} wk^{-1} to plots in the short *Spartina* zone. Total nitrogen content of both plants and soil, NH_4^+, NO_2^-, and NO_3^- concentrations, and salinity in the soil were measured monthly in the fertilized plots and in a control short *Spartina* area (Chalmers et al., 1976). Rates of denitrification, nitrogen fixation, algal primary productivity, and ATP concentrations were also measured.

Sludge additions resulted in increased plant biomass (Haines, 1979a) and in higher total nitrogen and ammonia concentrations in the soil (Chalmers, 1979). Nitrogen fixation rates did not differ between control and treated plots (Hanson, 1977a), and benthic algal productivity was not affected (Whitney and Darley, pers. comm.), nor was soil microbial ATP (Christian, 1976). Denitrification was inhibited (Sherr, 1977). At the end of the study we could account for only half of the applied nitrogen, most of that being found in the top 5 cm of soil. The amount present as plant biomass or lost as detritus was less than 5% of the amount added. The 44% lost was assumed to have been washed out of the study area by tides (Chalmers, 1979). Twenty months after the last sludge application, nitrogen content of surface soil and plant biomass in the fertilized area was higher than just outside the fertilized zone, although the percent nitrogen in the surface soil had decreased from 0.87% to 0.61%, compared to 0.38% for soil in the control area (Haines, 1979a).

Any nitrogen washed out of the fertilized area probably was rapidly assimilated by estuarine phytoplankton or microheterotrophs degrading nitrogen-poor detritus. The long-term effects of nitrogen additions to estuarine water are not known, but, like the marsh, estuarine water has an upper limit on the quantity of excess nitrogen its flora can assimilate. The organic production may become too great, creating temporary anoxic conditions in water that is usually oxygenated.

8.3. Net Flux of Phosphorus and Nitrogen

The major flux of phosphorus in and out of the salt marsh and its estuarine waters is in the form of dissolved phosphate. So far as we can tell, there is a net balance of input and output of phosphorus, with a considerable surplus relative to biological requirements present at all times, both in the sediments and in the tidal water. The concentration in the estuarine water tends to be slightly below the point of equilibrium with phosphate sorbed on marsh clays of 1 μM P l^{-1}, while water running off the marsh tends to exceed the clay equilibrium value. This suggests that there is active remineralization on the surface of the sediments and in the uppermost part of the sediments. The excess phosphate which is remineralized in the sediments is utilized by both autotrophic and heterotrophic microorganisms in the estuarine water, and much of it probably moves back through the marsh as living or nonliving particulate organic phosphorus.

Simple experiments with white clay as a marker demonstrate that the salt marsh at Sapelo Island is a depositional environment, receiving an appreciable daily input of sediment, including both organic fallout from the *Spartina* stands in the marsh and deposition of suspended organic particles carried into the marsh with the tide. Conversion of simple observations of deposition to time-averaged estimates of the rate of deposition and accumulation of elements is exceedingly difficult. The best estimates involve the use of biologically conservative radioactive isotopes (Delaune et al., 1978). This has not been done at Sapelo Island, but where it has been done there is evidence that there is some depletion of phosphorus from the soil with time. This may be, in part, the result of utilization by rooted plants and, in part, the loss of phosphate-rich interstitial water as the sediment is compacted. Superimposed on the long-term geological process of sedimentation, by which phosphorus is accumulating in the marsh, are a number of short-term biological recycling processes, some of which turn over phosphorus pools in the sediment and water in a matter of days. From the point of view of ecosystem processes, the marsh is in steady state with respect to the cycling of phosphorus. The studies of Flax Pond marsh showed a seasonal shift from intake of phosphate by the marsh in the colder seasons to intake of organic phosphorus and output of phosphate in the warmer seasons (Woodwell and Whitney, 1977), with the net annual balance being the input of organic phosphorus and the release of phosphate. While this is possibly true for the Sapelo Island marshes as well, the biological recycling is somewhat obscured by the equilibria with clay minerals and the presence of vigorous tidal flushing. Certainly there is no reason to believe that the Sapelo Island marshes are a source of phosphorus for the estuary or the ocean, nor is there reason to believe that the production of phosphate in the marsh has a significant influence on biological processes in the estuary or the coastal waters. Recycling of phosphorus in the coastal waters is rapid (Pomeroy, 1960), and the short-term cycling of phosphorus there appears to be independent of inputs, both from the marshes and from the ocean.

The biological processes of greatest magnitude in the cycle of nitrogen are fixation, denitrification, and autotrophic reduction of available nitrogen to organic nitrogen. Existing data are insufficient to determine whether nitrogen fixation and denitrification are equal for the marsh as a whole or whether one exceeds the other. The available information suggests that denitrification exceeds nitrogen fixation (Table 8.3), but the reverse could be true. In either case it is probable that the system is kept near a steady state by tidal exchanges of nitrate with the ocean and the Altamaha River. Both nitrogen fixation and denitrification vary with the seasons. The observed export of nitrate from the marsh suggests net fixation during the summer.

While there is significant nitrogen fixation in the marsh, there is not, so far as we know, significant fixation in the subtidal parts of the estuary or in the coastal waters. Denitrification does occur there in the anaerobic sediments, so the estuarine and coastal environment is a sink for biologically available nitrogen. If the salt marsh is also a

Table 8.3. A provisional Nitrogen Budget for the Duplin River Watershed at Sapelo Island[a]

Flux Components	Nitrogen Flux g N m^{-2} yr^{-1}
Inputs	
Rain	0.3
Sedimentation	3.3
Nitrogen fixation	14.8
Tidal exchange	46.6
Losses	
Denitrification	65.
Internal Cycles	
Primary production	70.
Soil remineralization	70.

[a]Tidal exchange is depicted as seeking a steady state, offsetting differences in nitrogen fixation and denitrification.

net sink for biologically available nitrogen, this means that the system can remain in steady state only through inputs of nitrate from the Altamaha River and the ocean by way of the denitrifying coastal zone. In this respect the cycle of nitrogen in the marshes of Sapelo Island differs from that described by Valiela and Teal (1979, 1979a) for Great Sippewissett marsh on Cape Cod, Massachusetts. The latter is a relatively small marsh adjacent to uplands, from which there is a significant flow of ground water. This flow brings in substantial amounts of nitrate, which percolate through the marsh peat, and causes net export of nitrogen as particulate organic nitrogen, ammonia, and nitrate. The Sapelo Island marshes might also export ammonia during the summer, were it not for the hydrographic isolation of most of the marshes. Ammonia lost from the marshes is utilized by phytoplankton within the three tidal segments of the Duplin River, and in that way much of it may be recycled through food webs associated with the marsh and tidal creeks.

A feature which both the Georgia and Massachusetts estuaries appear to share is the excess of denitrification over nitrogen fixation in the estuary and coastal zone as a whole. Denitrification in the anaerobic sediments of the marsh and tidal creeks influences not only the nitrogen budget of the marsh but also that of the estuary. This is reflected in the low ratio of available nitrogen to phosphorus. Although the Redfield ratio should not be applied as a standard, the departure from it to values as low as 2:1 is clearly indicative of nitrogen depletion. That depletion is a direct result of processes in the extensive anaerobic sediments of the marshes, estuaries, and coastal waters. The input of nitrate from the Altamaha River and from the ocean is probably of major importance. Although there is rapid recycling within the marsh and the estuarine water, it appears that a substantial input of nitrogen is necessary to balance the losses from denitrification.

Comparative data now exist on nitrogen budgets for several salt-marsh ecosystems, and these serve to illuminate the differences that exist in the flow of this biologically important element. A nitrogen budget for Great Sippewissett marsh reported by Valiela and Teal (1979, 1979a) shows nitrogen transport into the marsh by ground water, rain, and nitrogen fixation, mainly bacterial. Losses occur through tidal export, denitrification, and a small amount of volitilization. The large input of nitrogen from ground water, and a lesser amount from nitrogen fixation, is lost–approximately one-third to denitrification and two-thirds to tidal export. Overall, the marsh seems reasonably in balance with respect to nitrogen, as the 974 kg loss per year could easily be within the various measurement errors involved in such an accounting.

The long-term study of Flax Pond salt marsh, a small (57 ha) diked marsh with a single outlet on Long Island Sound, New York (Woodwell et al., 1979), found the annual tidal nitrogen flux in this marsh to be substantially lower than the flux in Great Sippewissett marsh. For the full year Woodwell et al. estimated a net loss from the marsh of 0.9 g N m^{-2} yr^{-1}, but because this estimate carried a standard error of 1.2, we must regard the marsh as more or less in balance. There was however, enough nitrogen fixation to account for the tidal losses of ammonium (2.1 g m^{-2} yr^{-1}). They also estimated the transport of organic nitrogen to Long Island Sound on an annual basis, but the amount, prorated over the entire marsh, would not exceed 2 to 3 g N m^{-2} yr^{-1}, far below the 24 g N m^{-2} yr^{-1} exported by the tides from Great Sippewissett marsh.

Nitrogen transfers within the Barataria Bay marshes have not been intensively studied. The nitrogen model of Hopkinson and Day (1977) is based mainly on information from other marsh systems. They report that the standing stocks predicted by the model agreed well with field measurements, but they do not discuss the overall input-output budget of the marsh.

The Sapelo Island marshes have high rates of nitrogen fixation, but, like the Great Sippewissett marsh, this is substantially exceeded by losses to denitrification. There is no appreciable ground water input to the marshes and only 0.3 g N M^{-2} yr^{-1} is added from rain. The marsh sediments approximately 3 g N m^{-2} yr^{-1}. Thus, unlike the Great Sippewissett and, to a lesser degree, the Flax Pond marshes, the Sapelo Island ecosystem does not export, but must import, a significant quantity of nitrogen each year, most likely as nitrate from seawater and the Altamaha River. Judging from the observed rates of primary production, however, the marginal balance of the nitrogen budget is less significant as a limiting factor in production than is the intensity of solar radiation, temperature, and overall climate.

9. A Model View of the Marsh

R.G. WIEGERT, R.R. CHRISTIAN, and R.L. WETZEL

And what if behind me to westward the wall of the woods
stands high?
The world lies east: how ample, the marsh and the sea
and the sky!

Sidney Lanier, "The Marshes of Glynn"

This view of the salt marsh ecosystem is a model, poetic to be sure, but a model nonetheless. Indeed, it anticipates the view of the marsh as an ecosystem, a group of components in interaction. In this chapter we address the nature of this interaction between marsh and sea and sky—more prosaically, between the soil, the water, and the air. In particular, we show how a simulation model of the salt marsh began as a summary of existing knowledge, was used to suggest new avenues of profitable research, was revised as the research progressed, and, finally, was employed to formulate hypotheses explaining both the dynamics of the marsh and its interaction with the estuary and the nearshore ocean.

9.1. What is a Model?

A model is an abstraction of reality. It should not distort reality, but, rather, should present the relevant aspects and delete the others. For Sidney Lanier, the poet, the reality of the marsh was the sense of space, of freedom, inherent in its connection with the limitless ocean and sky. The westward forest was a constraining, almost a forbidding boundary. Our ecological view of reality differs from this only in degree. Standing in the marsh one faces almost instinctively, not landward, but toward the creeks, the estuary, and the sea. From the sea come the tides, and to the marsh the tides are life.

These tides, the topography of the marshes, and the organisms all combine to produce a close coupling between marsh, estuary, sea, and atmosphere. In contrast, the interaction between marsh and land is at best tenuous, at least over the short run. We wish to explore the qualitative and quantitative nature of these interconnections by means of our model view of the marsh—our view, in other words, of ecological reality.

Models are sometimes defined as descriptions or analogies used to help visualize something that cannot be observed directly in its entirety. Another definition treats a model as a set of data, postulates, and inferences presented as a mathematical description of an entity or state of affairs. Both of these definitions have utility for the ecologist. An ecosystem cannot be observed in its entirety any more than can an electron. The latter is so small that our most perfect means of perception would seriously alter the behavior under observation. In practice, if not in theory, the large size and complexity of ecosystems pose an analogous set of problems. In both cases, resort is made to models based on explanatory theories; these in turn are developed from experiment and observation of parts of the system, of its overall behavior, and of its effect on other connected systems. In the preceding chapters we have presented summaries of such observations and experiments. Here we try to put these into the framework of an ecosystem model and "observe" the salt marsh ecosystem vicariously through the "eye" of the model.

9.1.1 Kinds of Models

The kind of model used deserves some comment. We consider only models of ecosystems. A system is defined as a collection of parts, displaying interactions between the parts and exhibiting whole system or unitary behavior (Miller, 1965). The ecosystem is a special class of general systems in which at least some of the parts comprise living organisms, usually populations. Thus the ecosystem typically has components (state variables), flows of matter/energy (fluxes), and flows of information (influential controls).

The simplest models are conceptual only (Rigler, 1975). They are general statements; they cannot explain and, thus, cannot be used alone to construct testable hypotheses.* For example, the verbal poetic model with which this chapter was begun expresses, or induces, an emotional, aesthetic feeling about the salt-marsh system, much the same feeling evoked by photographs or paintings, visual analogues of the verbal model. But such a model does not predict; it is a static model, portraying a view of reality that is valid only for a particular instant in time, that time when the data or impressions on which the model is based were gathered. The box and arrow diagram of Figure 9.1 is also a static model. But in this case, because the pathways of flow between components are illustrated, the statement made by the model is somewhat more explicit and thus easier to disprove. For example, the verbal poetic model elicits a subjective interpretation or view of the marsh from each reader—there can be no one "right" view. On the other hand, if a picture were labeled as a salt marsh when in reality

*Ecosystem models themselves are not usually explanatory scientific hypotheses or theories, but, if constructed in a certain way, they can be used to generate testable hypotheses.

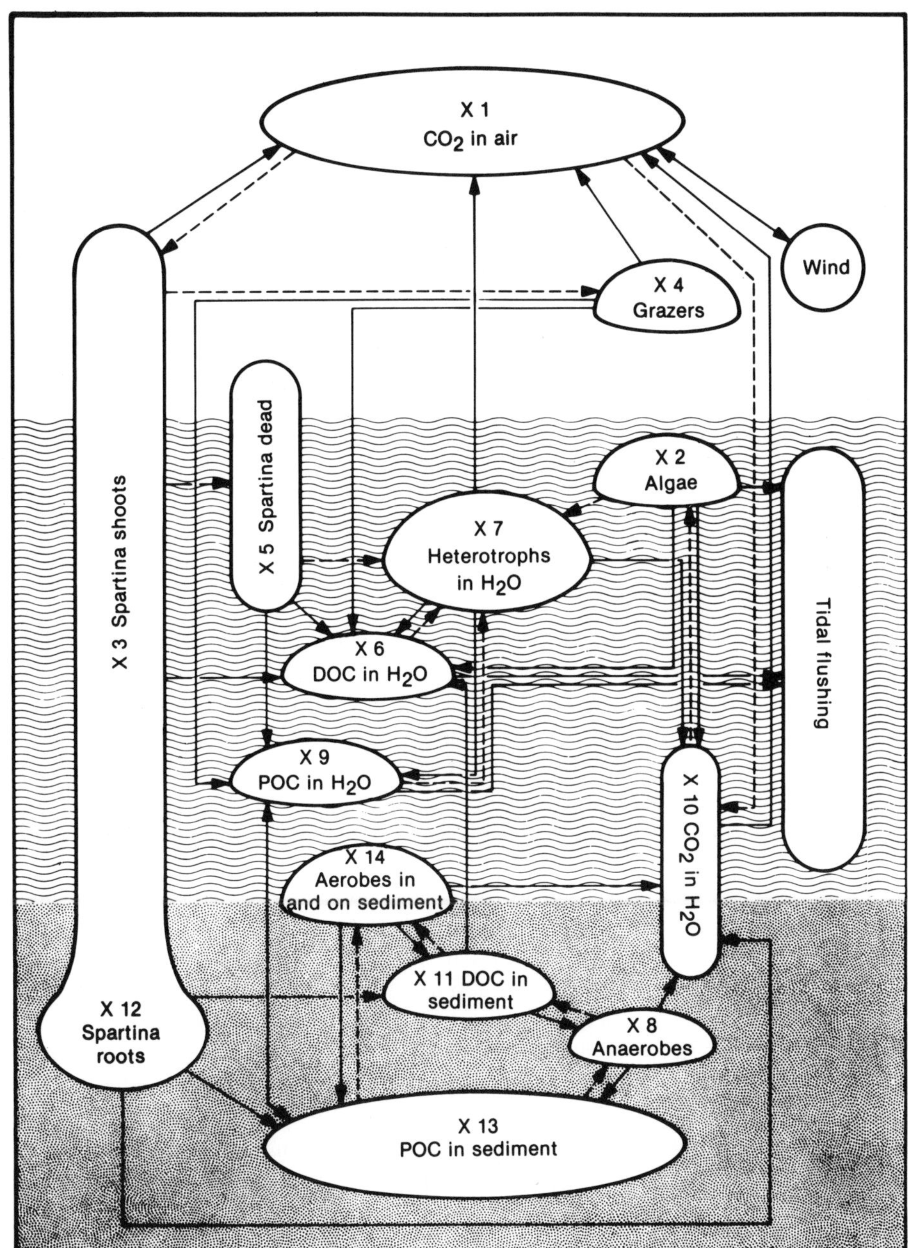

Figure 9.1. Model of carbon flux in the salt marsh, showing the 14 compartments and their relation to air, water, and soil. Adapted from Wiegert and Wetzel (1979). Reprinted from *Marsh-Estuarine Systems Simulation*, Edited by Richard F. Dame, Number 8 in the Belle W. Baruch Library in Marine Science, by permission of the University of South Carolina Press. Copyright © University of South Carolina 1979.

it was taken of a freshwater reed swamp, that visual model could be disproved by a knowledgeable viewer. Because the model in Figure 9.1 makes even more explicit statements about the number of components and the pathways of flow in a salt marsh, it could be disproved perhaps even more easily; certainly its statement is more precise. But none of these models explains much, if anything. They are, as we said earlier, simply statements of fact or of interpretation, akin to the statement, or law, of nature that, viewed from the Earth, the sun rises each morning in the east. A single morning without the appearance of the sun would, of course, falsify the statement. But a statement is not a hypothesis. A proper example of the latter might read: Each morning the sun rises in the east *because* the sun is moving around the stationary Earth from east to west. The ultimate falsification of this explanatory hypothesis did nothing to disturb the truth of the statement about the invariable easterly rising sun. For a model to become capable of generating testable hypotheses of this type, we must add statements about function to the structural diagram of Figure 9.1, thus permitting the prediction of behavior.

9.1.2 Structure vs. Function

Structure and function are used commonly in the ecological literature with an implied meaning; the words are seldom defined. As an example of the universal acceptance of these terms, a widely used and authoritative textbook (Odum, 1971) defines ecology as "the study of the structure and function of nature," but nowhere are the words structure and function defined; they do not even appear in the index. Ecologists seem to designate structure as the material "things" in the ecosystem and function as what these things do. With reference to Figure 9.1, the boxes *and* their occupants at a particular time are regarded as structure; the pathways of matter or energy flow and influence and the flows themselves, along with the regulatory aspects, are regarded as functional attributes. These are certainly reasonable, utilitarian definitions. Trees, animals, and rocks are the structures in an ecosystem, just as parks, people, and buildings are structures in a city. But another set of definitions, one more explicit and specific, is applied to structure and function in the field of systems science (Ashby, 1956). In models of systems, structure is given an abstract rather than a utilitarian meaning. Structure here refers to the boxes and arrows–pathways of matter or energy and information flux–of Figure 9.1; whereas, function refers to the occupants of the box, the actual matter or energy transferred and the form of the informational controls (Wiegert, 1980). A comparison of the use of the terms in systems science versus their use in ecological literature is given in Figure 9.2.

The trophic structure of–as distinct from the structures in–the ecosystem comprises the various possible components, and the associated pathways of transfer for matter or energy or information. The function of the ecosystem is defined by the characteristics of the particular occupants of these niches, that is, their standing stocks, their rates of matter or energy acquisition and transfer, and their modes of interaction. In the same way, the political structure of a city, viewed in the abstract, is independent of the personalities of the people actually occupying the offices of mayor and city council. Different cities sharing a mayoral system of municipal government will share similar political structures, but will experience differences in function. Because the

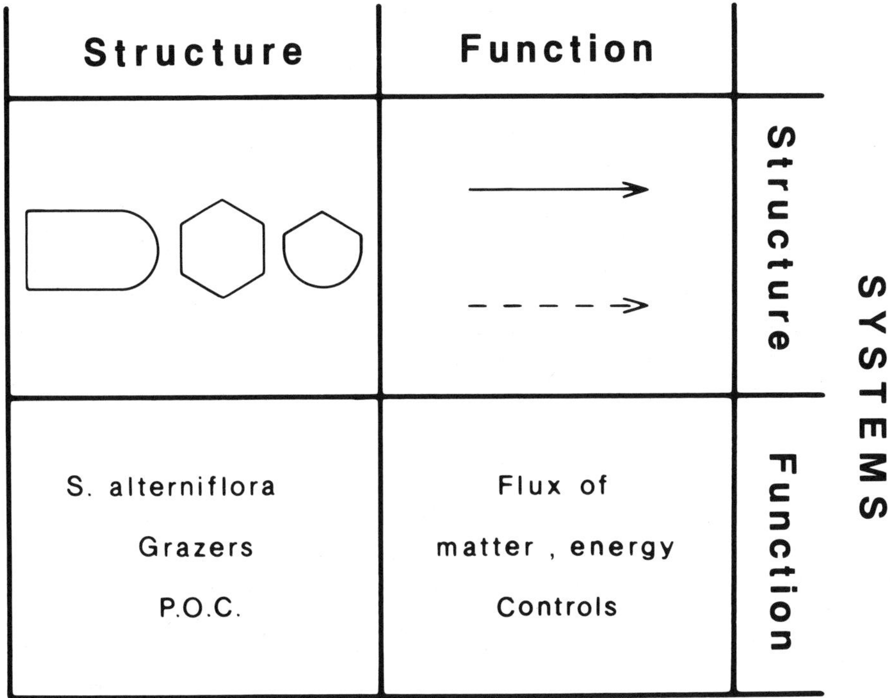

Figure 9.2. Differing views of structure and function prevalent in ecology and in systems science. The quadrants illustrate, clockwise from the upper left, the compartments (autotroph, heterotroph, or abiotic storage); the potential pathways of matter, energy, or information flow; the fluxes of matter or energy and control imposed by information transfer; and, last, the actual occupants of compartments (species, species groups, or categories of dead material).

utility of separating the abstract structural from the concrete functional attributes of ecosystems serves well in modeling, throughout this chapter we use the definitions of systems science illustrated in Figue 9.2 rather than those of ecology.

In general system structure is a more conservative characteristic than system function. The interaction of structure and function produces system behavior; that is, the interaction produces change in fluxes and thus in component sizes. Models that incorporate both structure and function are not static, but dynamic. If they are also constructed realistically so that the structure and function are biologically feasible, such models will generate or suggest usable, testable hypotheses.

9.2. The Salt-Marsh Model

The first step in model construction is to define the structure of the model, that is, to specify the variables of state and connect these with pathways of transfer. Chapter 1 illustrated and discussed the three distinct regions or divisions of the salt marsh, divisions based on processes occurring in the soil, water, and air (Figure 1.5). For our preliminary simulation model, we decided to subdivide the salt-marsh ecosystem to a greater and more useful degree. Our decision was based on an evaluation of the existing knowledge of the biological components and physical attributes of the marsh. This subdivision resulted in the flow diagram of Figure 9.1, a diagram showing the pathways of material transfer within the ecosystem and between the system and its surroundings. However, such a flow diagram is both an incomplete and imprecise representation of the structure of the ecosystem: incomplete because information flows are not shown, imprecise because the categories of compartment—autotroph, heterotroph, or passive storage—are indicated only by including the names of specific components, i.e. by including some of the functional attributes of the system. A better view of the structure of our first generation salt-marsh model is shown in Figure 9.3, where both material flows and informational transfers are shown explicitly and the shape of the box together with its connecting material flows show its trophic position.

9.2.1 Structural Attributes

Transfers of active material or energy include active feeding, passive absorption, egestion, and mortality, as well as purely physical transfers such as diffusion or sedimentation. There were 39 fluxes in the preliminary model (MRSH1V1), not counting the reversible transfer of CO_2 from air to the surroundings. Twenty-seven were entirely controlled by the donor compartment and 12 were controlled by either the donor compartment, the recipient compartment, or both, depending on the density of carbon, or the standing stock of each. Of course, the terms *donor-* and *recipient-*controlled refer only to control of carbon flux exerted in response to changes in the pools of carbon, whether living (biotic) or nonliving (abiotic). Wherever necessary, such factors as temperature, light, and nutrients other than carbon are included as implicit controls operating through seasonal changes in the maximum permitted rates of carbon transfer or transformation.

Donor-controlled flows of carbon, those whose magnitude is largely or entirely a function of the donor compartment plus any abiotic factors, include respired CO_2, egested waste, and nonpredatory mortality. All of these are the result of metabolic processes within the organisms from which the flows originate.

Ingestion (grazing, predation, etc.) is the most common and important example of a flow of carbon that is affected by either (or both) the component serving as the food resource (donor) and the component doing the feeding (recipient). Fluctuation in available food can change the rate of feeding. Similarly, changes in the density of the feeding population can change the rate of specific feeding through diversion of feeding activity to other behaviors, for example, to intraspecific interactions or "squabbling."

Any flow entering an abiotic pool of carbon (feces, respired CO_2, dead material) is usually a donor-controlled flow because of the lack of any dynamic capability in the abiotic component. The component itself cannot grow, and there is no within-component interaction. Most flows into a biotic component represent an active uptake and thus are potentially controllable by change in either the donor or the recipient.

In Figure 9.3 the compartments are represented by symbols showing whether they are biotic (hexagons) or abiotic (tanks) autotrophs and heterotrophs. To simplify the

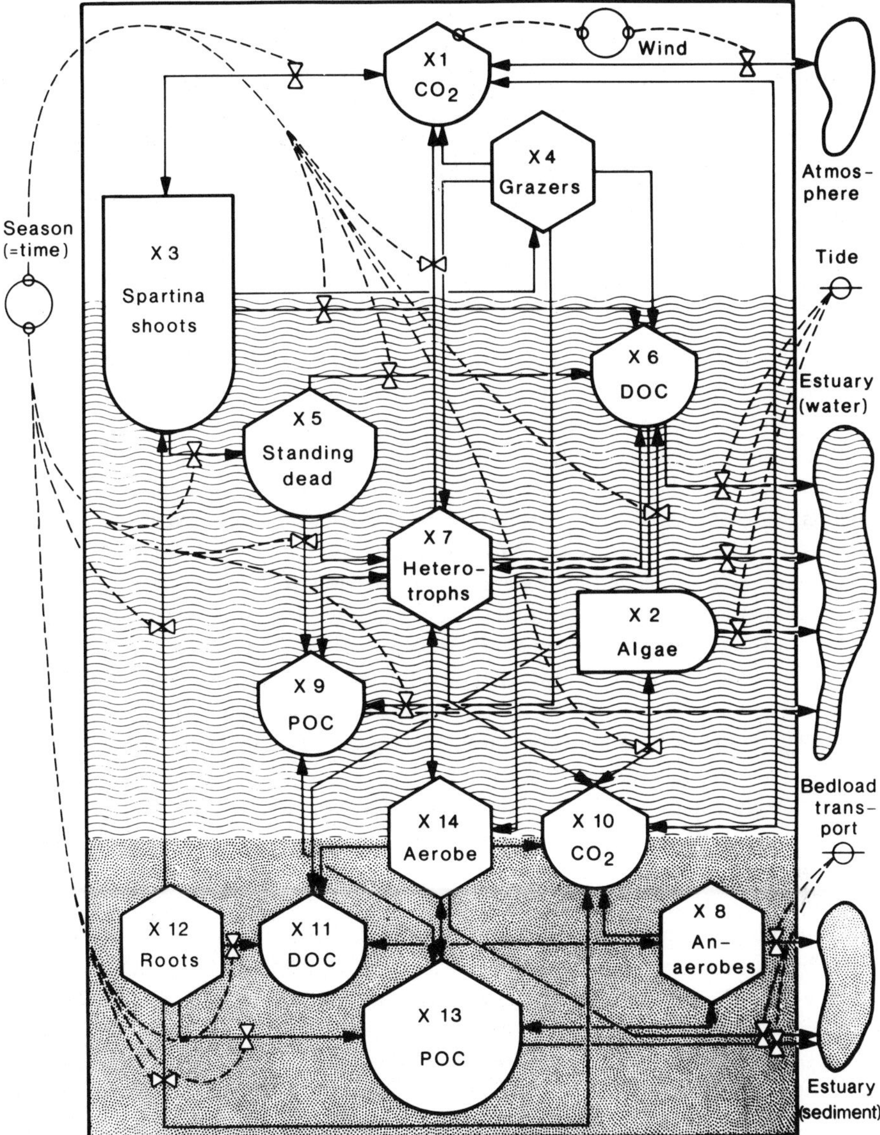

Figure 9.3. Influence and flow diagram of version 6 of the salt-marsh carbon model. Symbols are from Forrester (1961) and Odum (1971). Solid lines are carbon flows; dashed lines are flows of information. Bullets are autotrophs; hexagons are heterotrophs; tanks are abiotic storages. Circles are auxiliary variables supplying information but not simulated. Sources of information are indicated by small circles, material source-sinks as small "clouds." Wave and sediment boundaries indicate the demarcation between the three primary divisions of the marsh.

diagram, the only information flows shown as controls are those not involving scarcity of a resource or limitation by space. In the case of any flux to a *biotic* component, both of these factors may be assumed to be potential controlling factors, depending on the densities of donor and recipient. In the case of a flow to an *abiotic* component, control by the standing stock of the donor and no influence by the recipient are assumed. The symbolism in Figure 9.3 is based on the notational schemes of Forrester (1961) and H. T. Odum (1971).

9.2.2 Functional Components

The first step in the functional characterization of the model was selection of the functional components. Most was known about the ecology of *Spartina*, least about those processes and organisms, particularly macroheterotrophs, associated with the water. We agreed on a preliminary model that would portray the dynamics of 14 components, or variables of state. Seven of these were biotic and seven were abiotic (Table 9.1). The standing stock of each was expressed as g C m^{-2}, and carbon was used as the conserved measure of flow in all versions of the model. The models, as revised during the progress of the research, were named by generation and version. The first-generation model in its various revised versions occupies the major attention of this chapter, but we do discuss briefly the planned changes in Marsh Models 2 and 3.

In Marsh Model 1, version 1 (MRSH1V1), the carbon dioxide of the air (X1) served as the source or sink for carbon. During simulations, this compartment was usually held constant; the sum of input or output during any period was a measure of the net fixation or loss of carbon from the system.

The algae designation (X2) encompassed both phytoplankton and benthic algae, the latter dominated by diatoms, but including blue-green algae. The combination reflects our initial lack of information about the ecology of the two groups and it caused difficulties in interpreting model predictions of behavior by this compartment.

As the dominant plant in the system, smooth cordgrass (*S. alterniflora*) was represented by three state variables: shoots (X3), standing dead material (X5), and roots plus rhizomes (X12). Since the standing dead material included the associated detritus community, it was not entirely abiotic, although in MRSH1V1 it was given no capability for growth except via mortality of *Spartina* shoots. *Spartina* roots and rhizomes were entirely within the soil.

Grazers of *Spartina* shoots composed compartment (X4) which included the abundant homoptera species and the salt-marsh grasshopper, *Orchelimum*, plus their predators and parasites. Although this compartment represented literally all of the nonaquatic organisms of the marsh, the parameters used were based entirely on the data from the two major groups of herbivores.

Dissolved organic carbon (DOC) in the water (X6) included both labile and refractory compounds. The same was true of particulate organic carbon (POC) in the water (X9), although the latter included only nonliving particles.

The heterogenous group of organisms categorized as heterotrophs in the water (X7) comprised forms as diverse as aerobic bacteria, fish, and blue crabs, but the model parameters of X7 largely represented those of heterotrophic microorganisms. As in the

Table 9.1. The 14 Components of the First-Generation Salt-Marsh Ecosystem Model.

Symbol	Abbreviation	Name
X1	CO_2-air	Atmospheric carbon dioxide
X2	Algae	Benthic algae + phytoplankton
X3	S. shoots	Living shoots of *S. alterniflora*
X4	Grazers	Consumers of living *Spartina* + predators
X5	S. dead	*Spartina* standing-dead + wrack[a]
X6	DOC-H_2O	Dissolved organic carbon
X7	Heterotrophs-H_2O	All aerobic heterotrophs–bacteria to fish
X8	Anaerobes-soil	All anaerobic organisms in the soil
X9	POC-H_2O	Abiotic particulate organic carbon
X10	CO_2-H_2O	Carbon dioxide in creek, marsh, and interstitial water
X11	DOC-soil	Dissolved organic carbon in the interstitial water of the soil
X12	*Spartina* roots	Roots and rhizomes of *S. alterniflora*
X13	POC-soil	Abiotic particulate organic carbon in the soil
X14	Aerobes-soil	All aerobic heterotrophs in or on the soil

[a]Wrack comprises the large rafts and windrows of uncomminuted *Spartina* stems.

case of the algae (X2), and the grazers (X4), this degree of condensation in the model was caused by lack of knowledge. Detailed submodels of this and a few other condensed state variables will be discussed later in this chapter.

Anaerobic (X8) and aerobic (X14) microorganisms in the soil were important compartments; the former was particularly important because of the large storage of carbon in the soil of the marsh and because much of the nutrient cycling of the marsh takes place in this zone. The anaerobic organisms are dispersed throughout the soil below the aerobic zone. The aerobic microorganisms are in the aerobic zone, the few millimeters at the surface, and around the roots of *Spartina*, but this aerobic component also included such noninsect invertebrates as fiddler crabs and snails, invertebrates which are found in or on the marsh soils, not in the water.

Although soils underlie the marsh and sediments are found in the tidal creeks (Chapter 1), the first-generation model did not distinguish between the two areas, so sediment/soil can be viewed as referring to the substrate, prorated over the entire ecosystem, whether creek or marsh.

Gaseous carbon dioxide and methane in the water (X10) were used as a source or sink in much the same way as CO_2 in the air (X1), but were not held constant, since there was exchange with the air by diffusion.

Dissolved organic carbon (X11) in the interstitial water and particulate organic carbon (X13) in the soil were the analogues of compartments X6 and X9 in the water. They also contained both labile and refractory organic compounds.

9.2.3 Functional Parameters

The living parts of the ecosystem comprise populations of organisms interacting with each other and with their abiotic environment. We wished our model view of this system to encompass at least some of these capabilities for change, change in numbers, change in feeding rates, change in feeding preference. To the extent that a model does represent these in a realistic manner, it has enhanced value as a tool for predicting the future state of the ecosystem. The dynamic capability of a model stems from the relationship between the parameters, the state variables, and the functional form of the controls of the parameter values. The overall equation representing any particular flow of carbon is simply a mathematical rephrasing of the verbal statement of how these three separate parts of the system are interrelated. In this section and the next we give the essence of this verbal statement by defining parameters and functional controls. Perhaps giving the mathematical statement first will clarify the ensuing explanation. We begin with the basic equations for rates of carbon transfer through ingestion, physical transfer (diffusion, sedimentation), respiration (including anaerobic CO_2 and methane production), excretion, nonpredatory mortaility (any grazing or predation is of course represented as ingestion by the predator or grazer), and egestion. Although short and simplified, this list contains all the parameters and relationships necessary to describe realistically the basic flows of carbon in any ecosystem. The symbol F_{ij} simply stands for the flow of carbon (per unit area or volume, per unit time) from any donor compartment i to any recipient compartment j.

Ingestion: $$F_{ij} = \pi_{ij} \cdot \tau_{ij} \cdot x_j \cdot f_{ij} \cdot f_{jj} \tag{9.1}$$

Physical transport: $$F_{ij} = \tau_{ij} \cdot x_i \tag{9.2}$$

Respiration: $$F_{ij} = \rho_i \cdot x_i \tag{9.3}$$

Excretion: $$F_{ij} = \eta_i \cdot x_i \tag{9.4}$$

Nonpredatory mortality: $$F_{ij} = \mu_i \cdot x_i \tag{9.5}$$

Egestion: $$F_{ij} = F_{hi} \cdot \epsilon_{hi} \tag{9.6}$$

Equation 9.1 must be controlled because it realistically represents the rate of feeding as the product of a specific rate and the standing stock of the consumer, or recipient. If not controlled, the amount of x_j increases exponentially without limit. The symbol f is used to designate a control function, either keyed to the amount of a resource, the donor, as f_{ij} or to the amount of space available to x_j as f_{jj}. These control functions are discussed in Section 9.2.4. We now need only know that they must each take a value of $0 \leqslant f \leqslant 1.0$. Equations 9.2 through 9.5 are simple linear relationships specifying the flow as a function of the product of a rate times a standing stock of donor. Clearly, these are oversimplifications of the complex respiratory, excretory, and mortality processes occurring in nature. But adding complexity to these equations to handle specific real flows does not involve changing the basic relationship given by Equations 9.2 through 9.5. It only means making the realized rate a function of many other factors, not the constant implied by the equations. Equation 9.6, egestion, simply states in symbolic form the generalization that egestion usually is some fraction of the particular food ingested, thus the ϵ_{hi} implies a different parameter value ϵ for each particular pathway for ingestion of carbon.

The parameters of a realistic i.e. explanatory, simulation model must have a clear biological definition; they should not be mathematical artifacts inserted to provide a best-fitting equation (Wiegert, 1975). The salt-marsh models, and the Equations 9.1 through 9.6, employ 10 different parameters. Four are used in the feedback controls shown and discussed in Section 9.2.4. The units of the parameters vary. All rates are specific rates; that is, they represent the rate of carbon transferred or transformed per unit of carbon in the standing stock per unit of time. The carbon units cancel, leaving the specific rate with units time^{-1}. The inverse of the specific rate is the delay time, giving the length of time necessary to transfer or transform an amount of carbon equal to the standing stock. Parameters specifying threshold levels have units of carbon per unit area. Parameters specifying fractions (egestion, feeding preference, etc.) or threshold ratios (some control parameter) are dimensionless.

The feeding rate of any population or a physically controlled transfer between two abiotic state variables, such as sedimentation or diffusion, are both represented by the parameter τ (tau). For a population, τ is the maximum specific rate of ingestion, or the maximum specific rate of logarithmic growth divided by the assimilation efficiency plus the specific rate of loss sustained under the conditions conducive to maximum growth. Used to describe physical movements of material, τ is simply the specific rate of transfer.

Oxidative metabolism and fermentation result in carbon loss via CO_2, and methanogenesis causes loss of carbon via CH_4 (Chapters 6, 7). These carbon fluxes are governed by the specific rate ρ (rho). For a given physiological state, CO_2 or CH_4 production is constant per organism. Thus these fluxes from a biotic to an abiotic component are donor-controlled.

The specific rate of carbon loss by biotic components via the excretion of dissolved carbon compounds, usually nitrogenous waste products, is represented by η (eta). Physiological mortality, or the loss of carbon to nonpredatory death or the loss of only parts of the plant or animal is μ (mu). This parameter expresses the minimum level assumed by this loss under optimum conditions for population growth.

For a given donor or food resource, the egestion factor ϵ (epsilon) indicates the proportion of carbon ingested that is egested by a population. When a given recipient population ingests from more than one donor, some method is necessary to apportion ingestion. Otherwise, when all food resources were optimally available, the sum of all realized specific rates of ingestion could exceed the largest maximum rate of ingestion. This is biologically impossible, so feeding preferences (parameters $\leqslant$ 1.0) are necessary for each pathway in such a multiple resource situation. The basic feeding preference value is computed based on the abundance of a resource in the diet when all resources are optimally available, that is, all resources equal or exceed the satiation threshold. In each iteration of the simulation model, a realized feeding preference π (pi) is computed based on the abundance of a resource relative to the abundance of all other resources of a consumer (Wiegert and Wetzel, 1979). Parameters specifying fractions (egestion, feeding preference, etc.) or threshold ratios (some control parameters) are dimensionless. In nature, populations seldom ingest and grow at the maximum physiologically permissible rate. Instead, the standing stock fluctuates more or less widely around some long-term mean value. Fluxes of material (in this instance, carbon) are controlled, and the realized rate of ingestion, for example, will be something less than the maximum value τ.

9.2.4 Control Functions

With the parameters identified we now turn to the way in which changes in the density of the components affect the realized rates of ingestion. In a donor-recipient controlled interaction, the realized rate of ingestion can depend on either the standing stock of the donor or that of the recipient. The donor or the recipient densities at which effects are first apparent are the thresholds at which control begins or ends. With two populations involved, each with a maximum and minimum threshold, there are at least four parameters that must be included in these control functions. The simplest control functions are linear equations.

Consider first control by scarcity of a resource, or donor.

$$f_{ij} = \left\{ 1 - \left(\frac{\alpha_{ij} - x_i}{\alpha_{ij} - \gamma_{ij}} \right)_+ \right\}_+ \tag{9.7}$$

where the subscript + restricts the value inside the parentheses or brackets to be $\geqslant 0$.

The satiation threshold of population i is α_{ij}, below which the donor is perceived as scarce by the recipient; that is, the recipient population cannot ingest at the maximum physiological rate, τ. For densities of the donor equal to or exceeding α_{ij}, $f_{ij} = 1$, and food scarcity exerts no control (see Equation 9.1). There is a minimum density of the donor population (γ_{ij}), at or below which the donor is no longer available as food for population j. This is the refuge level; when $x_i \leqslant \gamma_{ij}$, $f_{ij} = 0$ and no ingestion of x_i is possible for x_j.

Control by scarcity of space is due to some direct intracompartmental interference with feeding and reproduction that is related to density of the recipient x_j.

$$f_{jj} = \left\{ 1 - \left(1 - \frac{\eta_j + \mu_j + \rho_j}{\tau_{ij}(1-\epsilon_{ij})} \right) \left(\frac{x_j - \alpha_{jj}}{\gamma_{jj} - \alpha_{jj}} \right)_+ \right\}_+ \tag{9.8}$$

This control function is also linear; the effect of an added unit of x_j on the ingestion by x_j does not change with density of x_j. But Equation 9.8 is more complex because of the extra term involving η, μ, ρ, and ϵ. This is necessary because when x_j reaches the density γ_{jj}, f_{jj} must be a value that will permit just enough ingestion for maintenance, thus γ_{jj} is the carrying capacity. Inspection of Equations 9.1 and 9.8 will show that, under these conditions and assuming no scarcity of resources:

$$F_{ij} = x_j \cdot \frac{\eta_j + \mu_j + \rho_j}{1 - \epsilon_{ij}}$$

Inclusion of nonpredatory mortality in the numerator of this correction is done on the assumption that its value is minimal for the genotype and constant. If μ is itself an important control variable, it should not be included in the maintenance suite of parameters.

The parameter α_{jj} in Equation 9.8 is the threshold response density. Whenever x_j exceeds α_{jj}, space is limiting. As the density of x_j increases towards γ_{jj}, limitation increases linearly.

The maximum density of the recipient population is γ_{jj}, at or above which the deterimental effects of intraspecific interference competition become so pronounced as to render further growth impossible. This threshold is the carrying capacity of the system for the recipient population.

Generally the separate effects of scarcity of food, donor control, and intraspecific interference, recipient control, are multiplicative. Thus, if prey were so scarce that the normal searching effort produced only half the maximum ingestion and if predators were so dense that squabbling among them took one-half the time normally devoted to searching, then the realized rate of ingestion should be the maximum (τ) multiplied by the product 0.5×0.5, or a realized rate of ingestion of $.25\tau$. We mentioned above the simple linear nature of Equations 9.7 and 9.8 in which a unit change in either donor or recipient produces a corresponding unit change in the realized rate of ingestion for any density of donor between γ_{ij} and α_{ij} and for any recipient density between the limits α_{jj} and γ_{jj}. The basic equation for ingestion (Equation 9.1) is, of course, always nonlinear except in the case of uncontrolled growth. But sometimes the nature of the ecological interactions demands a nonlinear control function as well. The searching behavior of a predator may change with prey density, or intraspecific interference may increase more or less rapidly with increase in predator density. Examples of these cases in the salt-marsh model, as well as in theoretical models of populations and ecosystems, are given in several of the papers cited in this chapter.

9.2.5 Testing the Salt-Marsh Model

The functional forms of the negative feedback* controls in a model represent hypotheses about how the biological populations affect each other. Thus simulations with the model can be compared with actual experiments. Failure of the predicted results to conform to the measured responses can result from either of two causes. First, the parameter values may be in error. If they are not, or if the degree of sensitivity of a parameter (how much changes in it affect a given variable) is low, then a postulated functional relationship may itself be incorrect. When this sort of error is identified in a simulation model, a hypothesis is falsified, and it must be replaced with an alternative hypothesis, another functional relationship.

These two important constraints must be constantly kept in mind. First, the error in the functional form of the feedback control must be clear: one should not arbitrarily decide that the functional form is wrong simply because the compartment it

---

*Negative feedback refers to a stabilizing form of control in which an increase in a variable causes a decrease in the rate of input, thus eventually bringing the variable within some specified limits.

controls may not be "behaving" properly. Second, the alternative functional form, or hypothesis, must have a reasonable ecological basis. Ideally, replacing a function would only proceed from observational or experimental data disproving one functional form and directly suggesting the alternative. For example, suppose a feedback term initially assumed a unit decrease in ingestion for a unit decrease in a particular resource, once the latter dropped below the satiation level. Subsequently, the dynamics of the consumer in question appeared unrealistic and, at this point experiments relating ingestion rate to resource availability were made. The results then showed a nonlinear response of population ingestion to decreasing availability of the resource. In this case, the data failed to support the first hypothesis and directly led to the new functional form. Although this ideal situation may seldom be achieved, substitution of alternative hypotheses about control can be done on the basis of data from the literature, or on the basis of ecological experience. Although the degree of objectivity decreases, circularity of reasoning can be avoided if the new functional form chosen can be justified apart from a) any particular effect it will have on the dynamic behavior of the compartment it controls most directly or b) any effect it will have on the model as a whole.

9.3. First-Generation Models

Six different versions of the salt-marsh model were employed in our research, all of them modifications of the basic plan presented in Section 9.2. All retained the same structural and functional compartments, that is, all have 14 compartments and the same occupants given in Table 9.1. Relatively minor structural changes were made in pathways, but some major changes were made in certain functional parameter values and control functions.

9.3.1 Marsh Model 1, Version 1 (MRSH1V1)

This first or preliminary model of the salt marsh was begun in 1973 and drew upon the information available from prior research in the Sapelo Island marshes as well as work on *Spartina* marshes in other areas. Specifically, the model was constructed to help answer three questions (Wiegert et al., 1975):

1) Based on present knowledge of the ecology of the coastal Georgia salt marsh, is this ecosystem a potential source of organic carbon to the offshore waters, or is the marsh instead a sink, extracting and storing or degrading organic matter produced offshore?

2) Which groups of organisms are most responsible for the production and the degradation of organic carbon compounds in this system?

3) Which of the many parameters contributing to the functional characteristics of the marsh, as reflected in our initial model, are potentially important but as yet poorly known? As a corollary, how might research effort be allocated efficiently to improve the predictive accuracy of the model regarding seasonal changes and responses to perturbations?

We could hardly expect any of the predictions from the initial model to be beyond question. Indeed, the entire philosophy of constructing models based on preliminary or scanty data is that the model might be used not as the final word, but as a tool for exposing gaps in existing knowledge quickly and efficiently, and for pinpointing "sensitive" areas for further research. As a case in point, our initial estimates for the 81 parameter values contained in Marsh Model 1, version 1 (MRSH1V1) were based approximately one-third on hard data from studies at Sapelo, one-third on values computed from literature reporting studies on areas outside of the Georgia marsh ecosystem, and one-third on guesses by those most familiar with the particular compartment or flow involved.

The structural and functional attributes of MRSH1V1 followed closely the descriptions of Section 9.1 with but two major exceptions. The feedback control functions representing the effect of scarcity of a material nutrient and the effects of competition for space were used additively instead of multiplicatively as discussed in Section 9.2.4. Secondly, for those microorganisms feeding on particulate organic carbon, the feedback control representing scarcity of a material resource incorporated a *ratio* of donor to recipient at the response threshold (Wiegert et al., 1975; Christian and Wetzel, 1978).

The Georgia salt marshes have long been regarded as a significant producer of organic matter. Prior to 1973 the evidence for this concept had been based on a few studies of input-output measurements of organic matter from tidal creeks, among which the most often cited was probably Odum and de la Cruz (1967), and on calculations derived from the static energy-flow model of the marsh compiled by Teal (1962), a model showing a net excess of production over respiration. A more detailed discussion of this point is given later; we show here only the prediction of the preliminary model, simulated for several years. Two kinds of simulations were run. In the first, the marsh was regarded as an ecosystem under glass, and no exchange of carbon with the surroundings was permitted. This simulation experiment caused violent seasonal fluctuations in the CO_2 content of the air, suggesting the rather obvious conclusions of a) insufficient CO_2 in the atmosphere immediately above the marsh to sustain the summer demand by *Spartina* and algae, and/or b) rates of CO_2 regeneration by organic carbon degradation insufficient to meet these demands. The second type of simulation was more realistic, allowing CO_2 in the air to exchange freely with the infinite source-sink of atmospheric CO_2, thus maintaining virtually a constant density of CO_2 in compartment X1. The results of this simulation are in Figure 9.4. The constant seasonal cycle of *Spartina* carbon is contrasted with the rapid buildup of organic carbon as POC in the soil and DOC in the water. The marsh is depicted by the simulation model as a definite source for carbon. Deliberate alteration of thresholds and parameter values in the model sufficient to counteract this build up and cause a net loss of carbon, that is, to depict the marsh as a sink, required rates of respiration and degradation that were unrealistically high. Either the marsh was truly a source of organic carbon or our preliminary model was so badly flawed structurally that it bore virtually no resemblance to the real ecosystem, at least with respect to carbon input-output. Because we had some confidence in the model, we proceeded on the assumption of a correct model prediction.

The possible fate of this excess carbon was an entirely different matter. The marsh soil was not appreciably increasing in carbon content, certainly at nowhere near the rate indicated in Figure 9.4. Furthermore, the levels of DOC in the water shown by the model were several orders of magnitude greater than the measured values. The preliminary model was simply too crude to resolve this question.

We turned to the next questions posed for the model. Which of the 81 parameter values are sensitive in terms of a change in the parameter value affecting the normal annual variation in standing stock of one or more of the compartments? For this sensitivity analysis we used the open system simulation as the standard against which to compare our perturbations. The simplest of analyses was made, in which each of the 81 parameters was varied by itself and the last year of the simulation compared with the standard, to see which compartments had been affected and how much. In general, a parameter was first doubled in value, then halved, unless such changes produced extinction(s), in which case more moderate perturbations were tried. In a few instances the parameter was only increased 10%, and occasionally a very small initial value of the parameter made a fivefold increase advisable.

Of the 81 parameters in the model, perturbations of only 21 produced a change more than 5% in standing stock of any compartment at the fifth year.* These parameters we considered important or "sensitive." Some of these sensitive parameters, for instance those associated with *Spartina* shoots, affected many compartments of the model, while others affected only a few. Most of those that were sensitive by this measure were greatly so, affecting two or more compartments in a major way. Indeed, if the criterion of sensitivity required a change of 25% or more in the fifth-year standing stock, 19 of the 21 would still remain sensitive.

The rates and thresholds associated with *Spartina*, algae, algal consumers, and the microbial decomposers composed the bulk of the sensitive rates. These were in general the parameters about which least was known. *Spartina* shoots, roots, and rhizomes constitute the bulk of the living biomass of the marsh and are heavily connected to other compartments (Figure 9.3.). The overall connectivity of MRSH1V1 is high—20.4% of the 196 possible pathways. The algae composed only a fraction of the standing stock, but they had a very rapid turnover and a high excretion rate. The microbes acting on DOC and POC were important because of the rapid turnover of DOC and the large amount of POC in the marsh soils and sediments. Thus the preliminary marsh model provided some clear predictions against which we could compare our initial concepts of the marsh and its dynamics. Many of the research projects of the three years that followed had as primary or secondary objectives the clarification of model function and particularly the revision of sensitive parameters.

Two revisions of MRSH1V1 were made. The first, MRSH1V2, differed from MRSH1V1 only in the provision of a rate coefficient for simulating tidal export from those compartments that were contributing to the marked buildup of organic carbon compounds in the DOC-water (X11) and POC-soil (X13) compartments. Retention of these mechanisms, plus additional changes in parameter values and feedback controls, produced a version of the model (MRSH1V3) that was more realistic and was deemed suitable for testing against the results from certain field experiments.

*The figure is mistakenly listed as 0.5% in Wiegert et al. (1975).

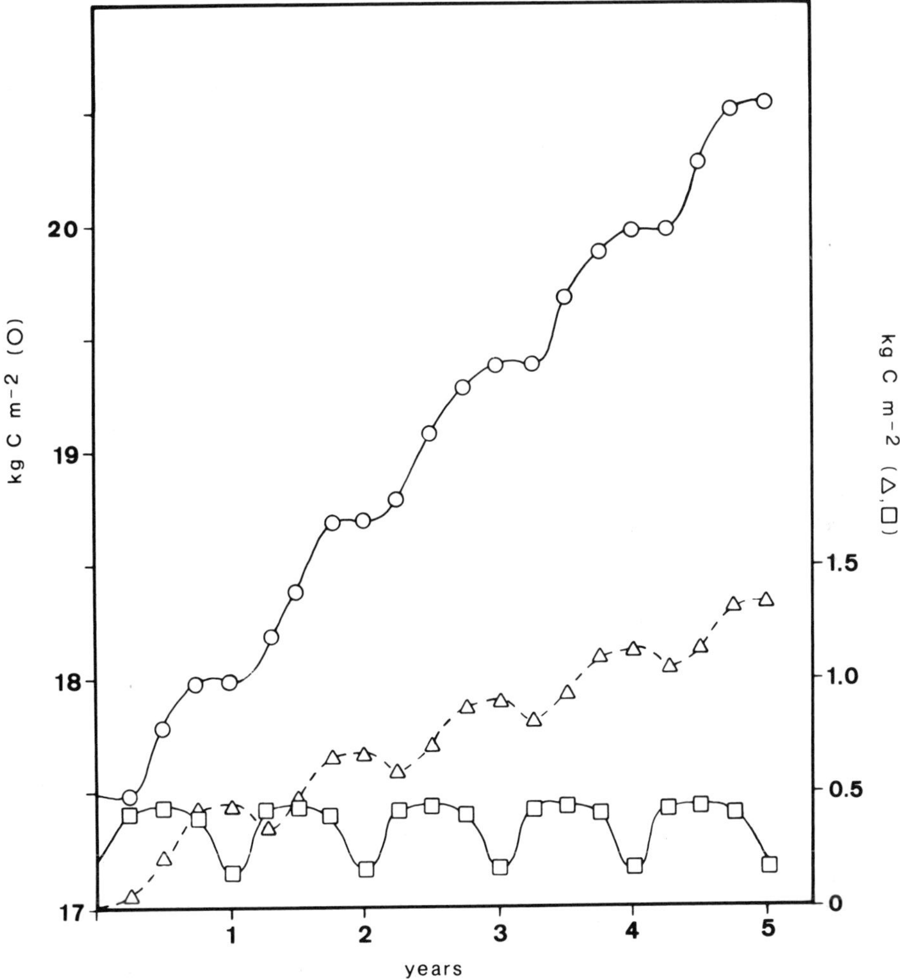

Figure 9.4. Simulations with MRSH1V1. Open system (CO_2 in air replenished). ○ POC in soil and sediments, △ DOC in water, □ carbon shoots of *Spartina.* Both scales are the same, but the scale for organic carbon in soil and sediments starts at 17, not 0. Adapted from Wiegert et al. (1975), with change to linear scaling.

9.3.2 Marsh Model 1, Version 3 (MRSH1V3)

The frequent feedback between the field and laboratory research and the continuing development of the model during the three years after the construction of MRSH1V1 culminated in a model (MRSH1V3) which could simulate quite well the seasonal dynamics of all 14 compartments (Wiegert and Wetzel, 1979). By comparing model output to field measurements of standing stock, a certain confidence in the model was developed. Although not validated, the model was, to a degree, corroborated in the sense of Caswell (1976). Wiegert and Wetzel (1979) compare this version in detail with

MRSH1V1 and discuss the experiment performed with version 3. Here we briefly review the changes that were made and the corroborative experiments.

The most useful and fundamental improvement of version 3 over version 1 is the capability of simulating tidal export of water-borne organic carbon. Three compartments in MRSH1V3 are subject to the action of this export coefficient (TIDE): algae (X2), DOC in water (X6), and POC in water (X9). Not all algae are suspended in water; a certain proportion are phytoplankton but, of the remaining benthic algae, only a small fraction is carried into the water column by tidal turbulence. Thus a resuspension coefficient (BENTH) had to be employed to simulate this movement of the benthic algae. The addition of these tidal exports, plus the addition of a reversible transfer explicitly intended to monitor CO_2 exchange with the surroundings via a direct pathway from heterotrophs in water (X7) to CO_2 in air, brought the total fluxes in MRSH1V3 from 39 to 45 and brought the total parameters from 81 to 91. Feedback controls in version 3 were linked multiplicatively rather than additively as they had been in MRSH1V1. Although this is fundamentally a very important conceptual modification, its practical significance was nil as far as the normal simulations of marsh dynamics by MRSH1V3 was concerned, because compartments in this marsh model are not severely limited by more than one factor at a time. Numerical solution of version 3, as of version 1, was the inexpensive, yet satisfactory Euler method, with a 0.1 day integration interval (Wiegert and Wetzel, 1974).

Direct comparison of the predictions of model versions 1 and 3 with field data is made in Table 9.2. The simulation data of Table 9.2 are taken from the last year of a five-year run, by which time a constant annual cycle had been achieved by all compartments other than those showing a continual accretion of carbon; neither model had any stochastic elements. Although the data of Table 9.2 hardly justify any rigorous statistical comparison, the generally closer agreement between field data and the predictions of MRSH1V3 is clear. The greatest deficiency of version 1 was the failure to provide a means of exporting the excess organic carbon. This not only changed the compartments where the accumulation was taking place, but also affected other compartments. MRSH1V3, by simulating the removal of these accumulations, preserved a more normal relationship between all compartments. We emphasize the artificial nature of this simulation—the hydrologic picture is far from the simple washout simulated by the coefficient TIDE—but, from the viewpoint of the marsh, the excess material is "gone."

In every case where the average annual value predicted by version 3 differed greatly from the prediction of version 1, the former was in closer agreement with the measured values. This initial corroboration of MRSH1V3 was then extended by comparing the predictions from model perturbation experiments with expected or measured results from the field. Five experimental manipulations of the model were made:

1) winter grazing on the shoots of *Spartina* was added, based on studies of the herbivorous insects in the marsh;

2) the effect of increased grazing on the algae was examined;

3) harvesting all aboveground living *Spartina* in two successive years was designed as a model mimic of an existing field experiment;

4) the effect of restricting transport of carbon between shoots and root-rhizomes was compared with a field experiment in clipping and root pruning;

5) the effect of varying the tidal export coefficient was evaluated.

Table 9.2. Annual Mean Standing Crop (g C m^{-2}) of the 14 State Variables. Predictions of MRSH1V1 and MRSH1V3 Compared with Field Measurements.

		MRSH1V1		MRSH1V3		Field Data[a]	
Compartment	Abbreviation	Annual Average	Annual Change	Annual Average	Annual Change	Annual Average	Annual Change
X1	CO_2-air	782		875		875	
X2	Algae	15		.5		1	
X3	*Spartina* shoots	125		118		135	
X4	Grazers	.5		.4		1	
X5	Dead *Spartina*	98		115		130	
X6	DOC-H_2O	995	+225	4		5.6	
X7	Heterotrophs-H_2O	10		12		7.5-30[b]	
X8	Anaerobes-soil	25	+1	44		30-45	
X9	POC-H_2O	240		8		9	
X10	CO_2-H_2O	46		35		?	
X11	DOC-soil	.8		36		26	
X12	*Spartina* roots	137		475		450	
X13	POC-soil	20,291	+589	17,733	+26	18,200	+50[c]
X14	Aerobes-soil	10		9		3-6	

[a] From Wiegert and Wetzel (1979).
[b] From Imberger et al. (in ms.)
[c] Based on mean annual increase in sediment of 1 mm.

When expressed per unit area and compared to other grasslands (Wiegert and Evans, 1967), the amount of carbon ingested by the arthropod grazers on the marsh is large; but, because of the luxuriant production of *Spartina*, the percentage of net annual primary production ingested by primary consumers is low (Chapter 5). However, the major grazing pressure exerted in the simulations with MRSH1V1 was in the summer, when the standing stock of *Spartina* was very high. Furthermore, in the sensitivity analysis of MRSH1V1, *Spartina* was shown to respond drastically to rather small (10%) decreases in rate of grass photosynthesis or to equally small increases in losses of carbon. From Teal (1962) we knew that the winter populations of planthoppers were high (Chapter 5). Reasoning that these might exert a moderate-to-high grazing pressure during that season, the winter mortality of grazers was relaxed and the populations allowed to remain high. We wished to ascertain if grazing, although still relatively low in terms of total net production consumed, could be controlling *Spartina* production. The result was a doubling of both the average annual standing crop and of the carbon ingested by the grazers with but negligible effect on *Spartina.* The annual net production was reduced by only 16 g C x m^{-2} x yr^{-1}, or 0.1%. We concluded that grazing, at the levels normally encountered in the salt marsh, has little direct effect on the productivity of the system. Even indirect effects, within the frame of model structure, function, and time employed (five years), were not seen. However, these qualifications are important, for, since nutrients other than carbon are not explicit in the model, possible impacts of grazing on nutrient scarcity and recycling cannot be simulated.

Simulation experiments on the algal compartment were difficult to interpret because phytoplankton and benthic algae were combined. This artificial simplification was maintained on the strength of the idea (Ragotzkie, 1959) that the phytoplankton had a vanishingly small role in the productivity of the estuary. We now know this to be far from the truth (Chapter 3), and the second-generation model will be modified accordingly. But when the perturbation experiments with MRSH1V3 were made, we assumed the effects were representative only of the benthic algae, primarily diatoms. Furthermore, the parameter values governing the dynamics of the algae compartment (X2) in MRSH1V3 were obtained from data on the benthic algae. The productivity of the latter, prorated over the entire marsh, may exceed 10% of the net production by *Spartina* (Chapter 3). Because of the potentially high turnover rate of algae and the possibility of crowding and nutrient limitation, the effects of increased grazing cannot *a priori* be assumed detrimental. Increased cropping could increase turnover by lowering the standing stock, thus freeing the population from limits by crowding, or it could increase nutrient regeneration rates. Clearly, increasing the cropping rate without limit can have a detrimental effect on net production. Thus, the question is: Where is the natural population on the curve relating net production to ingestion by grazers? Fenchel and Kofoed (1976) found field populations of snails that were stimulating algal production, as compared to net production in the absence of grazers, at some field densities, but decreasing it at others. Figure 9.5 represents a common phenomenon in some predator-prey relationships. Pace (1977) found that the mud snail, *Ilyanassa* (*Nassarius*) *obsoleta*, depressed the net production of the benthic algae; that is, the normal field population densities placed grazing pressure to the right of the netral line in Figure 9.5.

By relaxing the availability thresholds of benthic algae to algal grazers (X7), both algal standing crop and productivity of the algae were decreased, a result predictable

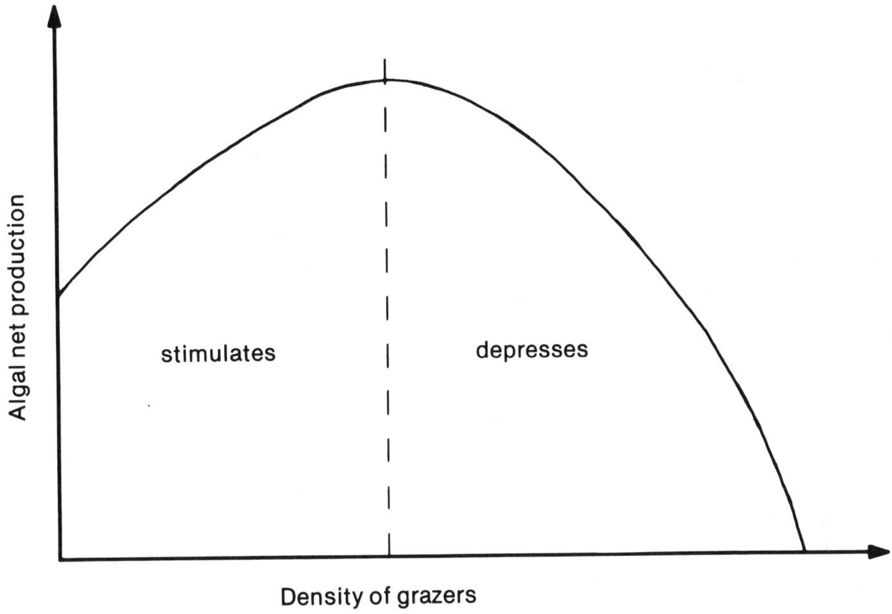

Figure 9.5. Diagram illustrating the possible dual stimulating and depressant effect of grazers on net production of benthic algae.

from the finding by Pace (1977) that nominal densities of snails were already having a detrimental effect. At least the model predicted the right directional effect. However, in addition to the complication of a combined algal compartment, we also note the extremely heterogenous nature of the algal grazer component (X7). Because this compartment feeds from several others in the model system, any experiments of this type must be interpreted with caution. Situations of this nature were the major prompters for the aerobic submodel currently under development (Section 9.4).

S. alterniflora is seldom grazed or harvested for forage, in contrast to the historical practices with *S. patens* which dominates the high marsh farther north along the Atlantic Coast. One practical reason for this is probably the nature of the soil which, even in the high marshes, is flooded often enough to remain muddy and easily disturbed. There may, however, be other reasons. A major probability is that *S. alterniflora* simply cannot withstand periodic cutting and still maintain itself, at least not in the face of competition from other salt-tolerant plants.

Pomeroy and Reimold (unpubl.) performed an experiment several years ago in which all shoots of *S. alterniflora* were cut from large areas of marsh in June of two successive years. To simulate this experiment, we introduced appropriate statements into the model to remove all accumulated shoot biomass on June 30 of two successive years. Otherwise the simulation proceeded normally, except that it was stopped at the end of six, not five, years. The results of this model experiment were of interest for two reasons. First, although the perturbation, in a modeling sense, was extremely simple, the results were striking, and, in one respect, unexpected. Second, the results, compared with the field experiment, very clearly show both the potential and the limitations of predictions by models.

The results based on the one field experiment are not complete, but the general results are clear. Two annual harvests of *S. alterniflora* reduced the standing crop of roots and rhizomes drastically. Total net production and standing stocks recovered slowly over several years in some areas, but *Spartina* was replaced by *Salicornia* in many areas.

The predictions of the simulation experiment are presented in Table 9.3. Clipping had an impressive effect on the vegetation, reducing the January 1 standing stock in the third year, one and a half years after the first clipping, to less than 10% of the nominal value. But note the unexpected result for roots and rhizomes. By the third year they had already begun to recover slightly. This predicted effect would merit investigation to see if it is observed in the field. However, net primary production still decreased to ∿ one-fifth of normal and full recovery required three and a half years after cessation of clipping.

The model is obviously limited to predicting effects on the compartments within it and cannot predict species replacement. That capability would entail much research on the potential invaders. But given the drastic reduction in standing stock and productivity, both that observed in the field experiment and that predicted by the model, it is easy to foresee some competitive replacement. Because the clipping experiment was acute, not chronic, if competitive replacement does not occur, *Spartina* recovery seems probable. Ascertaining conditions under which one or the other of these alternatives occurs would be an interesting topic for research.

A smaller-scale experiment that involved simultaneous clipping and root pruning was begun by a group headed by Robert Christian in 1975 (Chapters 6, 7). Small plots (0.1 to 0.15 m^2), with all *Spartina* material removed from the soil were set up. Periodically the plots were clipped to remove new growth. In addition, the roots and rhizomes were clipped to a depth of 15 cm, and a permanent plastic shield was inserted to prevent lateral growth into the plot. The researchers wished to test the idea that the major microbial activity in the top few centimeters of the soil was directly sustained and influenced by the carbon compounds translocated down to the roots and rhizomes by the productive shoots. This experiment provided another ideal opportunity to test the predictive capabilities of the model, so an analogous simulation was run. Specifically, we looked at the effects of this treatment on 1) standing stock of roots and rhizomes of *Spartina* (X12), 2) standing stock of soil microbes (X8 and X14), and 3) total carbon transformation, or metabolic activity, by the soil microbes.

For 12 months the results were surprisingly negative (Christian et al., 1978; Chapters 6, 7). Microbial standing stocks, measured as ATP, did not decline; metabolic activity, measured as energy charge, was unchanged. But standing stock of roots and rhizomes declined. After 18 months, only ATP was measured; it had declined by 40 to 50% in the clipped and pruned plots.

The model simulation produced results similar to the field experiment. Standing stocks of X8 and X14 were unchanged (Wiegert and Wetzel, 1979), as were carbon fluxes. In the model simulation, the standing stock of roots and rhizomes disappeared rapidly and was almost gone by the end of 18 months. The carbon necessary for the maintenance of this "business-as-usual" situation came from a decrease in the large POC-soil (X13) standing stock. Thus the model, simplified and crude as it was, presented the *Spartina* shoot-root-rhizome complex as primarily a long-term supplier of organic carbon to the soil, with very little other direct or immediate influence on the

Table 9.3. Simulated Effects (MRSH1V3) of Harvesting *S. alterniflora* Shoots for 2 Consecutive Years

January 1 of year	Standing Stocks of Shoots	Standing Stocks of Roots and Rhizomes	Annual Net Primary Production
1	54 clip[b]	345	1573
2	11 clip[b]	34	880
3	5	62	353
4	31	130	603
5	52	310	1447
6	54	342	1573

[a]Modified from Wiegert and Wetzel (1979).
[b]"Harvested" on June 30 in each of first two years. "Clip" means all aboveground or shoot biomass of *Spartina* was removed from X3 on that date and added to X5 (dead *Spartina*).

processes occurring there. We were pleased, but also somewhat surprised, to find the predictions of the model so fully supported by direct experimentation in the field. We usually tend to regard simplicity and lack of direct connection as a reflection of our ignorance or inability in model construction rather than as a property of the real system.

9.3.3 Varying Tidal Export (MRSH1V3)

Beginning with the preliminary model, which was used to evaluate the salt-marsh ecosystem as a source or sink of carbon, the question of carbon transport and fate has had an important effect on our research. The first version, MRSH1V1, labeled the marsh a source of organic and a sink for inorganic carbon. Because no evidence of large-scale accumulation of organic carbon in the marsh was available, the tidal export coefficient, TIDE, was employed in MRSH1V3 to rid the system of the excess carbon. The addition of this tidal export term, plus certain other refinements, improved the ability of the model to predict accurately the dynamic behavior of the marsh components. So dramatic was this improvement that an evaluation and revision of TIDE iself became an important goal of the research. As a first step in this direction, we sought to evaluate TIDE through simulation experiments with the MRSH1V3 model itself (Wiegert and Wetzel, 1979). We simply changed TIDE from the nominal value of 0.25 per day export of all suspended POC, DOC and algae to 0, 0.125% and 0.50 and made a five-year simulation run with each value. First we examined the effect on both the total carbon exported and the total carbon balance in the marsh. The results are presented in Table 9.4. Clearly, the zero tidal export, the situation in MRSH1V1, does not fit with reality. But there is surprisingly little difference between the remaining three, despite what would appear to be a rather drastic variation in the value of TIDE. Despite its importance, the latter does not seem to be a particularly sensitive parameter. As we shall see presently, that lack of sensitivity is related to other problems with the crude manner in which hydrologic events are modeled in MRSH1V3. But at the time these simulation results were first examined, we could certainly not choose among the

Table 9.4. Simulated Effects of Varying the Tidal Exchange in a Coastal Georgia Salt Marsh[a]

Exchange (% day^{-1})	Amount of Carbon Exported[b] (g C m^{-2} yr^{-1})	Change in the Carbon Balance of the Marsh[b] (g C m^{-2} yr^{-1})
0	0	1012
12.5	1025	53.4
25	1074	26.1
50	1033	17.4

[a]From Wiegert and Wetzel (1979).
[b]Values are those of the final year of a 5-year simulation.

values of TIDE based on total carbon exported alone. Considering the net annual change in total carbon in the marsh, the nominal value of TIDE as .25, and the increased value of .5, seemed to give depositions closest to the available geological estimates. But the immense carbon load in the soil and sediments (Table 9.2) and the relatively infinitesimal rate of annual accretion render a determination of the true value of TIDE by this comparison impossible.

Instead, we decided to examine the effects of a changing TIDE on the seasonal variation in standing stocks of those water-borne components directly affected by export, namely algae (X2), DOC (X6), and POC (X9). This produced some potentially helpful contrasts. Both DOC and POC are passive abiotic components. The only effect of progressively larger daily export rates was to lower the average standing stock and thus maintain the same annual export. Thus, for both of these variables there was an inverse relationship between the value of TIDE and the average standing stock throughout the year. Unfortunately, the only large difference was between zero export and all the other rates. The field data on seasonal variation in standing stocks of DOC and POC were, and are, scarce and variable. Certainly they were not sufficient to choose between the most realistic of the various values of TIDE (Wiegert and Wetzel, 1979), although the extremely low values for a TIDE of .5 (2 to 8 g C m^{-2} for POC and 1.5 to 3 g C m^{-2} for DOC) were generally below the range of field measurements. The different TIDE regimes did produce some clearly different seasonal dynamics in the standing stock of algae. Again, however, we did not have adequate measurements from the field with which to compare the predictions from the model. We thus concluded that 1) the marsh does produce an excess of carbon over degradation in place, 2) some way of getting rid of this excess was necessary to produce realistic simulations of marsh dynamics, and 3) the mechanism of transport was unknown but was simulated adequately by a tidal export of 12 to 50%, probably 12 to 25% of the water-borne carbon per day.

Attempts to measure directly the import-export of organic carbon by means of current meters and time sampling had met with little success in the Sapelo Island marshes, primarily because of difficulties in estimating water transport. The best, and most often quoted, data set showing a definite positive net annual export of organic carbon out of a tidal creek (Odum and de la Cruz, 1967) turned out, when the export was prorated over the watershed drained by the creek, to be on the order of 100 g C m^{-2}

yr^{-1}. This is barely 10% of the excess carbon that had to be accounted for. Also, the very real question arose at this time about what part, if any, of this excess carbon even reached the lower Duplin River let alone the sea (Section 4.1.3). Out of discussions in late 1976 concerning these frustrations was born the 1977 study of the hydrology of the Duplin River.

9.3.4 Marsh Model 1, Version 6 (MRSH1V6)

An interdisciplinary project, headed by Jörg Imberger, investigated the hydrologic mechanisms responsible for transport into and out of the Duplin River marshes. The purely hydrologic aspects have been presented in Chapter 2. The time-course measurements of carbon and nutrients have been covered in other chapters, and the integrated study is described fully in Imberger et al. (in ms.). Here we wish to briefly recount the role of the model in this study and the changes necessitated by our evolving view of carbon transport mechanisms.

Marsh Model 1, version 6 (MRSH1V6) differed from the previous versions only with respect to the mechanisms of carbon export, plus a few changes in other parameters as a result of the continuing acquisition of data from other studies, most notably in the parameters governing the dynamics of the algae. Although limited to only a small part of the model, the hydrologic changes are fundamental and change the structure of the model in a major way.

The hydrographic study reached several conclusions that were important in modeling the salt-marsh estuary:

1) The Duplin River is characterized by three distinct tidal segments, of which only the third, the one at the head of the river, is entirely pushed up onto the marsh at high tide. Some of the second segment is moved laterally onto the marsh along the middle of the river, and as well part of the first segment enters the marsh along the lower section of the river.

2) During normal (nonstorm) periods, exchange between the upper and lower segments is primarily by eddy diffusion, and the rate is on the order of 16% of the water per day. This slow mixing is a major reason why the differences between the concentrations of substance in each of the water masses can arise, and remain for days.

3. Turbulent mixing in the river and tidal creek suspends POC from the creek bottom. Some of this material is subsequently carried onto the marsh, dropped, and not picked up by the ebb tide, because flow on the marsh proper is slow and virtually laminar.

4. This daily accretion of POC on the marsh is interrupted by periodic storms in which wind action, or more commonly, erosion by heavy rain on the exposed marsh moves material into the creeks.

5. The fresh rainwater, if it falls fast enough and long enough, will not only erode material and transport it to the creeks and river, but can flush over the top of more saline water in the Duplin River and carry material out into Doboy Sound. Each year, on the average, four to six such storms occur over the Duplin River watershed.

The longitudinal mixing component of the Duplin River water has only a very minor influence on the dynamics of rapidly changing components. Such internal fluxes within a tidal segment as the movement of POC from sediment to water and back or the release and uptake of DOC far outweigh the longitudinal flux. Nevertheless, whatever export takes place has to be regulated by longitudinal mixing, with very labile DOC being assimilated by microorganisms, while the more refractory forms go on their way to the estuary and thence the sea (Hanson and Snyder, 1980). Much of the required export must be taking place during storms and must involve scouring of the refractory carbon components, namely lignocellulose-rich POC.

The simple, spatially homogenous, first-generation model, MRSH1V3, could not of course, represent adequately the expected patchy distribution of the variables caused by the imbalance between internal cycling and longitudinal mixing. However, several aspects—the diffusive drain relevant to the DOC, the sedimentation and resuspension of POC during a tidal cycle, and the effects of severe storms on carbon transport—did lend themselves to inclusion in the model (Imberger et al., in press). Many changes in MRSH1V3 were introduced, along with the parameter changes mentioned earlier, to form versions 4, 5 and 6. Here we consider the final version, MRSH1V6, which contained the four alterations described below.

Alteration I. The flux of a substance (Equation 2.1) is given by:

$$Q = EA\ \delta C/\delta X \tag{9.9}$$

where: Q = mass x time^{-1}

E = longitudinal diffusion coefficient = 41 m^2 x time^{-1}

A = cross sectional area (m^2) of river

C = concentration (g m^{-3})

X = length of river segment (m)

The term $\delta C/\delta X$ is estimated over the distance, approximately 10 km, from the mouth of the Duplin River to the head of the river, just before it splits into tidal creeks in the marsh as:

$$\frac{C - C_m}{L} \tag{9.10}$$

where: C = concentration at the head of the river estimated by the model

C_m = concentration measured at the river mouth

L = distance from mouth to head of river.

Substituting (9.10) into (9.9), canceling units, and using appropriate values of A and L gives:

$$\text{Flux (g day}^{-1}) = 0.16\ (C - C_m).$$

Concentrations* used must be those for materials for which the diffusion time (T_d) is short compared to the characteristic time for biological cycling. In the case of DOC, a constant 6 mg l^{-1} was assumed labile, and thus the difference ($C - C_m$) is computed only for the refractory portion.

The fraction 0.16 is between the values 0.125 and 0.25 that were considered the most realistic values for TIDE in MRSH1V3, but in that earlier version TIDE was multiplied by the concentration (C) at the head of the river. In MRSH1V6 we multiply the tidal transport term by $C - C_m$. Thus the estimates of tidal export predicted for normal tidal conditions by MRSH1V6 are considerably lower than those predicted by MRSH1V3.

Alteration II. The cycling of the POC from the creek beds to the marsh required a re-specification of compartments X9 and X13. At peak flood tide the concentration reaches 30 g POC m^{-2}. Because this concentration was entirely determined by the degree of turbulence in the creeks, the concentration in compartment X9 was fixed at 30 g m^{-2} for 6 hr day^{-1} and at 7 g C m^{-2} corresponding to the slack tide value, for the remaining 18 hours. These concentrations were used to determine the internal transfer, or biological utilization, and any excess was returned to compartment X13.

Alteration III. To simulate the effect of storms, we decided to completely empty the water-borne compartments X6 and X9, and to remove 50% and 33% of compartments X2 and X7 respectively. The 50% is based on the relative proportion of compartment X2 (algae) composed of the phytoplankton as opposed to the benthic algae on the surface of the marsh. Compartment X7 (heterotrophs) comprises all the larger motile organisms in the water—fish, shrimp, crabs— as well as the smaller planktonic organisms. The 33% is a conservative estimate of the proportion in this compartment of the smaller, less motile forms which are susceptible to washout along with the POC, DOC, and phytoplankton. On the day following the simulated storm, POC in the water was replenished by resuspension from POC in the sediment.

Inspection of the past five years of weather information from Sapelo revealed about four such storms occurring per year. If the rain falls at low tide, very severe erosion of the marsh surface and sloughing of the creek sides occurs. Therefore we simulated the effects of two categories of storm, a weak storm where 30 g POC m^{-2} is removed from the system by displacement of water in the creeks and Duplin River, and a strong storm where an additional 30 g POC m^{-2} is removed as a result of surface scouring, *Spartina* raft removal, and the increased hydrologic action caused by high tides and winds. The probability of the rain falling at low tide was 25%, since the water is very low or the marsh is exposed, on average, about one-fourth of the time. We simulated four storms per year, three of them weak and one strong.

Alteration IV. Our current view of cycling and transport of carbon in the salt-marsh system envisages a large deposition of POC on the surface of the marsh each day caused by suspension on the incoming tide and sedimentation at slack high water.

---

*In all versions of the marsh model, concentrations are reported per m^2. For water-borne components, this is equivalent to g m^{-3}, because the mean water depth prorated over the entire marsh at high tide is approximately one meter.

Little, if any, of this material is removed from the marsh at high tide, most of it accumulating until a storm scours the marsh. Intertidal creeks are scoured daily because the ebb has higher velocities than the flood tide. Aerobic decomposition of POC may be important in the sediments of these creeks, perhaps far more important than aerobic uptake of POC. Because of this and because the aerobic microbial heterotrophs in the water form only a small fraction (5% or less) of the total heterotrophs in X7, we decided to change the carbon flow pathway, giving aerobic microbes in and on the sediments (X14) the capability of taking DOC from the water; thus, in Figure 9.3, X6 → X14. This gave a more realistic representation of where the major uptake of DOC occurs and permitted the simulation of various rates of carbon incorporation by aerobic microbes and thence into food chains, culminating in the large motile heterotrophs–fishes, shrimp and crabs. Many of these organisms may leave the marsh and constitute a carbon export that has not hitherto been a part of the model simulations.

With these modifications, five-year simulations with MRSH1V6 were made–four storms per year, three weak and one strong. The nominal run differed mainly from MRSH1V3 with respect to the standing stock of algae, DOC and POC, and tidal export (Figure 9.6). Total export with MRSH1V6 was 586 g C m^{-2} compared to 1074 g C m^{-2} with MRSH1V3. This decreased drain on the components–algae, DOC and POC– supplying the bulk of the export caused the maintenance of a higher algal standing stock during the low winter growth period and a much higher annual DOC concentration (Figure 9.6). The mean POC concentration of version 6 was also higher (13.1 g C m^{-3}) than the value from version 3, but this was the result of the direct constraint on POC in the water and would be unaffected by any rate of export. No other components or carbon flows predicted by MRSH1V6 differed in any significant manner from those obtained from MRSH1V3. However, the diffusive and storm-caused export, together with degradation on the marsh, were not enough to account for all of the carbon. At the end of five years, MRSH1V6 predicted organic carbon accumulation at a rate of 207 g C m^{-2} yr^{-1}. The data on primary production and carbon degradation in the sediments are extensive and reliable (Wiegert, 1979). The overall long-term accumulation of sediment in the marsh is approximately one millimeter per year, the total accumulation very likely representing less than 50 g C m^{-2} yr^{-1}. Thus, conservatively, we have to account for an excess of 150 to 200 g C m^{-2} yr^{-1} which is seemingly neither deposited, degraded in the marsh, nor transported out by the normal diffusive exchange abetted by a regime of four catastrophic events per year.

There are three obvious mechanisms whereby this excess carbon might leave the Duplin River ecosystem. The first is via bedload transport of sediment, but measurement of sediment transport along the Duplin River by Gary Zarillo (*pers. comm.*) shows a relatively small loss of organic carbon. A second possibility is either that there are more than four storms of sufficient magnitude to replace the water at the head of the Duplin River or that our estimate of the amount of material removed per storm was too low. There were no data available concerning the latter point. But four such storms per year on the Georgia coast is a conservative number and the additional 30 mg l^{-1}, equivalent to 30 g C m^{-2}, assumed taken out by a strong storm is also on the conservative side. The third mechanism whereby the excess carbon could be removed from the marsh is via incorporation into food chains beginning with microbial utilization of POC and DOC.

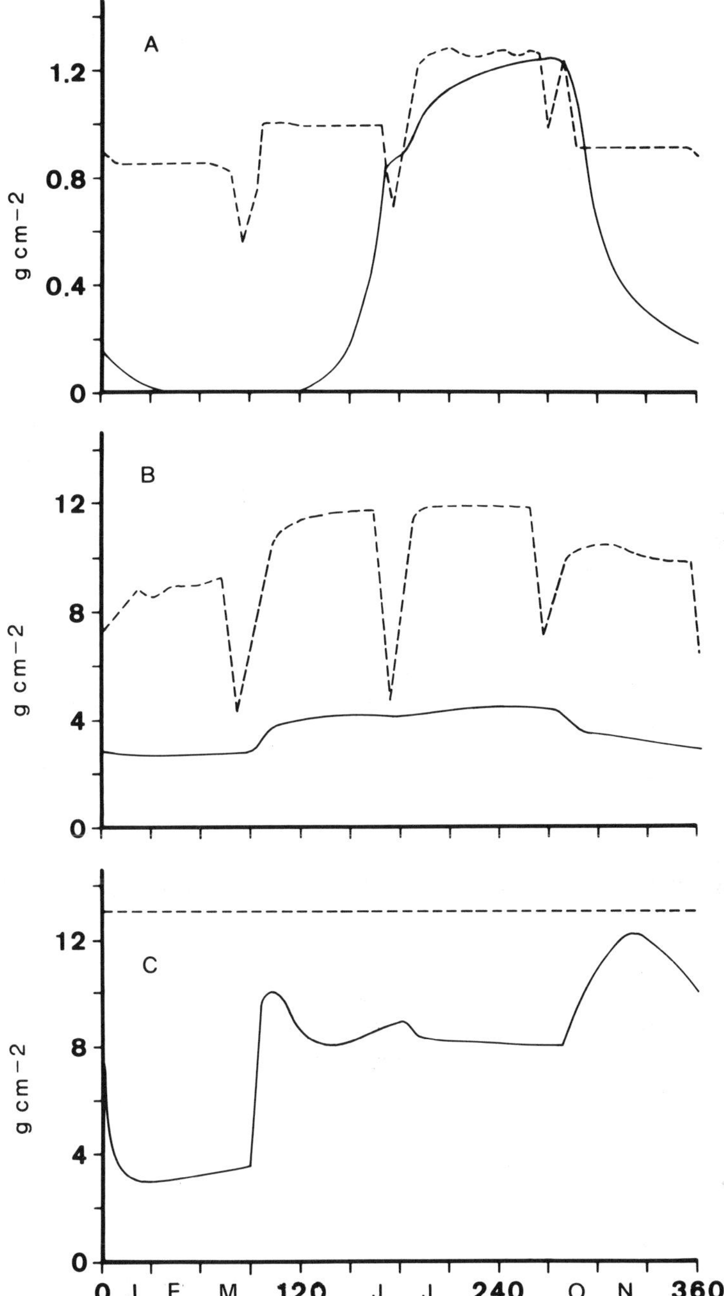

Figure 9.6. Seasonal variation in the simulated standing stocks of algae (a), dissolved organic carbon (b), and nonliving particulate organic carbon (c) in water of the marsh, tidal creeks, and the Duplin River. All values are prorated per m^2 for the entire marsh, including creeks, and are equivalent to concentrations per m^3 of water.

Mechanisms to simulate the second and third alternatives were incorporated into the model, and a five-year simulation of each was run. The modifications were as follows:

1. The storm regime was changed from four to six per year and all of them were strong; that is, in addition to flushing the contents of X6 and X9, an additional 30 g C m^{-2} was picked up from X13 and removed during each simulated storm.

2. A pathway (EMIGRAT),* a specific rate of loss equal to 3% per day of X14 and .15% (3 X .05) of X7, was provided. Most of this incorporated material does not "emigrate," but is respired, as discussed above.

Neither of these changes affected the simulated net primary production of *Spartina* or of the algae. In the model, processes immediately affecting the accumulation or loss of dead plant material were largely decoupled from the process of primary production. This would seem to be true for the marsh as well, at least within the five-year time sequence under consideration. The model had no provision for spatial separation of marsh and tidal creek–one-half of the respiration by aerobic microbes (X14) was assigned to sediment (i.e., marsh surface) and one-half to water (i.e., the bottoms of the subtidal creeks and rivers).

Increasing the hydrologic action to six storms per year, all capable of the additional specified scouring and removal from X13, virtually balanced the carbon of the marsh ecosystem, leaving only 16 g C m^{-2} as the annual accumulation. The annual increment of the marsh sediments is probably somewhat greater than this; therefore, the model prediction was that some regime of storm removal between four and six catastrophic events per year *could* be sufficient to balance the system.

Setting EMIGRAT equal to 0.03 also produced a balanced system in which the 29-gram annual increment in C m^{-2} is probably closer to the annual increment in carbon added to the marsh itself. Balance was obtained under this hypothesis by maintaining a high level of degradation so that total respiration in sediments, air, and water accounted for 55% of the total net production, identical to the percentage under the no-storm regime.

Thus the final contribution of the first generation salt-marsh model was to leave us three possible hypotheses that would explain the total carbon flux of the marsh. A new cycle of model development and research is called for, with some of the obvious deficiencies, such as spatial heterogeneity, explicit nutrient cycles other than carbon, and detailed submodels of carbon flow, yet to be explored. A start has been made in certain of these areas, and we discuss them in the final section, in the context of modeling and model state-of-the-art for salt marshes in general. A number of marsh research programs have not used models as a guide or optimization tool. As a consequence, simulation models of salt-marsh ecosystems or processes are few. Each existing model has certain unique features and objectives, and a brief comparison of their most salient features will round out our perspectives of models and their uses in ecological investigations of salt marshes.

---

*We included the pathway 0700 as emigration because we are currently testing this parameter as a direct movement of macrofauna rather than as a food chain transfer via bacteria.

9.4. A Man-Environment Model

A model of the salt marsh surrounding Barataria Bay in Louisiana (Hopkinson and Day, 1977; Wiegert, 1979a) was constructed for the same reasons as the Sapelo model, to summarize existing information and to gain insight into the operation of a natural ecosystem. Its evolution, however, took a somewhat different course, in part because of the clear importance of the Barataria Bay marshes to man and in part because of the equally clear impact of such man-induced perturbations to the marsh as dredging, flood prevention, and draining. The model was oriented to predict the impact of perturbation and to assist in the construction of larger scale man-environment models (Day, Hopkinson and Loesch, 1977). The distinction is only one of degree, but, in any case, more germane for our consideration is a comparison of the two marsh systems as viewed through the "eyes" of their models.

Hopkinson and Day (1977) wished to model the productivity and hydrology of the ecosystem in as realistic a manner as possible. Their model was thus *a priori* in approach, more explanatory than empirical. Their large-scale model of the ecosystem coupled the marsh with the offshore systems and incorporated, from the beginning, certain aspects of human interference. In Barataria Bay, *S. alterniflora* dominates the vegetation and fixes two to three times as much energy as the phytoplankton, perhaps slightly less dominance than in the Sapelo marshes, but certainly close. The water and the marsh immediately adjacent to the shore are more productive than the areas further inland. Because of the much smaller size of the Sapelo marshes there is no comparison with this phenomenon; the inland or upriver, marshes of the Georgia coast are dominated by species other than *S. alterniflora.* In contrast to the Sapelo marshes, the Barataria Bay ecosystem is very strongly affected by the land, being a product of the interaction of climate, physical gradients, and the all-important delta formation by the Mississippi River.

The Barataria Bay model included both carbon and nitrogen fluxes, and a number of ecologically realistic controls were employed. For example, the specific rate of net photosynthesis was determined by temperature, available light, and the level of nutrients; this rate was then multiplied by the standing stock of vegetation to obtain the instantaneous flux. Annual variation in sea level and temperature governed the transformation of dead standing *Spartina* into detritus. A potentially interesting area for perturbation simulations was not realized by their model because the transfer from the marsh plants and the detrital community to the marsh fauna was represented as a linear, donor-determined and donor-controlled flux. Thus variations in marsh fauna standing stocks could have no direct effect on their food.

Hopkinson and Day performed a sensitivity analysis of their model and found temperature to be by far the most important variable in the control of primary production. The Sapelo carbon model did not have temperature as an explicit variable, but changing the value of the maximum specific rate of photosynthesis seasonally (Wiegert et al., 1975) did implicitly recognize the importance of temperature (Figure 9.3). Nutrients and light in the Hopkinson-Day model were limiting only during blooms of phytoplankton. Sea level was instrumental in determining the timing and magnitude of flushing of organic matter into the estuary.

Simulations with the model included runs where primary production was eliminated alternately from the marsh and from the water. The elimination of production in the water had the greatest effect on the aquatic fauna despite the smaller standing stock and net production of the phytoplankton, probably because nitrogen cycling by the phytoplankton was much faster than that by the marsh plants. The phytoplankton not only produced organic carbon, but also helped enrich the organic food resources of the aquatic fauna. The aquatic fauna and not the marsh fauna is of major economic interest to man (Chapter 10). Barataria Bay, for example, accounts for almost half of Louisiana's commercial fisheries catch. Hopkinson and Day feel that the modeling effort gave them insights into the operation of the ecosystem that they could have obtained in no other way. Specifically, some new areas of research were suggested by results from the model, implying development of new hypotheses on the basis of model simulations. These new areas include:

1. efforts to find out why soil nitrogen is such a poor predictor of productivity in the marsh;
2. the importance of sea-level variation and the need for more concentration on hydrologic studies;
3. the measurements of the degree of variability in some of the more sensitive coefficients to permit inclusion of variable coefficients in modeling.

9.5. A Successional Model

Interest in the nominal dynamics and response to perturbation by ecosystems leads to development of large-scale compartmental efforts, such as the Sapelo and Barataria Bay models of the *a priori*, explanatory type. Interest in the processes of marsh formation leads to models of succession. In particular, if a true management model is desired, *a priori* model development will often yield to *a posteriori* models that are mainly empirical; that is, the models are designed to conform to a number of observations of real systems undergoing successional development, and the resulting correlative model can be used with considerable confidence in predicting the course of succession in another area for which a series of simple, required measurements has been made.

Zieman and Odum (1977) developed such a model (reviewed in Wiegert, 1979a). Their model was of ecological succession and production in two areas of salt marsh along Chesapeake Bay. The marsh area consisted of three zones: a low marsh area fringing the tidal creek, an ecotone, and a high marsh area where the incidence of tidal flooding was markedly less frequent. Mean tidal amplitude at these two sites was 1 meter. Salinity ranged from 0 to 15 $^{0}/_{0}$. Four species of vascular plants were sampled: in the low marsh, the dominant cordgrass, *S. alterniflora*, in the high marsh, *S. patens* and *Distichilis spicata*, and a minor component, *Aster tenuifolius.* All four species occurred in the ecotone.

Concern over the disposition of dredge spoil by the Corps of Engineers provided the initial stimulus for the research of Zieman and Odum. If a dredge spoil bank could be constructed in such a manner as to enhance the probability of a *Spartina* marsh developing on it, then a large amount of time and dredging effort could be saved, and increased benefits from marshland could be realized. The successional model was used to predict the conditions and time sequence necessary for the development of a salt marsh and the kind of spoil bank that must be constructed if a marsh is to form.

The model was unique in that a "point" concept of each species was constructed and coupled with the models of others to give simulated dynamics in both space and time. For example, each species was modeled as a number of spatial compartments. The model was empirical, developed directly from time-series data on colonization and growth, correlated with measurement of the physical and biological regulating factors. The resulting model was site specific. The selection of parameters for inclusion in the model was based solely on the degree of contribution to prediction obtained. By studying intensively two different sites and surveying many additional sites outside the Chesapeake Bay area, Zieman and Odum were able to obtain some idea of the range of applicability of the model.

Controls in the usual sense were not used. Total biomass increase in a given species of plant was a function of salinity, light, temperature, tide, and pH. In addition, available iron exerted an influence, and at any particular point elevation was a factor because of its relation to tidal inundation. Growth of *S. alterniflora* was correlated positively with number of tidal inundations and height, although the analysis did not extend seaward far enough to detect the presumed negative effect of too-frequent submersion. Growth was inversely related to sediment interstitial salinity. Data were obtained by continuous recording of variables correlated with measurements of standing stock and growth rate. Elevation and organic content were the site characters most important in determining the spatial limits of *S. alterniflora* marsh. Accurate prediction of growth and succession became more difficult as the elevation gradient was traversed from low marsh landward.

This type of model is an example of the proper and valuable use of correlation analyses to develop a management tool, one that can be used by persons with little training in either ecology or modeling. Indeed, our knowledge of succession in salt marshes is still so rudimentary that only the correlation type model is possible at present. The Zieman-Odum model would be a good point from which to develop a more fundamental explanatory model of succession. Randerson (in press) describes a successional model of salt marsh on the Norfolk coast of England; but this is a different kind of marsh and is dominated by forbs.

9.6. Future Directions

As our knowledge of salt marshes and their ecological relationships grows, so must our model view, because some very real constraints operate on the size and complexity of models. None can even approach a one-on-one representation of the real world, nor is this needed or desirable. The value of the model, as stated at the beginning of this chapter, is to simplify without losing the essence of what is to be predicted. Some of the constraints are purely technical—the capacity and operating cost of computers, for instance. Some relaxation of these constraints may be expected, but our ability to conceive of and to construct *good* ecosystem models is yet a long way from the technological asymptote. Our limitation at present is more often the ability of the ecological modelers to understand and work with their creation. There are a number of ways this limitation may be overcome. The limitation can be surmounted by the use of submodels or of separate models for spatially distinct parts of the system.

9.6.1 Submodels

We often know much more about certain biological groups in an ecosystem than we do about others. Or we may wish to examine the response of particular species or groups in certain trophic categories and be willing to view condensed responses in others. Hence, one may develop submodels, which are simply detailed models of a particular segement of an ecosystem. As part of the ecosystem modeling effort, the submodel may have its inputs and outputs tied directly to appropriate points in the larger model. In this sense, the marsh model of Hopkinson and Day (1977) could be considered a submodel of the larger man-environment model (Day, Hopkinson and Loesch, 1977).

Perhaps more commonly, the submodel will be used by itself to simulate the dynamics of a smaller segment of the ecosystem, a segment temporarily disconnected from the larger system so that the changing components of the submodel do not affect the main system over the term of the simulation. There is a close analogy between the use of the submodel in ecology and the use of the partial differential equation in the physical sciences.

Three submodels have been developed out of the Sapelo marsh modeling effort. Two of these submodels are still under development, but work based on the other has been published. They are: 1) a submodel of the dynamics of growth and death in *S. alterniflora*, compartments X3, X5 and X12 (Hayes and Gallagher, unpublished) 2) a submodel of the heterotrophs in the water and aerobic microorganisms in the sediment, compartments X7 and X14 (Wiegert and Wetzel, unpublished) and 3) a simple submodel of the microorganism, DOC, POC trophic perturbation in the water. (Compartments X6, DOC; X9, POC and X14, the aerobic microorganisms (Christian and Wetzel, 1978).

The submodel of *Spartina* was suggested by the wealth of information available on the ecology of *S. alterniflora*, the complexity of the life cycle in the sense of the separation of shoots and root-rhizomes, and the transformation into standing dead before the detritus (POC) category. The submodel is currently being revised to include data on the physiological ecology of *Spartina*. The submodel of heterotrophs arose as a response to the central importance of the heterotrophic component in the water (X7) and the recognition of this compartment as so heterogeneous as to be virtually impossible to simulate realistically. During development of the submodel, the necessity of incorporating the aerobic microorganisms was recognized and this led to the development of the simpler submodel published by Christian and Wetzel (1978). This published submodel serves as our prototype for extracting a small portion of an ecosystem model and using it to explore the consequences of perturbations that would be impossible to explore in the context of the larger ecosystem model.

Christian and Wetzel take an alternative modeling approach regarding the interactions of *Spartina* detritus, microorganisms, and animal consumers in the water column. This model is of the *a priori* type and it is general in that the parameters are unrelated to a specific site, but they are biologically realistic. It is small, with only five compartments and seven fluxes (Table 9.5). Controlling terms were similar to, or extensions of, those previously described for the Sapelo Island salt-marsh carbon flux model. Details of the equations are found in Christian and Wetzel (1978).

Table 9.5. Characteristics of a General Model of Detrital Substrate, Microbe, and Consumer Interactions.

State Variable	Flux from State Variable to:	Controls of Flows
Detrital substrate POC	Microbes	Biochemical composition of POC; relation of particle size to microbe density
Microbes	CO_2 Consumers	Respiratory requirements of microbes
DOC	Microbes	Density of microbes
Outside of system	DOC	Constant
CO_2	$\frac{2}{M}$	$\frac{2}{M}$
Consumers	CO_2	Respiratory requirements of consumers
	Out of system	Assimilation abilities of consumers

The impetus for the model was the lack of understanding of the dynamics of the detritus food web of *Spartina* in estuarine waters. Christian and Wetzel (1978) reviewed the then extant information on *Spartina* detritus and found that a minimum number of conclusions could be drawn. The existing view (Darnell, 1961) stated that microbes coated detritus particles and could be grazed by animals. However, few bacteria are usually seen in detritus, and the ability to support grazer energetics is suspect. The model was designed to determine under what conditions organic substrates could support microbial growth and under what conditions microbes could support growth of grazers. This was done through a series of manipulations of parameters, including alterations of 1) proportions of microbial uptake from POC and DOC, 2) of microbial ingestion rate, 3) of grazing pressure on microbes, 4) of microbial respiration, and 5) of spatial imitation. The model was analyzed with respect to which parameter values produced steady-state conditions during simulation of 360 days.

The model was sensitive to alterations in all parameters but spatial limitation. Optimum stability was characterized by a microbial compartment that 1) shows a high preference for dissolved substrate, 2) has the potential for rapid turnover, 3) has better than a 50% growth yield, and 4) is not grazed by consumers. However, the conditions of stability were not comparable to those found in the field. Stability was always accompanied by a large density of microbes on particles. We have inferred that the detrital food web in nature is, in fact, not stable. Either few bacteria ever attach and grow on particles, or rapid growth does occur, only to be followed by overgrazing. If the latter were true, the microbial link between *Spartina* as a substrate and grazers would be through microscale pulses of activity, rather than continuous low-level feeding.

9.6.2 Separate Models

The alternative to development of submodels is to simplify the total model by restricting its scope to areas of relative physical uniformity or homogeneity. Thus an alternative to the submodel route that we are exploring for the Sapelo model is to model separately different spatial areas, that is, to develop models of the creek bank, creek water, and high marsh as separate and unequal, or weighted, models. Similarly, different flows can be separately but simultaneously modeled; for example, a nitrogen model is currently under development that will run concurrently with and couple directly with the carbon model (Wiegert, 1979a). In this way, a high degree of reality, complexity, and predictive power can eventually be achieved with a minimum of confusion regarding the structure, function, and purpose of any individual model.

10. The Salt-Marsh Ecosystem: A Synthesis

R.G. WIEGERT and L.R. POMEROY

We began in Chapter 1 with an introduction to the salt marsh as an ecological system and progressively dissected that system into component parts and processes, examining interactions, impacts, and controls. We started to put the parts together again when discussing the carbon-flow model. We complete the process of unification in this final chapter, concentrating in particular on the mechanisms by which salt marshes are connected to the larger systems abutting and affecting them—the land and its rivers on the one side, and the sea on the other.

Salt marshes are many faceted. These extensions of an estuary are salt deserts. Yet they are also aquatic grasslands, where snails and grasshoppers coexist on a single stem of *Spartina.* They are tidal watersheds of high ion strength and low Reynolds number. As fermentation systems, one-third or more of their primary production appears to be transformed anaerobically. To some ecologists they are mixtures of several ecosystems, complex ecotones which are at once wet to dry, aerobic to anaerobic, and salt to saltier to fresh water. It is not surprising then, that salt marshes, even single marshes, have been perceived and studied at a number of levels and from various perspectives. We have tried to bring together these disparate views through the unifying device of systems modeling and by an interdisciplinary approach to research. Because of the high proportion of anaerobic metabolism in the salt marsh, we placed more emphasis on microbial ecology than is perhaps typical of an ecosystem study. The interactions between the microbial components and the more readily perceived macroorganisms are vital to our understanding of the marsh. These are not the same everywhere. Not only do they vary in different parts of the marshes of Sapelo Island, but they may be even more different in marshes in other parts of the world. Therefore, we have tried whenever possible to compare our findings with those of investigators studying other marshes.

Marshes, particularly those dominated by *Spartina*, are very similar in their gross appearance, tempting us to assume salt marshes are all much alike structurally and functionally in the sense these words are defined in Chapter 9. However, even the limited basis for comparative studies which now exists shows that is not the case. This is surprising. We are usually correct in expecting similarity in physical appearance and in type of dominant producers to lead to trophic and metabolic similarities as well. Indeed, we as ecologists are so conditioned to this view that we emphasize differences and often neglect to emphasize similarities between ecosystems. Why then are salt marshes on different coasts so variable? We suggest the cause is the dominant role played by physical factors in determining both the consumer species present and the patterns and rates of material transport. Climate is obviously an important variable, but the major differences appear to be related to the geological and hydrographic environment. Such features as the character of the sediments, the tidal regime, and the rate of deposition will influence not only the primary production of the vascular plants, but the rates of microbial transformation of that production and the kinds of end products which emerge from the marsh. Thus, in this synthesis we must first examine the relationship of diversity and stability in the salt marsh.

10.1. Diversity and Stability

Whether we use the standard of species richness or of equitability, we regard salt marshes as low-diversity ecosystems. Yet the truth of this statement depends both on the groups we are considering as well as on the spatial scale involved. Considering only the macroflora, the *Spartina* marsh is a virtual monoculture. Even when the numerous consumers and saprophages are included, the diversity in the salt marsh becomes only moderately high. The more than 100 insect species might not be expected because of the apparent dearth of niches in a system with, at most, a few higher plants. But what if the benthic algae are considered? Then we have a system in which hundreds of species of plants may exist in a few square centimeters. Surely this is a high diversity. However, these producers cannot create numerous physical niches for organisms higher in the trophic scheme, and because the consumers are mostly far larger than the benthic algae, they cannot easily specialize in feeding. Thus, despite the high numbers of algal species, the *Spartina* marshes are truly low-diversity systems in terms of structural complexity of the food web.

According to some ecologists, low diversity equates with lack of stability. Yet salt marshes are remarkably stable in the several senses in which that word is commonly used. *Spartina*-dominated marshes persist in place for long periods, the "stayability" of Smith (1972). The marshes are resistant to many perturbations in the sense of both Holling (1973) and Webster et al. (1975). However, certain perturbations, such as restriction of tidal flow, may affect *Spartina*, and thus the remainder of the marsh, in a very drastic manner. Similarly, an opening made in a terrestrial community such as a tropical rain forest is commonly filled with any one of tens or even hundreds of species, but in the salt marshes a disturbed area will usually eventually be filled by the same species that occupied it when perturbed.

Smith (1972) reversed the diversity-stability dogma and suggested theoretical support for the necessity of stability in climate and physical environment as a prerequisite for the evolution of diversity. Huston (1979) completed the hypothesis circle by noting that in a system with many potential occupants of a few niches, only periodic disturbance setting back the successional sequence will keep one or a few dominants from taking over completely. The salt marsh, at least in the lower elevations, would seem to support the latter theory.

In the higher elevations of the marsh more opportunities exist for variation in replacement of species, and a greater diversity of higher plants is often encountered. We saw, for example, replacement of *S. alterniflora* by *Salicornia* in plots where the former species had been mowed at the height of the growing season in two successive years (Chapter 9). Moreover, when entirely new areas such as dredge spoil banks are created, the resulting vegetation pattern is often extremely heterogeneous and diverse. Such findings suggest that an important determinant of the low-diversity domination by *Spartina* lies in the physical and chemical characteristics of the soil. Single-dominant forests are also known from the tropics, and a similar reason is given for their existence (Richards, 1964; Whitmore, 1975).

10.1.1 Physical and Chemical Factors

Although tidal flows are the dominant aspect of marked and rapid changes in the physical environment of salt marshes, tidal flows by themselves are not perturbations in the same sense discussed above. They are, overall, regular and predictable, in the same sense that sunlight is predictable in its seasonal changes. However, as with sunlight, the vagaries of the weather exert a large and unpredictable effect on the magnitude of the tides. Storm-caused erosion of the marshes, erosion by rain, and movements of sediments by nearshore currents are some of the major physical factors tending to destabilize salt marshes. Despite this, marshes largely remain intact; once established, they can persist and thus protect the integrity of the soil for long times in the face of fairly high-energy surf (Chapter 1).

The movement of water by the tides may be another way in which the potential stabilizing mechanism of nutrients in the soils is maintained. Pomeroy (1975) documented the large reserve of plant nutrients in the soils of salt marshes and the routes by which they may be in equilibrium with the water. Phosphorus is never limiting in the marsh, and although nitrogen may be in short supply for *Spartina* and macroorganisms in the high marsh, elsewhere it rarely, if ever, limits primary production. Moreover, the effect of excess nutrients is rarely seen, since excesses are rapidly assimilated in the naturally eutrophic marshes and have little impact. A naturally eutrophic system is relatively resistant to perturbations, except those instituted by humans (Section 10.1.3).

If the physical and chemical characteristics of the soil are important attributes conferring stability, then any perturbations which leave those characteristics relatively unchanged should have little effect on the marsh and vice versa. In general this seems to be the case, because long-term or permanent change in the pattern of tidal flow does affect the vegetation and also the chemical and physical characteristics of the soil. Perturbations of a more biological nature apparently have lesser effect on vegetation and sod and, hence, on the marsh as a whole.

10.1.2 Biological Factors

Evidence for stability as the result of biological attributes or interactions in the salt-marsh ecosystem is more elusive. *Spartina* does have a large biomass of roots and rhizomes and thus has a relatively slow turnover. In fact, the biomass of these reserves is greatest in the high marsh where potential limits imposed by lack of nutrients are most common. We may speculate on this as an induced response by the plants, conferring a measure of stability in resisting short-term-fluctuations in the supply of nutrients. In any case there is a concurrent small, active standing stock of algal primary producers that is turning over very rapidly. Because these two groups are not only being transformed on a different time scale but are also, for the most part, being transformed by organisms in a different food web, there is little basis for competition.

The same sort of dichotomy may prevail in the marsh soil for *Spartina* itself. Unless some other limit is imposed, a significant production of DOC by the roots may be transformed rapidly by the microorganisms into new biomass. The copious production of new roots and rhizomes each year (Chapter 3), however, is consumed on a time scale equalizing or exceeding that for leaves and stalks. Christian et al. (1978) found much of the root and rhizome material remaining in their plots 12 months after pruning and periodic clipping.

Despite their low plant diversity, salt marshes are seldom, if ever, devastated by natural consumers, as are comparable low-diversity terrestrial ecosystems such as coniferous forests. The marshes of Sapelo Island are a particularly striking example of an untouched natural ecosystem. Thus the natural predator populations have not been reduced and may confer a degree of protection against introduced pests or wide fluctuations in numbers of native consumers. But the major factor protecting salt-marsh vegetation is, no doubt, the tidal inundation. The daily influx of water and the consequent soft ground eliminates most vertebrate grazers, a major group exploiting terrestrial grassland. The influx also forces the large terrestrial arthropods to feed and seek shelter in the upper leaves where they are both more vulnerable to predators and very crowded, conditions inimical to maintenance of high population densities. We conclude that the relative stability of salt-marsh ecosystems is largely due to the physical and chemical attributes of the system. This is true, whether the marshes are single-dominant communities of *S. alterniflora* of the Atlantic coast of North America or of the sedge, *Carex lyngbyei*, of the Pacific Coast of North America, or whether they are more diverse assemblages, such as the composite-dominated salt marshes of the North Sea coasts of Britain and the Netherlands.

10.2. Transport of Materials

The geologist's view of salt marshes as depositional environments (Frey and Basan, 1978) contrasts with the ecologist's view of them as sources of various materials for the estuary and even the ocean (Odum, 1961). In part, this conflict is an artifact of the kinds of materials, the system boundaries, and the time frame. For example, marshes *are* clearly sedimentary traps. They accumulate sediment and transform it into soil during their lifetime. Indeed, it is this accumulation of sediments which permits the establishment of salt marshes in the first place. But the marsh ecosystem also transforms CO_2 into organic matter, insoluble material into soluble, and vice versa. It is the net balance between these processes and the import-export activities that is of interest to the ecologist. That a marsh is accreting at a given moment says nothing, by itself, about whether or not it is an importer or exporter of carbon, of phosphorous, or of specific organic compounds. Similarly with boundaries and time frame, the marsh may export some materials to the creeks but not to the estuary, or to the estuary but not to the sea. Long periods of geological time, during which marshes build up sediment, may be interspersed with shorter periods of ecological consequence, during which they export materials. Nevertheless, there is a residual conflict to be resolved, a conflict that centers on the effect of marshes on the productivity of the estuarine and nearshore area. In other words, in what way and how tightly is the marsh coupled to the sea?

Salt marshes are very productive; the Sapelo Island marshes rank with the most productive communities on the planet. The question ultimately motivating much of the research done on salt marshes during the past 25 years has concerned the fate of this productivity. Put simply, we ask: Where does all the carbon go? Throughout this book we have tried to develop an answer to this question. The problem, discussed in a recent review (Nixon, 1980), is a scarcity of information on what is happening to the algal and vascular plant detritus in the creeks and estuaries, and finding a solution to that problem is difficult. Such studies are time-consuming and expensive, but carbon fluxes have now been measured in a number of marshes and, we now have some good alternative hypotheses to the overly simple and demonstrably incorrect idea that all the excess carbon is carried offshore by tidal flow. Any effort to answer the question posed must consider 1) the carbon budget of a marsh, 2) the probable routes of transformation of this carbon within the marsh, 3) the routes of export, 4) the importance of this export in the overall marsh-estuary-nearshore productivity, and 5) the effect of man-induced perturbations on these processes. It is worth repeating here our warning that there is no single answer to the question about the flow of materials that will be valid for all salt marshes. However, enough comparative information exists, gathered mostly from eastern North America, to suggest some possibilities.

10.2.1 Carbon Budgets

Marshes differ significantly in both the spatial and seasonal patterns and the magnitude and seasonal balance of carbon flows. Differences in seasonal temperatures, in flora and fauna, and in hydrographic features all interact to produce this variety among marshes.

Flax Pond marsh on Long Island is a hydrographic sediment trap because it is a diked area with only one relatively narrow channel communicating with Long Island Sound. Measurements of input-output rates of carbon in various forms had, over the years, fleshed out a view of the marsh as a net importer of organic carbon (Woodwell et al., 1977). There was considerable variation in the input-output data, however, and the total fluxes were large in proportion to the net changes. A recent discussion of the carbon budget in this marsh by Woodwell et al. (1977) reports the marsh losing approximately 2 g C m^{-2} yr^{-1} as CO_2 and 8 g as DOC, while accumulating 61 g as POC, for a net import of carbon amounting to 51 g. However, Woodwell et al. (1979) report measurements on movements of large pieces of *S. alterniflora* which show net losses in this category of 85 to 170 g C m^{-2} yr^{-1}. Although this is not a large export, its significance lies in the fact that this marsh is probably not importing carbon, and in this respect is similar to other marshes. The reason the Flax Pond system does not export small POC is related to its hydrology. Filter-feeding organisms in the "pond" remove large amounts of POC, particularly in summer, releasing ammonium and DOC. The POC entering the basin, if not actively removed, would tend toward sedimentation anyway, while the dissolved compounds can, of course, move out with the water.

Great Sippewissett marsh in Massachusetts exports organic carbon (Valiela et al., 1978; Valiela and Teal, 1979a). However, because of the ground water movement laterally through the soil and the resulting effect of that movement on sulfate reduction, an important fraction of the biomass moving into the creeks is bacterial. This biomass comprises both the sulfate reducers and the aerobic sulfur oxidizers that used the reduced sulfur compounds as an energy source for fixing CO_2 from the water (Howarth and Teal, 1980). The Sapelo Island marshes are unlikely to export sulfate-reducing bacteria and reduced sulfur compounds in the major way that Great Sippewissett Marsh does (Chapter 7). Neither is it clear that such export is the general case for marshes in northeastern North America, because such export requires both a permeable, peaty marsh soil and a significant movement of ground water from the land.

To what extent the Sapelo Island marshes are exporters of particulate organic carbon is not fully resolved. The work of Imberger et al. (in ms.), which covered only a brief time span, suggests that the upper estuary is usually either a sink or neutral with respect to particulate organic matter (Chapters 2 and 9) but is a net exporter of dissolved humic materials, silicate, living biomass, and, probably, refractory particulate organic matter. Only during the infrequent correspondence of heavy rain and low tide is a major export of POC seen. But the marsh itself, as distinct from the estuarine system, produces annually a large excess of organic carbon. The actual form and fate of the excess carbon remains an open question. In Chapters 4, 9 and 10 we have presented evidence and some hypotheses addressing this question. Our current hypothesis, from simulations with MRSH1V6 as described in Chapter 9, is summarized in Table 10.1. The predictions in the table have been checked against the field data. Those on net primary production of *Spartina* agree very well (Chapter 3), but the estimate for algae may be only one-half of the total net production by benthic diatoms and phytoplankton (Section 10.2.2). Degradation measured at the soil surface also agrees closely with the output from the model. However, we do not have very good measurements of total CO_2 production in the water. In particular, the measurements from creek sediments and those of respiration of the aquatic macrofauna may be low, and thus the prediction for water is regarded as very minimal. However, the rather large tidal export due to daily and storm-caused transport poses a problem, for the material, mainly *Spartina* carbon, does not show up in appreciable quantities in the lower estuary or sound (Chapter 4). The possible fate of the approximately 200 g C m^{-2} yr^{-1} is still unaccounted for (Chapter 9). The most likely possibility is that this carbon is transformed microbially in the water on the marsh or in the upper tidal creeks. This ultimately becomes the secondary production of macroorganisms. The resulting living biomass swims or crawls out of the system or is captured by commercial or sport fishermen

Table 10.1. Simulated Annual Carbon Budget for the Duplin River Marshes, Sapelo Island, Georgia

Source or Process	Net Balance[a] ($g\ C\ m^{-2}\ yr^{-1}$)	
Production		
S. alterniflora	1575	
Algae	131	
Total production		1706
Loss		
Respiration (CO_2 + CH_4)		
in soil	- 623	
in water	- 222	
in air	- 68	
Tidal movement	- 586	
Total loss		-1499
Net change		207
Sedimentation	29	
Unexplained[b]	178	

[a] All values prorated over the entire marsh system—soil, creeks, and Duplin River.
[b] See Chapters 4, 9, 10.

within the system. Thus we suggest that the important links between the marsh and the sea are the trophic relationships of the macroconsumers in the water covering the marsh at high tide and residing in the creeks at low tide.

10.2.2 Transformation of Organic Matter

We have emphasized the segregation of food webs in the salt marsh and estuary at Sapelo Island. The direct consumers of grass remove only a small fraction of the production, on the order of 5 to 10% (Chapter 5). The remainder of the leaves and stems of *Spartina* and other marsh plants goes into the detritus food web. The evidence from carbon isotopic ratios (Chapter 4) suggests that most *Spartina* detritus is transformed and assimilated into macroinvertebrate biomass in the marsh proper or in the smaller tidal creeks. On the other hand, production of phytoplankton is now perceived to be more substantial than we had believed. That production, together with the production of benthic diatoms on the intertidal soil and sediment surfaces, is substantial (Table 3.8). If the detritus must be converted to microbial biomass, at an overall efficiency of not more than 50%, then the organic carbon and energy available to macroconsumers from algae approaches more closely that available from detritus. This, too, is a departure from earlier thinking, but it is in agreement wth the distribution of carbon isotopic ratios in the invertebrate populations.

Yet to be considered is the belowground production of roots, rhizomes, and dissolved organic compounds, which yield an amount comparable to the total available from all aboveground sources. This material is produced in a soil which is a mosaic of aerobic and anaerobic zones but is predominantly anaerobic. For the most part, belowground production must enter the anaerobic food web of fermentation, DNO reduction, sulfate reduction, and methanogenesis. It is clear that these are coupled, the production of one depending on end products of another (Chapter 7). What is less clear is the limits to flow of energy through the overall anaerobic food web. The experimental evidence points to substrate limitation, at least in the short *Spartina*, which is the greatest part of the marsh. Methane, the only gaseous end product for which we have data, does in fact represent a small fraction of total production of the system.

Ultimately, carbon incorporated into both aboveground and belowground biomass in the marsh must be incorporated by multicellular organisms. It is here that the impact of the marshes on the offshore areas is probably greatest, for the creeks and marshes provide both food and shelter for these species. They can, in turn, move high-quality carbon out to the sea or, in some cases, directly to the ultimate predator, man.

10.2.3 Export of Biomass

Estuaries are the nursery grounds for many coastal and estuarine vertebrates and invertebrates, including several of commercial importance. In many instances, spawning occurs offshore in coastal waters, and the very young larvae make their way into the estuaries and marsh creeks. The mechanisms by which this is accomplished by organisms too small to swim effectively can be surmised but have not been confirmed. The larvae and juveniles remain in the salt marshes and creeks, usually during the summer, to feed and find refuge from predation. Other advantages of the nursery-ground life history include the availability of vacant niches, higher water temperatures in summer, and reduced salinity. Low salinity, which is preferred by younger fishes but not by older ones, apparently accounts for the vacant niches and the reduction in abundance of predators in estuaries (Tabb, 1966). Studies documenting the nursery-ground role of salt marshes and estuaries are reviewed by Gunter (1967) and Setzler (1977). According to Gunter, over 97% of the total commercial catch of the states around the Gulf of Mexico is in some way dependent upon the estuaries. Setzler's quantitative survey of ichthyoplankton in the Duplin River and Doboy Sound verified the use of the estuary by postlarval and juvenile fishes of the families Scianidae and Engraulidae.

Estuaries and associated salt marshes provide habitat not only for larval and juvenile fishes and shellfish, but also for organisms that spend most or all of their lives there. On the Georgia coast these include the commercial shrimp, *Penaeus setiferus* and *P. aztecus*, the blue crab, *Callinectes sapidus*, and the oyster, *Crassostrea virginica.* Oysters and blue crabs spend all of their non-planktonic existence in estuaries, except for the female blue crabs, which move offshore to spawn (Williams, 1965). White shrimp (*P. setiferus*) are offshore spawners which enter the estuary as juveniles, < 7 mm in size, and grow to > 50 mm in the creeks and salt marshes before moving to the estuaries and offshore (Anderson, 1970). *P. aztecus* is probably also an offshore spawner, but it uses the estuary as a feeding ground for much of its life history. There is a small but significant local fishery for these species in Georgia (Table 10.2).

The one active commercial fishery in the Duplin River itself is for the blue crab. One fisherman maintains a line of crab pots in the river and tends them daily through

Table 10.2. Georgia Landings of Fishes and Shellfish in 1974[a]

Catch	Metric Tons
Fishes	331
Blue crab	4605
Shrimp	3287
Oysters	30
Total	8253

[a]Reported by the National Marine Fisheries Service.

much of the year. He reports that the catch is large, and it appears to be a steady fishery, with less variation from year to year than many others, such as the shrimp fishery, which is near the northern limit of the range of the species and is therefore sensitive to severe winters. Sports fisheries also occur in the Duplin River. Of course, the economic benefits of the sport fishery are even more remote from the marsh than are those of the crab fishery, since they are realized primarily in the sale of boating and fishing equipment and in the rental of lodging for vacationers.

So the two opposing perceptions of the salt marsh, as wasteland to be exploited and as a productive area supporting fisheries within and beyond its boundaries, now vie for supporters. Owners of marshland–much of it is indeed in private ownership, in spite of its intertidal character–are not in a position to benefit directly from its productivity, since the mature stocks of fishes and shellfish are usually geographically far removed from the marsh nursery. Exploitation of a fishery will be accomplished by others. Therefore, despite the intrinsic value of marshlands, their highest value to owners continues to be as real estate, subject to "improvement." Our present view of the food web of the marsh and estuary suggests that the preservation of fisheries depends as much upon protection of the smaller tidal creeks as upon protection of the marsh and its *Spartina* production. Since most land-development procedures that would modify the marsh would fill in the smaller creeks, the two are intrinsically connected.

We think another, and perhaps more serious threat, to the marsh-supported fisheries is pollution, which can come from the land, the river, or the sea, and can take many forms. For example, many marshes are sprayed to control mosquitoes. Many insecticides can be as effective in reducing populations of marsh invertebrates, such as fiddler crabs, as they are in reducing populations of mosquitoes. Marshes can also be affected by oil spills. But by far the most significant pollutants are those brought to the estuaries and marshes by rivers.

10.3. Assimilation of Wastes

In addition to receiving many other materials from sources upstream, estuaries have traditionally been used as receiving waters for industrial and municipal wastes. The ability of an estuary to remain biologically viable is, in part, a function of its flushing rate (Ketchum, 1951), a rate determined by the morphology of the estuary and by the tidal regime (Stommel and Farmer, 1952). The types of insults from pollution which may affect estuaries and their adjoining marshes are toxins, both heavy metals and organic compounds, biodegradable organic materials that produce excess biological oxygen demand, and plant nutrients which cause eutrophication. If estuaries or marshes are used as the receiving waters for primary-treated sewage, they potentially receive all three of these pollutants. Effluent which has received secondary treatment will also contain all three to some extent.

Heavy metals from both rivers and the ocean move through estuaries and salt marshes naturally, but their flux may be accentuated by anthropogenic inputs, such as wastes from industries and from automotive exhausts. Metal ions are adsorbed by clays, although around 80% of the input moves through estuaries without being involved in clay sorption reactions (Dunstan et al., 1975). Clays are potentially an effective detoxifying mechanism of silty estuaries, such as those in southeastern North America. However, a fraction of the sorbed metals is subsequently incorporated into *Spartina* and mobilized into the detritus food web when the plants die. Therefore, an estuary which has become heavily polluted with toxic metals will continue to regenerate metals into the food web from the sediments for many years after the source of pollution has been terminated. At a concentration one or more orders of magnitude above natural concentrations, some metals will be toxic to *Spartina* (Dunstan et al., 1975), with probable adverse consequences for the stability of the ecosystem. Mercury is, to some extent, a special case, both because it naturally accumulates to rather high concentrations in predatory marine fishes and mammals and because it occurs in both inorganic and methylated forms in natural waters. Anthropogenic mercury finds its way into a biogeochemical pathway which leads readily into sport and commercial fisheries.

Humates chelate metal ions and are an effective detoxifying mechanism in all natural waters (Wood, 1980). Therefore, analyses of total concentration of metals by methods such as atomic absorption spectrometry not only do not tell us the potential toxicity of the water, but they do not even tell us the amount of metals available for exchange with clays and incorporation into food webs. Analytical observations of increased total concentration of any heavy metal in estuarine water would be cause for concern, but such data could not be used to predict with any precision the immediate effects on survival or food web uptake of that metal. A more meaningful analysis must involve a measure of the binding capacity of the natural assemblage of humates for the metals in question (Wood, 1980). Analysis of metal-binding capacity in natural waters is still in its infancy.

Biodegradable organic materials, such as sewage and some mill wastes, pose a more direct and immediate threat to estuaries and marshes. In the highly productive southeastern marshes, there is already a high rate of production of organic materials which are being oxidized by aerobic microorganisms in the water and surface sediments. Under normal conditions, the concentration of dissolved oxygen is below saturation in the tidal rivers and creeks (Frankenberg, 1975), and on warm summer nights these waterways may become totally oxygen depleted for several hours, depending on the phase of the tide. Many marsh and estuarine organisms have mechanisms for coping with brief interludes of oxygen deficit. Mullet jump into the air; the grass shrimp, *Palaemonetes*, climbs out of water into *Spartina*; mollusks close their shells and respire anaerobically. These defenses are effective only for a short time at summer temperatures, so a prolongation of the oxygen deficit for as little as a day will result in a fish kill. Therefore, relatively small increments of biological oxygen demand may have serious consequences.

Another source of oxygen demand is the sediments. Although there is a natural and rather substantial oxygen demand by the sediment surface (Chapter 6), disturbance of the sediments and exposure of their anaerobic portion will release a large, probably chemical, oxygen demand (Frankenberg and Westerfield, 1968). This occurs to a limited extent under natural conditions, but it can be a major problem associated with dredging for channel maintenance, land filling, or strip mining.

Eutrophication of estuaries by the anthropogenic introduction of excess plant nutrients can occur. However, the most sensitive estuaries are those which lack clay sediments and are not naturally rich in plant nutrients (Pomeroy et al., 1972). The estuaries of southeastern North America, such as the Duplin River, are naturally eutrophic, with high rates of phytoplankton production (Chapter 3). They have large amounts of clays to adsorb excess phosphate, a nutrient which is never limiting in any case. They have populations of denitrifying bacteria which would, in the long term, tend to reduce any excess input of available nitrogen. However, the immediate effect of a nitrogen input would probably be an increase in photosynthesis of phytoplankton, benthic algae, and even short *Spartina* (Chapter 3). This would ultimately increase biological oxygen demand, and it might result in extended anoxic conditions in tidal creeks in summer. In potentially polluted urban estuaries, where fish kills are a regular event, it is difficult to identify the cause of the kills in most instances, because conditions change so rapidly that one cannot examine the water containing the dead fish and determine what killed them. In 25 years of research at Sapelo Island we have never seen a fish kill in the estuary. Therefore, although the system appears to be rather delicately poised with respect to dissolved oxygen and biological oxygen demand, it is resistant to mortality from brief, naturally occurring episodes of oxygen depletion. Experience in many other estuaries suggests, however, that the introduction of either additional available nitrogen or additional degradable organic matter might be deleterious to the fauna of the tidal creeks.

10.4. Aesthetics

Salt marshes have value as landscapes, they protect and feed animals, and they may be important in other, as yet undiscovered, ways. But, above all, salt marshes should persist simply because they are there, an intrinsic piece of the nature mankind is only slowly learning to appreciate and to cherish. Most people prefer to enjoy the beauty of the vast landscapes at some distance. Few visitors to canyons walk through them, enduring the heat and thirst and dust. Even fewer climb to the tops of mountains. Yet for those who do, the hardships seem to enhance rather than diminish their appreciation of the whole. We would like to think that Sidney Lanier, who thought so highly of the marshes of Glynn, was involved, as we have been, in a sufficiently immediate way to experience biting flies and gnats, deep, soft mud, and sharp oyster shells. For having experienced these discomforts, one is sure to have seen the myriad fascinating scenes in the marsh—the grasshoppers diving under the water to escape, the secretive rail hunting for food, and the surface of the mud turning golden brown with the upward movement of diatoms at ebb tide. The interactions of organisms and populations, interactions which can be perceived only at close range, are the heart of the ecosystem, and, to those of us who study them, these interactions are among the most rewarding of the aesthetic experiences the salt marsh can offer.

References

Abdollahi, H., Nedwell, D. B. (1976) Seasonal temperature as a factor influencing bacterial sulfate reduction in a salt marsh sediment. Microb. Ecol. *5*, 73-79

Abram, J. W., Nedwell, D. B. (1978) Hydrogen as a substrate for methanogenesis and sulfate reduction in anaerobic salt marsh sediments. Arch Microbiol. *117*, 93-97

Abram, J. W., Nedwell, D. B. (1978a) Inhibition of methanogenesis by sulfate reducing bacteria competing for transferred hydrogen. Arch. Microbiol. *117*, 89-92

Admiraal, W. (1977) Salinity tolerance of benthic estuarine diatoms as tested with a rapid polarographic measurement of photosynthesis. Mar. Biol. *39*, 11-18

Anderson, C. E. (1974) A review of structure in several North Carolina salt marsh plants. In: Ecology of Halophytes. Reimold, R. J., Queen, W. H. (eds.). New York: Academic Press, 307-344

Anderson, W. W. (1970) Contribution to the life histories of several penaeid shrimps (Penaeidae) along the south Atlantic coast of the United States. U. S. Fish and Wildlife Service, Spec. Sci. Rep. *605*

Andrews, R., Coleman, D. C., Ellis, J. E., Singh, J. S. (1974) Energy flow relationships in a shortgrass prairie ecosystem. In: Proc. 1st Int. Congr. Ecology. Wageningen: Centre for Agricultural Publishing and Documentation, 22-28

Andrezejewska, L. (1967) Estimation of the effects of feeding of the sucking insect, *Cicadella viridis* L. (Homoptera, Auchenorrhyncha) on plants. In: Secondary Productivity of Terrestrial Ecosystems. Petrusewicz, K. (ed.). Krakow: Polish Acad. Sci., 791-805

Andrezejewska, L., Wojcik, Z. (1970) The influence of Acridoidea on the primary production of a meadow (field experiment). Ekologia Polska *18*, 89-109

Antlfinger, A. E. (1976) Photosynthetic and water-use strategies of three salt marsh succulents. M.S. Thesis, Univ. Georgia, Athens

Antlfinger, A. E., Dunn, E. L. (1979): Seasonal patterns of CO_2 and water vapor exchange of three salt-marsh succulents. Oecologia 43, 249-260

Arndt, C. H. (1914) Some insects of the between tides zone. Proc. Indiana Acad. Sci., 323-336

Ashby, W. R. (1956) An introduction to Cybernetics. London: Chapman and Hall

Atkinson, L. P., Hall, J. R. (1976) Methane production and distribution in a Georgia salt marsh. Estuarine Coastal Mar. Sci. *4*, 677-686

Atkinson, L. P., Richards, F. A. (1967) The occurrence and distribution of methane in the marine environment. Deep-Sea Res. *14*, 673-684

Aurand, D., Daiber, F. C. (1973) Nitrate and nitrite in the surface waters of two Delaware salt marshes. Chesapeake Sci. *14*, 105-111

Ausmus, B. S. (1973) The use of the ATP assay in terrestrial decomposition studies. Bull. Ecol. Res. Comm (Stockholm) *17*, 223-234

Axelrad, D. M., Moore, K. A., Bender, M. E. (1976) Nitrogen, phosphorus, and carbon flux in Chesapeake Bay marshes. Virginia Polytech Inst.-Virginia Water Resources Res. Center Bull. *70*

Azam, F., Hodson, R. E. (1977) Dissolved ATP in the sea and its utilization by marine bacteria. Nature *267*, 696-698

Baas-Becking, L. G. M., Wood, E. J. F. (1955) Biological processes in the estuarine environment. I. Ecology of the Sulfur Cycle. Kon. Acad. Wetenshap. (Amsterdam) Proc. in Physical Sci., Ser. B. 58, 160-181

Bahr, L. M. (1974) Aspects of the structure and function of the intertidal oyster reef community in Georgia. Ph.D. Thesis, Univ. Georgia, Athens

Bahr, L. M. (1976) Energetic aspects of the intertidal oyster reef community at Sapelo Island, Georgia. Ecology *57*, 121-131

Baker-Blocker, A., Donahue, T. M., Mancy, K. H. (1977) Methane flux from wetlands. Tellus *29*, 245-250

Baker-Dittus, A. M. (1978) Foraging patterns of three sympatric killifish. Copeia 1978, 383-389

Balch, W. E., Fox, G. E., Magrum, L. J., Woese, C. R., Wolfe, R. S. (1979) Methanogens: Reevaluation of a unique biological group. Microbiol Rev. *43*, 260-296

Balderston, W. L., Payne, W. J. (1976) Inhibition of methanogenesis in salt marsh sediments and whole suspensions of methanogenic bacteria by nitrogenous oxides. Appl. Environ. Microbiol. *32*, 254-260

Bancroft, K., Paul, E. A., Wiebe, W. J. (1976) The extraction and measurement of adenosine triphosphate from marine sediments. Limnol. Oceanogr. *21*, 473-480

Barnes, R. D. (1953) The ecological distribution of spiders in non-forest maritime communities at Beaufort, North Carolina. Ecol. Monogr. *23*, 315-337

Barnwell, F. H. (1966) Daily and tidal patterns of activity in individual fiddler crabs (Genus *Uca*) from the Woods Hole region. Biol. Bull. *130*, 1-17

Basan, P. B., Frey, R. W. (1977) Actual-palaeontology and neoichnology of salt marshes near Sapelo Island, Georgia. In: Trace Fossils 2. Crimes, T. P., Harper, J. C. (eds.). Geol. J. Special Issue *9*, 41-70

Belay, A., Fogg, G. E. (1978) Photoinhibition of photosynthesis in *Asterionella formosa* (Bacillariophyceae). J. Phycol. *14*, 341-347

Bent, A. C. (1968) Life histories of North American Cardinals, Grosbeaks, Buntings, Towhees, Finches, Sparrows, and Allies, Part 2. New York: Dover, pp.

Berk, S. G., Brownlee, D. E., Heinle, D. R., Kling, H. J., Colwell, R. R. (1977) Ciliates as a food source for marine planktonic copepods. Microb. Ecol. *4*, 27-40

Blum, U., Seneca, E. D., Stroud, L. M. (1978) Photosynthesis and respiration of *Spartina* and *Juncus* salt marshes in North Carolina: some models. Estuaries *1*, 228-238

Bollag, J. M., Czlenkowski, S. T. (1973) Inhibition of methane formation in soil by various nitrogen containing compounds. Soil. Biol. Biochem. *5*, 673-678

Borror, D. J., DeLong, D. M. (1964) An Introduction to the Study of Insects. New York: Holt, Rinehart and Winson

Bowling, C. C. (1972) Notes on the biology of the rice water weevil, *Lissorhoptrus oryzophilus.* Ann. Entomol. Soc. Amer. *65*, 990-991

Brenner, D., Valiela, I., Van Raalte, C. (1976) Grazing by *Talorchestia longicornis* on an algal mat of a New England salt marsh. J. Exp. Mar. Biol. Ecol. *22*, 161-169

Briggs, K. B., Tenore, K. R., Hanson, R. B. (1979) The role of microfauna in detrital utilization by the polychaete, *Nereis succinea* (Frey and Leuckart). J. Exp. Mar. Biol. Ecol. *36*, 225-234

Brinkhuis, B. H. (1976) The ecology of temperate salt-marsh fucoids. I. Occurrence and distribution of *Ascophyllum nodosum* Ecads. Mar. Biol. *34*, 325-338

Brinkhurst, R. O. (1963) Observations on wing polymorphism in the Heteroptera. Proc. Roy. Ent. Soc. Lond. A *38*, 15-22

Broome, S. W., Woodhouse, W. W., Jr., Seneca, E. D. (1975) The relationship of mineral nutrients to growth of *Spartina alterniflora* in North Carolina. II. The effects of N, P, and Fe fertilizers. Soil. Sci. Soc. Amer. Proc. *39*, 301-307

Brown, D. H., Gibby, C. E., Hickman, M. (1972) Photosynthetic rhythms in epipelic algal populations. Brit. Phycol. J. *7*, 37-44

Bryant, M. P., Campbell, L. L., Reddy, C. A., Crabill, M. R. (1977) Growth of *Desulfovibrio* in lactate or ethanol media low in sulfate in association with H_2-utilizing methanogenic bacteria. Appl. Environ. Microbiol. *33*, 1162-1169

Buechler, D. G., Dillon, R. D. (1974) Phosphorus regeneration in fresh-water paramecia. J. Protozool. *21*, 331-343

Bunker, S. M. (1979) Retention of various components of *Spartina alterniflora* detritus by the striped mullet, *Mugil cephalus.* M. S. Thesis, Univ. Georgia, Athens

Buresh, R. J., Patrick, W. H., Jr. (1978) Nitrate reduction to ammonium in anaerobic soil. Soil. Sci. Soc. Amer. J. *42*, 913-918

Burkholder, P. R. (1956) Studies on the nutritive value of *Spartina* grass growing in the marsh areas of coastal Georgia. Bull. Torrey Bot. Club *83*, 327-334

Burkholder, P. R., Bornside, G. H. (1956) Decomposition of marsh grass by aerobic bacteria. Bull. Torrey Bot. Club *84*, 366-383

Byrne, D. M. (1978) Life history of the spotfin killifish, *Fundulus luciae* (Pisces: Cyprinodontidae) in Fox Creek Marsh, Virginia. Estuaries *1*, 211-227

Cameron, G. N. (1972) Analysis of insect trophic diversity in two salt marsh communities. Ecology *53*, 58-73

Cappenberg, T. E. (1974) Interrelation between sulfate-reducing and methane-producing bacteria in bottom deposits of a fresh water lake. I. Field observations. II. Inhibition experiments. Antonie van Leewenhoek J. Microbiol. Serol. *40*, 285-306

Cappenberg, Th. E., Prins, R. A. (1974) Interrelations between sulfate-reducing and methane-producing bacteria in bottom deposits of a freshwater lake. III. Experiments with ^{14}C-labeled substrates. Antonie van Leeuwenhoek J. Microbiol. Serol. *40*, 457-469

Carriker, M. R. (1967) Ecology of estuarine benthic invertebrates. In: Estuaries. Lauff, G. H. (ed.). Washington: AAAS Publ. *83*, 442-487

Carritt, D. E., Goodgal, S. (1954) Sorption reactions and some ecological implications. Deep-Sea Res. *1*, 224-243

Carter, W. (1973) Insects in Relation to Plant Disease. New York: Wiley

Caswell, H. (1976) The validation problem. In: Systems Analysis and Simulation. Patten, B. C. (ed.). Ecology *4*, 313-325

Caswell, H., Reed, F., Stephenson, S. N., Werner, P. A. (1973) Photosynthetic pathways and selective herbivory: a hypothesis. Amer. Nat. *107*, 465-480

Cavari, B. Z., Phelps, G. (1977) Denitrification in Lake Kinneret in the presence of oxygen. Freshwat. Biol. *7*, 385-391

Chalmers, A. G. (1977) Pools of nitrogen in a Georgia salt marsh. Ph.D. Thesis, Univ. Georgia, Athens

Chalmers, A. G. (1979) The effects of fertilization on nitrogen distribution in a *Spartina alterniflora* salt marsh. Estuarine Coastal Mar. Sci. *8*, 327-337

Chalmers, A. G., Haines, E. B., Sherr, B. F. (1976) Capacity of a *Spartina* marsh to assimilate nitrogen from secondarily treated sewage. Tech. Completion Rep., USDI/OWRT Project A-057-Ga.

Chapman, F. A., Graham, D. (1974) The effect of light on the tricarboxylic acid cycle in green leaves. I. Relative rates of the cycle in dark and the light. Plant. Physiol. *53*, 879-885

Chapman, R. L. (1971) The macroscopic marine algae of Sapelo Island and other sites on the Georgia coast. Bull. Georgia Acad. Sci. *29*, 77-89

Chapman, V. J. (ed.) (1977) Wet Coastal Ecosystems. Amsterdam: Elsevier, 428 pp.

Chew, R. M. (1974) Consumers as regulators of ecosystems: an alternative to energetics. Ohio J. Sci. *74*, 359-370

Chollet, R., Ogren, W. L. (1975) Regulation of photorespiration in C_3 and C_4 species. Bot. Rev. *41*, 137-179

Christian, R. R. (1976) Regulation of a salt marsh soil microbial community: a field experimental approach. Ph.D. Thesis, Univ. Georgia, Athens

Christian, R. R., Hall, J. R. (1977) Experimental trends in sediment microbial heterotrophy: radioisotopic techniques and analyses. In: Ecology of Marine Benthos. Coull, B. C. (ed.). Columbia: Univ. S. Carolina, 67-88

Christian, R. R., Hansen J. A. (1980) Effects of tidal subsidy and soil drainage on anaerobic, microbial heterotrophic potential in a Georgia salt marsh. Abst. Ann. Meeting, ASM

Christian, R. R., Wetzel, R. L. (1978) Interactions between substrate microbes and consumers of *Spartina* "detritus" in estuaries. In: Estuarine Interactions. Wiley, M. (ed.). New York, Academic Press, 93-114

Christian, R. R., Wiebe, W. J. (1978) Anaerobic microbial community retabolism in *Spartina alterniflora* soils. Limnol. Oceanogr. *23*, 328-336

Christian, R. R., Wiebe, W. J. (1979) Three experimental regimes in the study of sediment microbial ecology. In: Methodology for Biomass Determinations and Microbial Activities in Sediments. Litchfield, C. D., Seyfried, P. L. (eds.). Amer. Soc. Testing Materials STP673, 148-155

Christian, R. R., Bancroft, K., Wiebe, W. J. (1975) Distribution of adenosine triphosphate in salt marsh sediments at Sapelo Island, Georgia. Soil. Sci. *119*, 89-97

Christian, R. R., Bancroft, K., Wiebe, W. J. (1978) Resistance of the microbial community within salt marsh soils to selected perturbations. Ecology *59*, 1200-1210

Claypool, G. E., Kaplan, I. R. (1974) The origin and distribution of methane in marine sediments. In: Natural Gases in Marine Sediments. Kaplan, I. R. (ed.). New York: Plenum, 99-139

Conger, P. S. (1943) Ebullition of gases from marsh and lake waters. Chesapeake Biol. Lab. Pub. 59.

Connell, J. H. (1961) The influence of interspecific competition and other factors on the distribution of the barnacle, *Chthamalus stillatus.* Ecology *42*, 710-723

Coulson, J. C., Whittaker, J. B. (1978) Ecology of moorland animals. In: Production Ecology of British Moors and Montane Grasslands. Heal, O. W., Perkins, D. F. (eds.). New York: Springer-Verlag, 52-93

Coupland, R. T., Van Dyne, G. M. (1979) Systems synthesis. In: Grassland Ecosystems of the World: Analysis of Grasslands and Their Uses. Coupland, R. T. (ed.). Cambridge: Cambridge Univ. Press, 97-106

Crane, J. (1975) Fiddler Crabs of the World, Ocypodidae: Genus *Uca.* Princeton: Princeton Univ.

Crawford, C. C., Hobbie, J. E., Webb, K. L. (1974) The utilization of dissolved free amino acids by estuarine microorganisms. Ecology *55*, 551-563

Cushing, D. H. (1975) Marine Ecology and Fisheries. Cambridge: Cambridge Univ.

Dahlberg, M. D. (1972) An ecological study of Georgia coastal fishes. Fishery Bull. *70*, 323-353

Dahlberg, M. D. (1975) Fishes of Georgia and Nearby States. Athens: Univ. Georgia.

Dahlberg, M. D., Heard, R. W., III (1969) Observations on Elasmobranchs from Georgia. Quart. J. Florida Acad. Sci. *32*, 21-25

Dahlberg, M. D., Odum, E. P. (1970) Annual cycles of Species occurrence, abundance, and diversity in Georgia estuarine fish populations. Amer. Midl. Nat. *83*, 382-392

Darley, W. M., Dunn, E. L., Holmes, K. S., Larew, H. G., III (1976) A ^{14}C method for measuring microalgal productivity in air. J. Exp. Mar. Biol. Ecol. *25*, 207-217

Darley, W. M., Montague, C. L., Plumley, F. G., Sage, W. W., Psalidas, A. T. (1981). Factors limiting edaphic algal biomass and productivity in a Georgia salt marsh. J. Phycol. (in press)

Darley, W. M., Ohlman, C. T., Wimpee, B. B. (1979) Utilization of dissolved organic carbon by natural populations of epibenthic salt marsh diatoms. J. Phycol. *15*, 1-5

Darnell, R. M. (1961) Trophic spectrum of an estuarine community based on studies of Lake Ponchartrain, Louisiana. Ecology *42*, 553-568

Darnell, R. M. (1967) Organic detritus in relation to the estuarine ecosystem. In: Estuaries. Lauff, G. H. (ed.). Washington: AAAS Publ. 83, 376-382

Davis, J. B., Yarbrough, H. F. (1966) Anaerobic oxidation of hydrocarbons by *Desulfovibrio desulfuricans*. Chem. Geol. *1*, 137-144

Davis, L. V. (1978) Class Insecta. In: An Annotated Checklist of the Biota of the Coastal Zone of South Carolina. Zingmark, R. G. (ed.). Columbia: Univ. S. Carolina, 186-220

Davis, L. V. Gray, I. E. (1966) Zonal and seasonal distribution of insects in North Carolina salt marshes. Ecol. Monogr. *36*, 275-295

Day, J. W., Hopkinson, C. S., Loesch, H. C. (1977) Modeling man and nature in southern Louisiana. In: Ecosystem Modeling in Theory and Practice. Day, J. W., Hall, C. A. S. (eds.). New York: Wiley Interscience, 381-392

de la Cruz, A. A., Hackney, C. T. (1977) Energy value, elemental composition, and productivity of belowground biomass of a *Juncus* tidal marsh. Ecology *58*, 1165-1170

Delaune, R. D., Patrick, W. H., Jr., Buresh, R. J. (1978) Sedimentation rates determined by ^{137}Cs dating in a rapidly accreting salt marsh. Nature *275*, 532-533

DeLong, D. M. (1965) Ecological aspects of North American leafhoppers and their role in agriculture. Bull. Ent. Soc. Amer. *11*, 9-20

De Niro, M. J., Epstein, S. (1978) Influence of diet on the distribution of carbon isotopes in animals. Geochim. Cosmochim. Acta. *42*, 495-506

Denno, R. F. (1976) Ecological significance of wing-polymorphism in Fulgoridae which inhabit tidal salt marshes. Ecol. Entomol. *1*, 257-266

Denno, R. F. (1977) Comparison of the assemblages of sap-feeding insects (Homoptera-Hemiptera) inhabiting two structurally different salt marsh grasses in the genus *Spartina*. Environ. Entomol. *6*, 359-372

Denno, R. F. (1978) The optimum population strategy for plant hoppers (Homoptera: Delphacidae) in stable marsh habitats. Can. Entomol. *110*, 135-142

Denno, R. F., Grissell, E. E. (1979) The adaptiveness of wing-dimorphism in the salt marsh-inhabiting plant hopper, *Prokelisia marginata* (Homoptera: Delphacidae). Ecology *60*, 221-236

Dixon, A. F. G. (1971) The role of aphids in wood formation. I. The effect of the sycamore aphid, *Drepanosiphum platanoides* (Schr.) (Aphididae), on the growth of sycamore, *Acer pseudoplanatus* (L.). J. Appl. Ecol. *8*, 165-179

Dodson, K. S. (1961) Determination of inorganic sulfate in studies on the enzymic and non-enzymic hydrolysis of carbohydrate and other sulfate esters. Biochem. J. *78*, 312-314

Duffey, E. (1962) A population study of spiders in limestone grassland. J. Animal Ecol. *31*, 571-599

Dunstan, W. M., Atkinson, L. P. (1976) Sources of new nitrogen for the South Atlantic bight. In: Estuarine Processes. Wiley, M. (ed.). New York: Academic Press Vol. 1, 69-78

Dunstan, W. M., Windom, H. L., McIntire, G. L. (1975) The role of *Spartina alterniflora* in the flow of lead, cadmium, and copper through the salt-marsh ecosystem. In: Mineral Cycling in Southeastern Ecosystems. Howell, G. F., Gentry, J. B., Smith, M. H. (eds.). USERDA Symp. Ser. Conf. 740513, 250-256

Eaton, J. W., Moss, B. (1966) The estimation of numbers and pigment content in epipelic algal populations. Limnol. Oceanogr. *11*, 584-595

Edwards, J. M., Frey, R. W. (1977) Substrate characteristics within a Holocene salt marsh, Sapelo Island, Georgia. Senckenbergiana Maritima *9*, 215-259

Emlen, J. M. (1973) Ecology: An Evolutionary Approach. Reading: Addison-Wesley

Engler, R. M., Patrick, W. H. Jr., (1974) Nitrate removal from floodwater overlying flooded soils and sediments. J. Environ. Qual. *3*, 409-413

Enhalt, D. H. (1974) The atmospheric cycle of methane. Tellus *26*, 58-70

Estrada, M., Valiela, I., Teal, J. M. (1974) Concentration and distribution of chlorophyll in fertilized plots in a Massachusetts salt marsh. J. Exp. Mar. Biol. Ecol. *14*, 47-56

Evans, F. C., Murdoch, W. W. (1968) Taxonomic composition, trophic structure and seasonal occurrence in a grassland insect community. J. Animal Ecol. *37*, 259-273

Fenchel, T. (1967) The ecology of marine microbenthos. I. The quantitative importance of ciliates as compared with metazoans in various types of sediments. Ophelia *4*, 121-137

Fenchel, T. (1969) The ecology of marine microbenthos. IV. Structure and function of the benthic ecosystem, its chemical and physical factors and the microfauna communities with special reference to the ciliated Protozoa. Ophelia *6*, 1-182

Fenchel, T., Kofoed, L. H. (1976) Evidence for exploitative intraspecific competition in mud snails (Hydrobiidae) Oikos *27*, 367-376

Ferguson, R. L., Murdoch, M. B. (1975) Microbial ATP and organic carbon in sediment of the Newport River estuary, North Carolina. In: Estuarine Research. Cronin, L. E. (ed.). New York: Academic Press, 229-250

Fischer, H. B., List, E. J., Koh, R. C. Y., Imberger, J., Brooks, N. H. (1979) Mixing in Inland and Coastal Waters. New York: Academic Press

Forrester, J. W. (1961) Industrial Dynamics. Cambridge: MIT Press, 81-85

Foster, W. A., Treherne, J. E. (1976) Insects of marine salt marshes: problems and adaptations. In: Marine Insects. Cheng, L. (ed.). New York: American Elsevier, 5-42

Foster, W. A., Treherne, J. E. (1978) Dispersal mechanisms in an intertidal aphid. J. Animal Ecol. *47*, 205-217

Fox, G. E., Magrum, L. J., Balch, W. E., Wolfe, R. S., Woese, C. R. (1977) Classification of methanogenic bacteria by 16s ribosomal RNA characterization. Proc. Nat'l. Acad. Sci. USA. *74*, 4537-4541

Frankenberg, D. (1975) Oxygen in a tidal river: low tide concentration correlates linearly with location. Estuarine Coastal Mar. Sci. *4*, 455-460

Frankenberg, D., Burbanck, W. D. (1963) A comparison of the physiology and ecology of the estuarine isopod, *Cyathura polita* in Massachusetts and Georgia. Biol. Bull. *125*, 81-95

Frankenberg, D., Smith, K. L. (1967) Coprophagy in marine animals. Limnol. Oceanogr. *12*, 443-450

Frankenberg, D., Westerfield, C. W., Jr. (1968) Oxygen demand and oxygen depletion capacity of sediments from Wassaw Sound, Georgia. Bull. Georgia Acad. Sci. *26*, 160-172

Freney, J. R., Denmead, O. T., Simpson, J. R. (1978) Soil as a source or sink for atmospheric nitrous oxide. Nature *273*, 530-532

Frey, R. W., Basan, P. B. (1978) Coastal salt marshes. In: Coastal Sedimentary Environments. Davis, R. A. (ed.). New York: Springer-Verlag, 101-169

Gaastra, P. (1959) Photosynthesis of crop plants as influenced by light, carbon dioxide, temperature, and stomatal diffusion resistance. Medelingen Landbou. Wagenigen *59*, 1-68

Gallagher, J. L. (1974) Sampling macro-organic matter profiles in salt marsh plant root zones. Soil. Sci. Amer. Proc. *38*, 154-155

Gallagher, J. L. (1975) Effect of an ammonium nitrate pulse on the growth and elemental composition of natural stands of *Spartina alterniflora* and *Juncus roemerianus*. Amer. J. Bot. *62*, 644-648

Gallagher, J. L. (1975a) The significance of the surface film in salt marsh plankton metabolism. Limnol. Oceanogr. *20*, 120-123

Gallagher, J. L. (1979) Growth and element compositional responses of *Sporobolus virginicus* (L.) Kunth to substrate salinity and nitrogen. Amer. Midl. Nat. *102*, 68-75

Gallagher, J. L., Daiber, F. C. (1973) Diel rhythms in edaphic community metabolism in a Delaware salt marsh. Ecology *54*, 1160-1163

Gallagher, J. L., Daiber, F. C. (1974) Primary production of edaphic communities in a Delaware salt marsh. Limnol. Oceanogr. *19*, 390-395

Gallagher, J. L., Pfeiffer, W. J. (1977) Aquatic metabolism of the standing dead plant communities in salt and brackish water marshes. Limnol. Oceanogr. *22*, 562-564

Gallagher, J. L., Plumley, F. G. (1979) Underground biomass profiles and productivity in Atlantic coastal marshes. Amer. J. Bot. *66*, 156-161

Gallagher, J. L., Pfeiffer, W. J., Pomeroy, L. R. (1976) Leaching and microbial utilization of dissolved organic matter from leaves of *Spartina alterniflora*. Esturarine Coastal Mar, Sci, *4*, 467-471

Gallagher, J. L., Plumley, F. G., Wolf, P. L. (1977) Underground biomass dynamics and substrate selective properties of the Atlantic coastal salt marsh plants. Tech. Rept. D-77-28 for Chief of Engineers, U. S. Army. U. S. Army Waterways Exper. Sta., Vicksburg, Mississippi.

Gallagher, J. L., Reimold, R. J., Linthurst, R. A., Pfeiffer, W. J. (1980) Aerial production, mortality, and mineral accumulation-export dynamics in *Spartina alterniflora* and *Juncus roemerianus* plant stands. Ecology *61*, 303-312

Gallagher, J. L., Reimold, R. J., Thompson, D. E. (1972) Remote sensing and salt marsh productivity. Proc. 38th Ann. Meeting Amer. Soc. Photogrammetry, 338-348

Gallagher, J. L., Reimold, R. J., Thompson, D. E. (1973) A comparison of four remote sensing media for assessing salt marsh primary productivity. In: Proc 8th Internat. Symp. Remote Sensing of Environment. Cook, J. (ed.), 1287-1295

Gamble, T. N., Betlach, M. R., Tiedje, J. M. (1977) Numerically dominant denitrifying bacteria from world soils. Appl. Environ. Microbiol. *33*, 926-939

Gardner, L. R. (1975) Runoff from an intertidal marsh during tidal exposure: regression curves and chemical characteristics. Limnol. Oceanogr. *20*, 81-89

Gardner, W. S., Hanson, R. B. (1979) Dissolved free amino acids in interstitial waters of Georgia salt marsh soils. Estuaries *2*, 113-118

Gessner, R. V., Goos, R. D. (1973) Fungi from decomposing *Spartina alterniflora*. Can. J. Bot. *51*, 51-55

Gessner, R. V., Goos, R. D., Sieburth, J. McN. (1979) The fungal microcosm of the internodes of *Spartina alterniflora*. Mar. Biol. *16*, 269-273

Gieskes, W. W. G., Kraay, G. W., Baars, M. A. (1979) Current ^{14}C methods for measuring primary production: gross underestimate in oceanic waters. Neth. J. Sea Res. *13*, 58-78

Giurgevich, J. R. (1977) Seasonal patterns of carbon metabolism and plant water relations of *Juncus roemerianus* and *Spartina alterniflora* in a Georgia salt marsh. Ph.D. Thesis, Univ. Georgia, Athens

Giurgevich, J. R., Dunn, E. L. (1978) Seasonal patterns of CO_2 and water vapor exchange of *Juncus roemerianus* Scheele in a Georgia salt marsh. Amer. J. Bot. *65*, 502-510

Giurgevich, J. R., Dunn, E. L. (1979) Seasonal patterns of CO_2 and water vapor exchange of the tall and short height forms of *Spartina alterniflora* in a Georgia salt marsh. Oecologia *43*, 139-156

Giurgevich, J. R., Dunn, E. L. (In press) Seasonal patterns of daily net photosynthesis, transpiration, and net primary productivity of *Juncus roemerianus* and *Spartina alterniflora* in a Georgia salt marsh. Oecologia

Goering, J. J. (1968) Denitrification in the oxygen minimum layer of the eastern tropical Pacific Ocean. Deep-Sea Res. *15*, 157-164

Golley, F. B. (1961) Energy values of ecological materials. Ecology *42*, 581-584

Gosselink, J. G., Odum, E. P., Pope, R. M. (1973) The value of the tidal marsh. Pub. LSU-SG-74-03 Baton Rouge: Center for Wetland Resources

Grigarick, A. A., Beards, G. W. (1965) Ovipositional habits of the rice water weevil in California as related to a greenhouse evaluation of seed treatments. J. Econ. Entomol. *58*, 1053-1056

Grineva, G. M. (1963) Alcohol fermentation and excretion by plants roots under anaerobic conditions. Fiziologiya Rastenii *10*, 361-369

Gunter, G. (1967) Some relationships of estuaries to the fisheries of the Gulf of Mexico. In: Estuaries. Lauff, G. H. (ed.). Washington: AAAS Publ 83, 621-638

Hails, J. R., Hoyt, J. H. (1969) An appraisal of the evolution of the lower Atlantic coastal plain of Georgia, USA. Trans. Inst. British Geogr. *46*, 53-68

Haines, B. L., Dunn, E. L. (1976) Growth and resource allocation responses of *Spartina alterniflora* Loisel to three levels of NH_4-N, Fe, and NaCl in solution culture. Bot Gaz. *137*, 224-230

Haines, E. B. (1974) Processes affecting production in Georgia coastal waters. Ph.D. Thesis, Duke Univ., Durham, North Carolina

Haines, E. B. (1976) Stable carbon isotope ratios in the biota, soils, and tidal water of a Georgia salt marsh. Estuarine Coastal Mar. Sci. *4*, 609-619

Haines, E. B. (1976a) Relation between the stable carbon isotope composition of fiddler crabs, plants, and soils in a salt marsh. Limnol. Oceanogr. *21*, 880-883

Haines, E. B. (1976b) Nitrogen content and acidity of rain on the Georgia coast. Water Res. Bull. 12, 1223-1231

Haines, E. B. (1977) The origins of detritus in Georgia salt marsh estuaries. Oikos. *29*, 254-260

Haines, E. B. (1979) Nitrogen pools in Georgia coastal waters. Estuaries 2, 34-39

Haines, E. B. (1979a) Growth dynamics of cordgrass, *Spartina alterniflora* Loisel., on control and sewage sludge fertilized plots in a Georgia salt marsh. Estuaries 2, 50-53

Haines, E. B., Hanson, R. B. (1979) Experimental degradation of detritus made from the salt marsh plants *Spartina alterniflora* Loisel, *Salicornia virginica* L. and *Juncus roemerianus* Scheele. J. Exp. Mar. Biol. Ecol. *40*, 27-40

Haines, E. B., Montague, C. L. (1979) Food sources of estuarine invertebrates analysed using $^{13}C/^{12}C$ ratios. Ecology *60*, 48-56

Haines, E. B., Chalmers, A. G., Hanson, R. B., Sherr, B. (1977) Nitrogen pools and fluxes in a Georgia salt marsh. In: Estuarine Processes. Wiley, M. (ed.). New York: Academic Press, Vol. 2, 241-254

Hale, M. G., Foy, C. L., Shay, F. J. (1971) Factors affecting root exudation. Adv. Agron. *23*, 89-109

Hale, S. S. (1975) The role of benthic communities in the nitrogen and phosphorus cycles of an estuary. In: Mineral Cycling in Southeastern Ecosystems. Howell, F. G., Gentry, J. B., Smith, M. H. (eds.). ERDA Symp. Ser. CONF-740513, 291-308

Hamilton, P. V. (1976) Predation on *Littorina irrorata* (Mollusca: Gastropoda) by *Callinectes sapidus* (Crustacea: Portunidae). Bull. Mar. Sci. *26*, 403-409

Hansen, J. A. (1979) Effects of physical factors on fermentation in salt marsh soils. M. S. Thesis, Univ. Georgia, Athens

Hanson, R. B. (1977) Comparison of nitrogen fixation activity in tall and short *Spartina alterniflora.* Appl. Environ. Microbiol. *33*, 569-602

Hanson, R. B. (1977a) Nitrogen fixation (acetylene reduction) in a salt marsh amended with sewage sludge and organic carbon and nitrogen compounds. Appl. Environ. Microbiol. *33*, 846-852

Hanson, R. B., Snyder, J. (1979) Microheterotrophic activity in a salt-marsh estuary, Sapelo Island, Georgia. Ecology *60*, 99-107

Hanson, R. B., Snyder, J. (1980) Glucose exchanges in a salt-marsh estuary: biological activity and chemical measurements. Limnol. Oceanogr. *25*, 633-642

Hanson, R. B., Wiebe, W. J. (1977) Heterotrophic activity associated with particulate size fractions in a *Spartina alterniflora* Loisel. salt marsh-estuary, Sapelo Island, Georgia, and the continental shelf waters. Mar. Biol. *42*, 321-330

Hargrave, B. T. (1976) The central role of invertebrate faeces in sediment decomposition. In: The Role of Terrestrial and Aquatic Organisms in Decomposition Processes. Anderson, J. M., Macfadyen, A. (eds.). Oxford: Blackwell, 301-321

Harper, J. L. (1977) Population Biology of Plants. New York: Academic Press

Harrington, R. W., Harrington, E. S. (1961) Food selection among fishes invading a high subtropical salt marsh from onset of flooding through the progress of a mosquito brood. Ecology *42*, 646-666

Harrington, R. W., Harrington, E. S. (1972) Food of female marsh killifish, *Fundulus confluentus* Goode and Bean, in Florida. Amer. Midl. Nat. *87*, 492-502

Harrison, W. G. (1978) Experimental measurements of nitrogen remineralization in coastal waters. Limnol. Oceanogr. *23*, 684-694

Haven, D. S., Morales-Alamo, R. (1966) Aspects of biodeposition by oysters and other invertebrate filter feeders. Limnol. Oceanogr. *11*, 487-498

Heard, R. W. (1975) Feeding habits of white catfish from a Georgia estuary. Florida Scientist *38*, 20-28

Heinle, D., Flemer, D. A. (1976) Flows of material between poorly flooded tidal marshes and an estuary. Mar. Biol. *35*, 359-373

Hochochka, P. W., Mustafa, T. (1972) Invertebrate facultative anaerobiosis. Science. *178*, 1056-1060

Hoese, H. D. (1971) Dolphin feeding out of water in a salt marsh. J. Mammalogy *52*, 222-223

Holling, C. S. (1973) Resilience and stability of ecological systems. Ann. Rev. Ecol. System. *4*, 1-23

Holm-Hansen, O. (1973) Determination of total microbial biomass by measurement of adenosine triphosphate. In: Estuarine Microbial Ecology. Stevenson, H. L. and Colwell, R. R. (eds.). University of South Carolina Press, Columbia, 73-89

Holm-Hansen, O., Booth, C. R. (1966) The measurement of adenosine triphosphate in the ocean and its ecological significance. Limnol. Oceanogr. *11*, 510-519

Hopkinson, C. S., Day, J. W. (1977) A model of the Barataria Bay salt marsh ecosystem. In: Ecosystem Modeling in Theory and Practice. Hall, C. A. S., Day, J. W. (eds.). New York: Wiley Interscience, 235-266

Hopkinson, C. S., Day, J. W., Jr., Gael, B. T. (1978) Respiration studies in a Louisiana salt marsh. An. Centro Mar. Limnol. Univ. Nat. Autón. Mexico *5*, 225-238

Howard, J. D., Remmer, G. H., Jewitt, J. L. (1975) Estuaries of the Georgia coast, USA: sedimentology and biology. VII. Hydrography and sediments of the Duplin River, Sapelo Island, Georgia. Senckenbergiana Maritima 7, 237-256

Howard, W. J. (1940) Wintering of the greater snow goose. Auk. *57*, 523-531

Howarth, R. W. (1979) Pyrite: Its rapid formation in a salt marsh and its importance to ecosystem metabolism. Science *203*, 49-51

Howarth, R. W., Teal, J. M. (1979) Sulfate reduction in a New England salt marsh. Limnol. Oceanogr. *24*, 999-1013

Howarth, R. W., Teal, J. M. (1980) Energy flow in a salt marsh ecosystem: the role of reduced inorganic sulfur compounds. Amer. Nat. 116, 862-872

Hoyt, J. H. (1967) Barrier island formation. Bull. Geol. Soc. Amer. *78*, 1125-1136

Hughes, E. H. (1980) Estuarine subtidal food webs analyzed with stable carbon isotope ratios. Masters Thesis, University of Georgia.

Humphreys, W. F. (1979) Production and respiration in animal populations. J. Animal Ecol. *48*, 427-453

Hunter, K. A., Liss, P. S. (1979) The surface charge of suspended particles in estuarine and coastal waters. Nature *282*, 823-825

Hustedt, F. (1955) Marine littoral diatoms of Beaufort, North Carolina. Bull. Duke Univ. Mar. Sta. *6*, 1-67

Huston, M. (1979) A general hypothesis of species diversity. Amer. Nat. *113*, 81-101

Imberger, J. (1977) On the validity of water quality models for lakes and reservoirs. Proc. 17th Cong. of IAHR. 6, 293-303

Imberger, J., Berman, T., Christian, R. R., Haines, E. B., Hanson, R. B., Pomeroy, L. R., Whitney, D. E., Wiebe, W. J., Wiegert, R. G. (in ms.) The influence of water motion on the spatial and temporal variability of chemical and biological substances in a salt marsh estuary.

Jackewicz, J. R. (1973) Energy utilization in the marsh crab, *Sesarma reticulatum* (Say). M. S. Thesis, Univ. Delaware, Newark

Jacobs, J. (1968) Animal behavior and water movement as co-determinants of plankton distribution in a tidal system. Sarsia *34*, 355-370

Jarvis, P. G. (1971) The estimation of resistances to carbon dioxide transfer. In: Plant Photosynthetic Production: Manual of Methods. Sestak, Z., Catsky, J., Jarvis, P. G. (eds.). Den Hague: Junk, 566-631

Jassby, A. D. (1975) An evaluation of ATP estimations of bacterial biomass in the presence of phytoplankton. Limnol. Oceanogr. *20*, 646-648

Jeffries, H. P. (1972) Fatty-acid ecology of a tidal marsh. Limnol. Oceanogr. *17*, 433-440

Jeffries, H. P. (1975) Diets of juvenile Atlantic menhaden (*Brevoortia tyrannus*) in three estuarine habitats as determined from fatty acid composition of gut contents. J. Fish. Res. Bd. Canada *32*, 587-592

Johannes, R. E. (1964) Phosphorus excretion and body size in marine animals: microzooplankton and nutrient regeneration. Science *146*, 923-924

Johannes, R. E., Satomi, M. (1966) Composition and nutritive value of fecal pellets of a marine crustacean. Limnol. Oceanogr. *11*, 191-197

Johannes, R. E., Satomi, M. (1967) Measuring organic matter retained by aquatic invertebrates. J. Fish. Res. Bd. Can *24*, 2467-2471

Johannes, R. E., Coward, S. J., Webb, K. L. (1969) Are dissolved amino acids an energy source for marine invertebrates? Comp. Biochem. Physiol. *29*, 282-288

Johnson, A. S., Hillistad, H. O., Shanholtzer, S. F., Shanholtzer, G. F. (1974) An Ecological Survey of the Coastal Region of Georgia. Washington: U. S. National Park Service Sci. Monogr. Ser. No. 3.

Johnston, R. F. (1956) Population structure in salt marsh song sparrows. I. Environment and annual cycle. Condor *58*, 24-44

Johnston, R. F. (1956a) Population structure in salt marsh song sparrows. II. Density, age structure, and maintenance. Condor *58*, 611-621

Joiris, C. (1977) On the role of heterotrophic bacteria in marine ecosystems: some problems. Helgo. wiss. Meeresunters. *30*, 611-621

Jones, W. J. (1975) Ecology of methanogenesis in salt marsh and marine sediments. M. S. Thesis, Clemson Univ., Clemson, South Carolina

Jφrgensen, B. B. (1977) Bacterial sulfate reduction within reduced microniches of oxidized marine sediments. Mar. Biol. *41*, 7-17

Jφrgensen, B. B. (1977a) The sulfur cycle of a coastal marine sediment (Limnfjorden, Denmark). Limnol. Oceanogr. *22*, 814-832

Joshi, M. M., Ibrahim, I. K. A., Hollis, J. P. (1975) Hydrogen sulfide: effects on the physiology of rice plants and relation to straighthead disease. Phytopathology *65*, 1165-1170

Kaczmarek, W., Kaszubiak, H., Pedziwilk, Z. (1976) The ATP content in soil microorganisms. Ekol. Pol. *24*, 299-406

Kaestner, A. (1970) Invertebrate Zoology. III. Crustacea. New York: Interscience

Kale, H. W., III (1964) Food of the long-billed marsh wren, *Telmatodytes palustris griseus* in the salt marshes of Sapelo Island, Georgia. Oriole *29*, 47-66

Kale, H. W. (1965) Ecology and Bioenergetics of the Long-Billed Marsh Wren, *Telmatodytes palustris griseus* (Brewster), in Georgia salt marshes. Cambridge: Nuttall Ornithol. Club. Publ. 5

Kale, H. W., Hyppio, P. A. (1966) Additions to the birds of Sapelo Island and vicinity. Oriole *31*, 1-11

Kaplan, W. A., Teal, J. M., Valiela, I. (1977) Denitrification in salt marsh sediments: Evidence for seasonal temperature selection among populations of denitrifiers. Microbial Ecol. *3*, 193-204

Karl, D. M., Haugsness, J. A., Campbell, L., Holm-Hansen, O. (1978) Adenine nucleotide extraction from multicellular organisms and beach sand: ATP recovery, energy charge ratios, and determination of carbon/ATP ratios. J. Exp. Mar. Biol. Ecol. *34*, 163-181

Keefe, C. W. (1972) Marsh production: A summary of the literature. Contrib. Mar. Sci. Univ. Texas *16*, 163-168

Ketchum, B. H. (1951) The exchange of fresh and salt waters in tidal estuaries. J. Mar. Res. *10*, 18-38

Khan, A. W., Trotter, T. M. (1978) Effects of sulfur-containing compounds on anaerobic degradation of cellulose to methane by mixed cultures obtained from sewage sludge. Appl. Environ. Microbiol. *35*, 1027-1034

King, G. M. (1978) The nature of methanogenesis in soils of a Georgia salt marsh. Ph.D. Thesis, Univ. Georgia, Athens

King, G. M., Wiebe, W. J. (1978) Methane release from soils of a Georgia salt marsh. Geochim. Cosmochim. Acta *42*, 343-348

King, G. M., Wiebe, W. J. (In press) Regulation of sulfate concentration on methanogenesis in salt marsh soil. Estuarine Coastal Mar. Sci.

Kjerfve, B. (1970) Volume transport, salinity distribution, and net circulation in the Duplin Estuary, Georgia. M. S. Thesis, Univ. Washington, Seattle

Kneib, R. T., Stiven, A. E. (1978) Growth, reproduction, and feeding of *Fundulus heteroclitus* (L.) on a North Carolina salt marsh. J. Exp. Mar. Biol. Ecol. *31*, 121-140

Koike, I., Hattori, A. (1978) Denitrification and ammonia formation in anaerobic coastal sediments. Appl. Environ. Microbiol. *35*, 278-282

Komatsu, Y., Tagaki, M., Yamaguchi, M. (1978) Participation of iron in denitrification in water-logged soil. Soil Biol. Biochem. *10*, 21-26

Kowalenko, C. G., Lowe, L. E. (1975) Evaluation of several extraction methods and of a closed incubation method of soil sulfur mineralization. Can J. Soil Sci. *55*, 1-8

Koyama, T. (1963) Gaseous metabolism in lake sediments and paddy soils and the production of methane and hydrogen. J. Geophys. Res. *68*, 3971-3973

Kraeuter, J. N. (1973) Pycnogonida from Georgia. J. Nat. Hist. *7*, 493-449

Kraeuter, J. N. (1976) Biodeposition by salt marsh invertebrates. Mar. Biol. *35*, 215-223

Kraeuter, J. N., Setzler, E. M. (1975) The seasonal cycle of Scyphozoa and Cubozoa in Georgia estuaries. Bull. Mar. Sci. *25*, 66-74

Kraeuter, J. N., Wolf, P. L. (1974) The relationship of marine macroinvertebrates to salt marsh plants. In: Ecology of Halophytes. Reimold, R. J., Queen, W. H. (eds.). New York: Academic Press, 449-462

Kuenzler, E. J. (1959) The phosphorus budget of a mussel population and its effect on a salt marsh ecosystem. Ph.D. Thesis, Univ. Georgia, Athens

Kuenzler, E. J. (1961) Structure and energy flow of a mussel population in a Georgia salt marsh. Limnol. Oceanogr. *6*, 191-204

Kuenzler, E. J. (1961a) Phosphorus budget of a mussel population. Limnol. Oceanogr. *6*, 400-415

Kuno, E. (1973) Population ecology of rice leafhoppers in Japan. Rev. Plant Production Res. *6*, 1-16

Kuschel, G. (1951) Revision de *Lissorhoptrus* Le Conte y genera vecinos de America. Rev. Chilena Ent. *1*, 23-74

Lee, J. J., Inman, D. L. (1975) The ecological role of consumers—an aggregated systems view. Ecology *56*, 1455-1458

Levinton, J. S. (1972) Stability and trophic structure in deposit-feeding and suspension-feeding communities. Amer. Nat. *106*, 472-486

Levinton, J. S., Lopez, G. R. (1977) A model of renewable resources and limitation of deposit-feeding benthic populations. Oecologia *31*, 177-190

Linthurst, R. A., Reimold, R. J. (1978) Estimated net aerial primary productivity for selected estuarine angiosperms in Maine, Delaware, and Georgia. Ecology *59*, 945-955.

Linthurst, R. A., Reimold, R. J. (1978) An evaluation of methods for estimating the net aerial primary productivity of estuarine angiosperms. J. appl. Ecol. *15*, 919-931

Lopez, G. R., Levinton, J. S. (1978) The availability of microorganisms attached to sediment particles as food for *Hydrobia ventrosa* Montague (Gastropoda: Prosobranchia). Oecologia *32*, 263-275

Lopez, G. R., Levinton, J. S., Slobodkin, L. B. (1977) The effect of grazing by the detritivore *Orchestia grillus* on *Spartina* litter and its associated microbial community. Oecologia *30*, 111-127

McCarthy, J. J., Taylor, W. R., Taft, J. L. (1975) The dynamics of nitrogen and phosphorus cycling in the open waters of the Chesapeake Bay. In: Marine Chemistry in the Coastal Environment. Church, T. M. (ed.). Washington: Amer. Chem Soc. Symp. Ser. No. 18, 664-681

McNaughton, S. J. (1979) Grazing as an optimization process: grass-ungulate relationships in the Serengeti. Amer. Nat. *113*, 691-703

Mah, R. A., Wood, D. M., Baresi, L., Glass, T. L. (1977) Biogenesis of methane. Ann. Rev. Microbiol. *31*, 309-341

Mangat, B. S., Levin, W. B., Bidwell, R. G. S. (1974) The extent of dark respiration in illuminated leaves and its control by ATP levels. Can J. Bot *52*, 673-681

Manjunath, T. M. (1977) A note on the oviposition in the macroptrous and bracypterous forms of the rice brown planthopper, *Nilaparvata lugens* Stal. (Homoptera, Delphacidae). Proc. Indian Acad. Sci., Sec. B *86*, 405-408.

Marcus, E., Marcus, E. (1967) Some opisthobranchs from Sapelo Island, Georgia, USA. Malacologia *6*, 199-222

Marples, T. G. (1966) A radionuclide study of arthropod food chains in a *Spartina* salt marsh estuary. Ecology *47*, 270-277

Marshall, J. T., Jr. (1948) Ecologic races of song sparrows in the San Francisco Bay region. I. Habitat and abundance. Condor. *50*, 193-215

Marshall, J. T., Jr. (1948a) Ecologic races of the song sparrow in the San Francisco Bay region. II. Geographic variation. Condor *50*, 233-256

Marshall, S. M., Orr, A. P. (1960) Feeding and nutrition. In: The Physiology of Crustacea. Waterman, T. H. (ed.). New York: Academic Press, Vol. 1, 227-258

Martens, C. S., Berner, R. A. (1977) Interstitial water chemistry of anoxic Long Island Sound sediments. I. Dissolved gases. Limnol. Oceanogr. *22*, 10-25

Martin, M. M., Martin, J. S. (1979) The distribution and origins of the cellulolytic enzymes of the higher termite, *Macrotermes natalensis*. Physiol. Zool. *52*, 11-21

Martof, B. S. (1963) Some observations on the herpetofauna of Sapelo Island, Georgia. Herpetologia *19*, 70-72

Mattson, W. J., Addy, N. D. (1975) Phytophagous insects as regulators of forest primary production. Science *190*, 515-522

Meanley, B. (1975) Birds and Marshes of the Chesapeake Bay Country. Cambridge, Maryland: Tidewater Publishers

Menzel, D. W., Vaccaro, R. F. (1964) The measurement of dissolved and particulate organic carbon in sea water. Limnol. Oceanogr. *9*, 138-142

Menzies, R. J., Frankenberg, D. (1966) Handbook of the Common Isopod Crustacea of Georgia. Athens: Univ. Georgia

Miles, P. W. (1968) Insect secretions in plants. Ann. Rev. Phytopath. *6*, 137-164

Miller, D. C. (1965) Studies on the systematics, ecology, and geographical distribution of certain fiddler crabs. Ph.D. Thesis, Duke University, Durham, North Carolina

Miller, J. G. (1965) Living systems: basic concepts. Behavioral Sci. *10*, 193-411

Mishima, J., Odum, E. P. (1963) Excretion rate of ^{65}Zn by *Littorina irrorata* in relation to temperature and body size. Limnol. Oceanogr. *8*, 39-44

Mitchell, J. E., Pfadt, R. E. (1974) A role of grasshoppers in a shortgrass prairie ecosystem. Environ. Entomol. *3*, 358-360

Montague, C. L. (1980) The net influence of the mud fiddler crab, *Uca pugnax*, on carbon flow through a Georgia salt marsh: the importance of work by macroorganisms to the metabolism of ecosystems. Ph.D. Thesis, Univ. Georgia, Athens

Morris, G. K., Walker, T. J. (1976) Calling songs of *Orchelimum* meadow katydids (Tettigoniidae) I. Mechanism, terminology, and geographic distribution. Canad. Entomol. *108*, 785-800

Mountfort, D. O., Asher, R. A. (1979) Effect of inorganic sulfide on the growth and metabolism of *Methanosarcina barkeri* Strain D. M. Appl. Environ. Microbiol. *37*, 670-675

Nasu, S. (1969) The Virus Diseases of the Rice Plant. Baltimore: Johns Hopkins

Neales, T. F., Incoll, L. D. (1968) The control of leaf photosynthesis rate by the level of assimilate concentration in the leaf: A review of the hypothesis. Bot. Rev. *34*, 107-125

Nedwell, D. B., Floodgate, G. D. (1972) Temperature induced changes in the formation of sulphide in a marine sediment. Mar. Biol. *14*, 18-24

Nestler, J. (1977) Interstitial salinity as a cause of ecospheric variation in *Spartina alterniflora*. Estuarine Coastal Mar. Sci. *5*, 707-714

Nestler, J. (1977a) A preliminary study of the sediment hydrography of a Georgia salt marsh using rhodamine WT as a tracer. Southeastern Geol. *18*, 265-271

Newell, R. C. (1965) The role of detritus in the nutrition of two marine deposit-feeders, the prosobranch, *Hydrobia ulvae* and the bivalve, *Macoma balthica*. Proc. Zool Soc. Lond. *144*, 25-45

Newell, R. C. (1976) Adaptations to intertidal life. In: Adaptations to Environment. Newell, R. C. (ed.). London: Butterworth, 1-82

Nixon, S. W. (1980) Between coastal marshes and coastal waters—A review of twenty years of speculation and research on the role of salt marshes in estuarine productivity and water chemistry. In: Estuarine and Wetlands Processes. Hamilton, P., Macdonald, K. (eds.). New York: Plenum

Nixon, S. W., Oviatt, C. A., Gerber, J., Lee, V. (1976) Diel metabolism and nutrient dynamics in a salt marsh embayment. Ecology *57*, 740-750

Oades, J. M., Jenkinson, D. S. (1979) Adenosine triphosphate content of the soil microbial biomass. Soil Biol. Biochem. *11*, 201-204

Odum, E. P. (1961) The role of tidal marshes in estuarine production. New York State Conservationist *15*, 12-15

Odum, E. P. (1971) Fundamentals of Ecology, 2nd Ed. Philadelphia: Saunders

Odum, E. P., de la Cruz, A. A. (1967) Particulate organic detritus in a Georgia salt marsh-estuarine ecosystem. In: Estuaries. Lauff, G. H. (ed.). Washington: AAAS Publ. 83, 383-388

Odum, E. P., Fanning, M. E. (1973) Comparison of the productivity of *Spartina alterniflora* and *Spartina cynosuroides* in Georgia coastal marshes. Bull. Georgia Acad. Sci. *31*, 1-12

Odum, E. P., Smalley, A. E. (1959) Comparison of population energy flow of a herbivorous and a deposit-feeding invertebrate in a salt-marsh ecosystem. Proc. Nat'l Acad. Sci. US *45*, 617-622

Odum, H. T. (1971) Environment, Power, and Society. New York: Wiley Interscience

Odum, W. E. (1968) The ecological significance of fine particle selection by striped mullet, *Mugil cephalus*. Limnol. Oceanogr. *13*, 92-98

Odum, W. E. (1968a) Mullet grazing on a dinoflagellate bloom. Chesapeake Sci. *9*, 202-204

Odum, W. E. (1970) Utilization of the direct grazing and plant detritus food chains by the striped mullet, *Mugil cephalus*. In: Marine Food Chains. Steele, J. H. (ed.). Berkeley: Univ. California, 222-240

Odum, W. E., Heald, E. J. (1972) Trophic analysis of an estuarine mangrove community. Bull. Mar. Sci. *22*, 671-738

Odum, W. E., Heald, E. J. (1975) The detritus based food web of an estuarine mangrove community. In: Estuarine Research. Cronin, L. E. (ed.). New York: Academic Press, Vol. 1, 265-286

Ogelsby, R. T., Christman, R. F., Driver, C. H. (1967) The biotransformation of lignin to humus—facts and postulates. Appl. Environ. Microbiol. *9*, 171-184

Oney, J. (1951) Food habits of the clapper rail in Georgia. J. Wildlife Management *15*, 106-107

Oremland, R. S. (1979) Methanogenic activity in plankton samples and fish intestines: a mechanism for *in situ* methanogenesis in oceanic surface waters. Limnol. Oceanogr. *24*, 1136-1141

Oremland, R. S., Taylor, B. F. (1978) Sulfate reduction and methanogenesis in marine sediments. Geochim. Cosmochim. Acta *42*, 209-214

Oshrain, R. L. (1977) Aspects of anaerobic sulfur metabolism in salt marsh soils. M. S. Thesis, Univ. Georgia, Athens

Ott, J., Schiemer, F. (1973) Respiration and anaerobiosis of free living nematodes from marine and limnetic sediments. Neth. J. Sea Res. *7*, 233-243

Oviatt, C. A., Nixon, S. W. (1976) Sediment resuspension and deposition in Narragansett Bay. Estuarine Coastal Mar. Sci. *3*, 201-217

Owen, D. F., Wiegert, R. G. (1976) Do consumers maximize plant fitness? Oikos *27*, 488-492

Owen, D. F., Wiegert, R. G. (1981) Mutualism between grasses and grazers: an evolutionary hypothesis. Oikos *36*, 376-378

Pace, M. L. (1977) The effect of macroconsumers grazing on the benthic microbial community of a salt marsh mud flat. M.S. Thesis, Univ. Georgia, Athens

Pace, M. L., Shimmel, S., Darley, W. M. (1979) The effect of grazing by a gastropod, *Nassarius obsoletus,* on the benthic microbial community of a salt marsh mud flat. Estuarine Coastal Mar. Res. *9*, 121-134

Palmer, J. D., Round, F. E. (1967) Persistent vertical-migration rhythms in benthic microflora. VI. The tidal and diurnal nature of this rhythm in the diatom, *Hantzchia virgata.* Biol. Bull. *132*, 45-55

Palmer, R. A. (1974) Studies on the blue crab (*Callinectes sapidus*) in Georgia. Georgia Game and Fish Div. Contrib. Ser. No. 29

Panganiban, A. T., Jr., Patt, T. E., Hart, W., Hanson, R. B. (1979) Oxidation of methane in the absence of oxygen in lake water samples. Appl. Environ. Microbiol. *37*, 303-309

Patrick, W. H., Jr., Delaune, R. D. (1976) Nitrogen utilization by *Spartina alterniflora* in a salt marsh in Barataria Bay, Louisiana. Estuarine Coastal Mar. Sci. *4*, 59-64

Patrick, W. H., Jr., Khalid, R. A. (1974) Phosphate release and absorption by soils and sediments: effect of aerobic and anaerobic conditions. Science *186*, 53-55

Payne, W. J. (1973) Reduction of nitrogenous oxides by microorganisms. Bacteriol. Rev. 37, 409-452

Perkins, D. F. (1978) The distribution and transfer of energy and nutrients in the *Agrostis-Festuca* grassland ecosystem. In: Production Ecology of British Moors and Montane Grasslands. Heal, O. W., Perkins, D. F. (eds.). New York: Springer-Verlag 375-395

Perkins, E. J. (1963) Penetration of light into littoral soils. J. Ecol. *51*, 687-692

Peterson, B. J., Howarth, R. W., Lipschutz, F., Ashendorf, D. (1980) Salt marsh detritus: An alternative interpretation of stable carbon isotope ratios and the fate of *Spartina alterniflora.* Oikos *34*, 173-177

Pfeiffer, W. J. (1974) Cattle egrets feeding in salt marsh. Oriole *39*, 44-45

Pomeroy, L. R. (1959) Algal productivity in salt marshes of Georgia. Limnol. Oceanogr. *4*, 386-397

Pomeroy, L. R. (1960) Residence time of dissolved phosphate in natural waters. Science *131*, 1731-1732

Pomeroy, L. R. (1970) The strategy of mineral cycling. Ann. Rev. Ecol. System. 1, 171-190

Pomeroy, L. R. (1975) Mineral cycling in marine ecosystems. In: Mineral Cycling in Southeastern Ecosystems. Howell, F. G., Gentry, J. B., Smith, M. H. (eds.). ERDA Symp. Ser. CONF-740513, 209-223

Pomeroy, L. R., Bush, F. M. (1959) Regeneration of phosphate by marine animals. Intern. Oceanogr. Congr. Preprints. Washington, AAAS, 893-894

Pomeroy, L. R., Deibel, D. (1980) Aggregation of organic matter by pelagic tunicates. Limnol. Oceanogr. *25*, 643-652

Pomeroy, L. R., Haskin, H. H., Ragotzkie, R. A. (1956) Observations on dinoflagellate blooms. Limnol. Oceanogr. *1*, 54-60

Pomeroy, L. R., Mathews, H. M., Min, H. S. (1963) Excretion of phosphate and soluble organic phosphorus compounds by zooplankton. Limnol. Oceanogr. *8*, 50-55

Pomeroy, L. R., Smith, E. E., Grant, C. M. (1965) The exchange of phosphate between estuarine water and sediments. Limnol. Oceanogr. *10*, 167-172

Pomeroy, L. R., Odum, E. P., Johannes, R. E., Roffman, B. (1966) Flux of ^{32}P and ^{65}Zn through a salt-marsh ecosystem. Proc. Symp. Disposal of Radioactive Wastes into Seas, Oceans, and Surface Waters. Vienna: IAEA., 177-188

Pomeroy, L. R., Johannes, R. E., Odum, E. P., Roffman, B. (1969) The phosphorus and zinc cycles and productivity of a salt marsh. In: Symposium on Radioecology. Nelson, D. J., Evans, F. C. (eds.). USAEC CONF-67503., 412-419

Pomeroy, L. R., Shenton, L. R., Jones, R. D. H., Reimold, R. J. (1972) Nutrient flux in estuaries. In: Nutrients and Eutrophication. Likens, G. E. (ed.). Amer. Soc. Limnol. Oceanogr. Spec. Symp. 1, 274-291

Pomeroy, L. R., Bancroft, K., Breed, J., Christian, R. R., Frankenberg, D., Hall, J. R., Maurer, L. G., Wiebe, W. J., Wiegert, R. G., Wetzel, R. L. (1977) Flux of organic matter through a salt marsh. In: Estuarine Processes. Wiley, M. (ed.). New York: Academic Press, Vol. 2, 270-279

Post, W. (1974) Functional analysis of space-related behavior in the seaside sparrow. Ecology *55*, 564-575

Poulet, S. A. (1976) Feeding of *Pseudocalanus minutus* on living and non-living particles. Mar. Biol. *34*, 117-125

Powell, E. N., Crenshaw, M. A., Rieger, R. M. (1979) Adaptations to sulfide in the meiofauna of the sulfide system. I. ^{35}S-sulfide accumulation and the presence of a sulfide detoxification system. J. Exp. Mar. Biol. Ecol. *37*, 57-76

Ragotzkie, R. A. (1959) Plankton productivity in estuarine waters of Georgia. Publ. Inst. Mar. Sci. Univ. Texas *6*, 146-158

Ragotzkie, R. A., Bryson, R. A. (1955) Hydrography of the Duplin River, Sapelo Island, Georgia. Bull. Mar. Sci. *5*, 297-314

Ragotzkie, R. A., Pomeroy, L. R. (1957) Life history of a dinoflagellate bloom. Limnol. Oceanogr. *2*, 62-69

Randerson, P. F. (In press) The ecology of the Wash. V.: A simulation model of salt marsh succession. J. Appl. Ecol.

Randolph, P. A., Randolph, J. C., Barlow, C. A. (1975) Age-specific energetics of the pea aphid, *Acyrthosiphon pisum.* Ecology *56*, 359-369

Reimold, R. J. (1972) The movement of phosphorus through the marsh cord grass, *Spartina alterniflora* Loisel. Limnol. Oceanogr. *17*, 606-611

Reimold, R. J., Daiber, F. C. (1970) Dissolved phosphorus concentrations in a natural salt-marsh of Delaware. Hydrobiologia *36*, 361-371

Reimold, R. J., Gallagher, J. L., Thompson, D. E. (1973) Remote sensing of tidal marsh. Photogram. Eng. *39*, 477-488

Reimold, R. J., Gallagher, J. L., Linthurst, R. A., Pfeiffer, W. J. (1975) Detritus production in coastal Georgia salt marshes. In: Estuarine Research. Cronin, L. E. (ed.). New York: Academic Press, 217-228

Reimold, R. J., Linthurst, R. A., Wolf, P. L. (1975a) Effects of grazing on a salt marsh. Biol. Conserv. *8*, 105-125

Reise, K., Ax, P. A. (1979) A meiofaunal "thiobios" limited to the anaerobic sulfide system of marine sand does not exist. Mar. Biol. *54*, 225-238

Rhoades, D. F., Cates, R. G. (1976) Toward a general theory of plant antiherbivore chemistry. In: Recent Advances in Phytochemistry. Wallace, J. W., Mansell, R. L. (eds.). New York: Plenum, Vol. 10, 168-213

Rich, P. H., Wetzel, R. G. (1978) Detritus in the lake ecosystem. Amer. Nat. *112*, 57-71

Richards, P. W. (1964) The Tropical Rain Forest. Cambridge: Cambridge U. Press

Rickards, W. L. (1968) Ecology and growth of juvenile tarpon, *Megalops atlantis*, in a Georgia salt marsh. Bull. Mar. Sci. *18*, 220-239

Ridgeway, S. H. (1972) Mammals of the Sea; Biology and Medicine. Springfield, Illinois: C. C. Thomas

Rigler, F. H. (1975) The concept of energy flow and nutrient flow between trophic levels. In: Unifying Concepts in Ecology. van Dobben, W. H., Lowe-McConnell, R. H. (eds.). Den Hague: Junk., 15-26

Roman, M. R., Tenore, K. R. (1978) Tidal resuspension in Buzzards Bay, Massachusetts. I. Seasonal changes in the resuspension of organic carbon and chlorophyll *a*. Estuarine Coastal Mar. Sci. *6*, 37-46

Rublee, P., Cammen, L., Hobbie, J. E. (1978) Bacteria in a North Carolina salt marsh: standing crop and importance in the decomposition of *Spartina alterniflora.* Univ. North Carolina Sea Grant Publ. UNC-SG-78-11.

Rudd, J. W. M., Hamilton, R. D. (1978) Methane cycling in a eutrophic shield lake and its effects on whole lake metabolism. Limnol. Oceanogr. *23*, 337-348

Satomi, M., Pomeroy, L. R. (1965) Respiration and phosphorus excretion in some marine populations. Ecology *46*, 877-881

Schaeffer, C. W., Levin, N. L., Milch, P. (1968) Death from dessication in the mud snail, *Nassarius obsoletus* (Say): effects of temperature. Nautilus *81*, 109-114

Schindler, D. W., Holmgren, S. K. (1971) Primary production and phytoplankton in the Fisheries Research Board Experimental Lakes Area, northwestern Ontario, and other low-carbonate waters, and a liquid scintillation method for determining ^{14}C activity in photosynthesis. J. Fish Res. Bd. Can. *28*, 189-202

Scott, J. A., French, N. R., Leetham, J. W. (1979) Patterns of consumption in grasslands. In: Perspectives in Grassland Ecology. French, N. R. (ed.). New York: Springer-Verlag, 89-105

Scranton, M. I., Brewer, P. G. (1977) Occurrence of methane in the near-surface of the western subtropical North Atlantic. Deep-Sea Res. *24*, 127-138

Sellner, K. G., Zingmark, R. G. (1976) Interpretations of the ^{14}C method of measuring the total annual production of phytoplankton in a South Carolina estuary. Bot. Mar. *19*, 119-125

Setzler, E. M. (1977) A quantitative study of the movement of larval and juvenile Scianidae and Engraulidae into the estuarine nursery grounds of Doboy Sound, Sapelo Island, Georgia. Ph.D. Thesis, Univ. Georgia, Athens

Shanholtzer, G. F. (1974) Relationship of vertebrates to salt marsh plants. In: Ecology of Halophytes. Reimold, R. J., Queen, W. H. (eds.). New York: Academic Press, 463-474

Shanholtzer, S. F. (1973) Energy flow, food habits, and population dynamics of *Uca pugnax* in a salt marsh system. Ph.D. Thesis, Univ. Georgia, Athens

Sharp, H. F., Jr. (1967) Food ecology of the rice rat, *Oryzomys palustris* (Harlin), in a Georgia salt marsh system. J. Mammal. *48*, 557-563

Sherr, B. F. (1977) The ecology of denitrifying bacteria in salt marsh soils—an experimental approach. Ph.D. Thesis, Univ. Georgia, Athens

Sherr, B. F., Payne, W. J. (1978) Effect of the *Spartina alterniflora* root—rhizome system on salt marsh soil denitrifying bacteria. Appl. Environ. Microbiol. *35*, 724-729

Sherr, B. F., Payne, W. J. (1979) Role of the salt marsh grass, *Spartina alterniflora* in the response of soil-denitrifying bacteria to glucose enrichment. Appl. Environ. Microbiol. *38*, 747-748

Shimmel, S. M. (1979) The effect of grazing on epibenthic algae in a Georgia salt marsh. M. S. Thesis, Univ. Georgia, Athens

Sibley, C. G. (1955) The response of salt-marsh birds to extremely high tides. Condor *57*, 241-242

Sieburth, J. McN. (1977) Convener's report on the informal session on biomass and productivity of microorganisms in planktonic ecosystems. Helgo. wiss. Meeresunters. 30, 697-704 (1977)

Siegfried, W. R. (1971) The food of the cattle egret. J. Appl. Ecol. *8*, 447-468

Sikora, J. P., Sikora, W. B., Erkenbrecher, C. W., Coull, B. C. (1977) Significance of ATP, carbon, and caloric content of meiobenthic nematodes in partitioning benthic biomass. Mar. Biol. *44*, 7-14

Simpson, K. W. (1976) Shore flies and brine flies (Diptera:Ephydridae). In: Marine Insects. Cheng, L. (ed.). New York: Elsevier, 465-495

Sinclair, A. R. E. (1975) The resource limitation of trophic levels in tropical grassland ecosystems. J. Animal Ecol. *44*, 497-520

Sivanesen, A., Manners, J. G. (1972) Bacteria of muds colonized by *Spartina townsendii* and their possible role in *Spartina* die-back. Plant and Soil *36*, 349-361

Skyring, G. W., Chambers, L. A. (1976) Biological sulfate reduction in carbonate sediments of a coral reef. Austral. J. Mar. Freshwater Res. *27*, 595-602

Skyring, G. W., Oshrain, R. L., Wiebe, W. J. (1979) Assessment of sulfate reduction rates in Georgia marshland soils. Geomicrobiology J. *1*, 389-400

Slater, J. A. (1977) The incidence and evolutionary significance of wing polymorphism in lygaeid bugs with particular reference to those of South Africa. Biotropica *9*, 217-229

Smalley, A. E. (1959) The growth cycle of *Spartina* and its relation to the insect populations in the marsh. In: Proc. Salt Marsh Conf. Ragotzkie, R. A., Teal, J. M., Pomeroy, L. R., Scott, D. C. (eds.). Athens: Univ. Georgia, 96-100

Smalley, A. E. (1959a) The role of two invertebrate populations, *Littorina irrorata* and *Orchelimum fidicinium*, in the energy flow of a salt marsh ecosystem. Ph.D. Thesis, Univ. Georgia, Athens

Smalley, A. E. (1960) Energy flow of a salt marsh grasshopper population. Ecology *41*, 672-677

Smith, F. E. (1972) Spatial heterogeneity, stability, and diversity in ecosystems. Trans. Conn. Acad. Arts Sci. *44*, 307-335

Smith, M. R., Mah, R. A. (1978) Growth and methanogenesis by *Methanosarcina* strain 227 on acetate and methanol. Appl. Environ. Microbiol. *36*, 870-879

Smith, T. J., III, Odum, W. E. (1981) The effects of grazing by snow geese on coastal salt marshes. Ecol. *62*, 98-106

Sooksai, S., Tugwell, P. (1978) Adult rice water weevil feeding symptoms: number of samples required, spatial and seasonal distribution in a rice field. J. Econ. Entomol. *71*, 145-148

Sφrenson, J. (1978) Capacity for denitrification and reduction of nitrate to ammonia in a coastal marine sediment. Appl. Environ. Microbiol. *35*, 301-305

Sorenson, J. (1978a) Occurrence of nitric and nitrous oxides in a coastal marine sediment. Appl. Environ. Microbiol. *36*, 809-813

Sottile, W. S. (1974) Studies of microbial production and utilization of dissolved organic carbon in a Georgia salt marsh-estuarine ecosystem. Ph.D. Thesis, Univ. Georgia, Athens

Southwood, T. R. E. (1961) A hormonal theory of the mechanism of wing polymorphism in Heteroptera. Proc. Roy. Ent. Soc. Lond. *26*, 63-66

Southwood, T. R. E. (1962) Migration of terrestrial arthropods in relation to habitat. Biol. Rev. *37*, 171-214

Stanley, D. W., Hobbie, J. E. (1977) Nitrogen recycling in the Chowan River. Univ. North Carolina, Water Resources Res. Inst. Report 121

Steeman-Nielsen, E. (1952) The use of radioactive (C^{14}) for measuring organic production in the sea. J. Cons. Int. Explor. Mer *18*, 117-140

Stiven, A. E., Kuenzler, E. J. (1979) The response of two salt marsh molluscs, *Littorina irrorata* and *Geukensia demissa*, to field manipulations of density and marsh grass. Ecol. Monogr. *49*, 151-171

Stommel, H., Farmer, H. G. (1952) On the nature of estuarine circulation. Woods Hole Oceanogr. Inst. Tech. Rep. 52-51, 52-53, 52-88

Strong, D. R., Jr., McCoy, E. D., Ray, J. E. (1977) Time and the number of herbivorous species: The pests of sugarcane. Ecology *58*, 167-175

Stroud, L. M. (1976) Net primary production of belowground material and carbohydrate patterns of two height forms of *Spartina alterniflora* in two North Carolina marshes. Ph.D. Thesis, North Carolina State Univ., Raleigh

Sullivan, M. J. (1975) Diatom communities from a Delaware salt marsh. J. Phycol. *11*, 384-390

Sullivan, M. J. (1976) Long term effects of manipulating light intensity and nutrient enrichment on the structure of a salt marsh diatom community. J. Phycol. *12*, 205-210

Sullivan, M. J., Daiber, F. C. (1974) Response in production of cord grass, *Spartina alterniflora*, to inorganic nitrogen and phosphorus fertilizer. Chesapeake Sci. *15*, 121-123

Sullivan, M. J., Daiber, F. C. (1975) Light, nitrogen, and phosphorus limitation of edaphic algae in a Delaware salt marsh. J. Exp. Mar. Biol. Ecol. *18*, 79-88

Sweet, M. H. (1964) The biology and ecology of the Rhyparochrominae of New England (Homoptera: Lygaeidae). I. Entomological Amer. *43*, 1-123

Tabatabi, M. A. (1974) Determination of sulfate in water samples. Sulphur Inst. J. *10*, 11-13

Tabb, D. C. (1966) The estuary as a habitat for spotted seatrout. In: A Symposium on Estuarine Fisheries. Smith, R. F., Swartz, A. H., Massmann, W. H. (eds.). Amer. Fish. Soc. Spec. Pub. 3

Tanner, W. F. (1960) Florida coastal classification. Trans. Gulf Coast Assoc. Geol. Soc. *10*, 259-266

Taylor, W. R. (1964) Light and photosynthesis in intertidal diatoms. Helgo. wiss. Meeresunters. *10*, 29-37

Teal, J. M. (1958) Distribution of fiddler crabs in Georgia salt marshes. Ecology *39*, 185-193

Teal, J. M. (1959) Respiration of crabs in Georgia salt marshes and its relation to their ecology. Physiol. Zool. *32*, 1-14

Teal, J. M. (1959a) Birds of Sapelo Island and vicinity. Oriole *24*, 1-14

Teal, J. M. (1962) Energy flow in the salt marsh ecosystem of Georgia. Ecology *43*, 614-624

Teal, J. M., Kanwisher, J. (1961) Gas exchange in a Georgia salt marsh. Limnol. Oceanogr. *6*, 388-399

Teal, J. M., Kanwisher, J. (1966) Gas transport in the marsh grass, *Spartina alterniflora*. J. Exper. Bot. *17*, 355-361

Teal, J. M., Teal, M. (1969) Life and Death of a Salt Marsh. Boston: Little, Brown

Teal, J. M., Wieser, W. (1966) The distribution and ecology of nematodes in a Georgia salt marsh. Limnol. Oceanogr. *11*, 217-222

Teal, J. M., Valiela, I., Berlo, D. (1979) Nitrogen fixation by rhizosphere and free-living bacteria in salt marsh sediments. Limnol. Oceanogr. *24*, 126-132

Tenore, K. R. (1977) Growth of the polychaete, *Capitella capitata*, cultured on various levels of detritus derived from different sources. Limnol. Oceanogr. *22*, 936-941

Tenore, K. R. (1977a) Utilization of aged detritus from different sources by the polychaete, *Capitella capitata*. Mar. Biol. *44*, 51-56

Tenore, K. R., Tietjen, J. H., Lee, J. J. (1977) Effect of meiofauna on incorporation of aged eelgrass, *Zostera marina*, detritus by the polychaete, *Nephthys incisa*. J. Fish. Res. Bd. Can. *34*, 563-567

Thayer, G. W. (1971) Phytoplankton production and the distribution of nutrients in a shallow unstratified estuarine system near Beaufort, N. C. Chesapeake Sci. *12*, 240-253

Theede, H., Ponat, A., Hiroka, K., Schleiper, C. (1969) Studies on the resistance of marine bottom invertebrates to oxygen-deficiency and hydrogen sulfide. Mar. Biol. *2*, 325-337

Thomas, J. P. (1966) Influence of the Altamaha River on primary production beyond the mouth of the river. M. S. Thesis, Univ. Georgia, Athens

Thomson, J. M. (1966) The grey mullets. Oceanogr. Mar. Biol. Ann. Rev. *4*, 301-335

Tiedje, J. M., Firestone, M. K., Smith, M. S., Betlach, M. R., Firestone, R. B. (1978) Short-term measurement of denitrification rates in soils using ^{13}N and acetylene inhibition methods. In: Proceedings in Life Sciences–Microbial Ecology. Loutit, M. W., Miles, J. A. R. (eds.). Berlin: Springer, 132-137

Tippins, H. H., Beshear, R. J. (1971) On the habitat of *Haliaspis spartinae* (Comstock) (Homoptera: Diaspididae). Ent. News *82*, 165

Tomkins, I. R. (1941) Notes on Macgillivary's seaside sparrow. Auk *58*, 38-51

Traczyk, T. (1968) Studies on the primary production in meadow community. Ekol. Pol. *16*, 59-100

Turner, R. E. (1976) Geographic variation in salt marsh macrophyte production: a review. Contrib. Mar. Sci. Univ. Texas. *20*, 47-68

Turner, R. E. (1978) Community plankton respiration in a salt marsh estuary and the importance of macrophytic leachates. Limnol. Oceanogr. *23*, 442-451

Ubben, M. S., Hanson, R. B. (1980) Tidal induced regulation of nitrogen fixation activity in a Georgia salt marsh, Sapelo Island, Georgia. Estuarine Coastal Mar. Sci. *10*, 445-453

Valiela, I., Babeic, D. F., Atherton, W., Seitzinger, S., Krebs, C. (1974) Some consequences of sexual dimorphism: Feeding in male and female fiddler crabs, *Uca pugnax* (Smith). Biol. Bull. *147*, 652-660

Valiela, I., Teal, J. M. (1974) Nutrient limitation in salt marsh vegetation. In: Ecology of Halophytes. Reimold, R. J., Queen, W. H. (eds.). New York: Academic Press, 547-563

Valiela, I., Teal, J. M. (1979) The nitrogen budget of a salt marsh ecosystem. Nature *280*, 652-656

Valiela, I., Teal, J. M. (1979a) Inputs, outputs, and interconversions of nitrogen in a salt marsh ecosystem. In: Ecological Processes in Coastal Environments. Jeffries, R. L., Davy, A. J. (eds.). Cambridge: Blackwell, 399-419.

Valiela, I., Teal, J. M., Volkmann, S., Shafer, D. (1978) Nutrient and particulate fluxes in a salt marsh ecosystem: tidal exchanges and inputs by precipitation and groundwater. Limnol. Oceanogr. *23*, 798-812

Valiela, I., Wright, J. E., Volkmann, S. B. (1977) Growth, production, and energy transformations in the salt-marsh killifish, *Fundulus heteroclitus*. Mar. Biol. *40*, 135-144

Van Dolah, R. F. (1978) Factors regulating the distribution and population dynamics of the amphipod, *Gammarus palustris*, in an intertidal salt marsh community. Ecol. Monogr. *48*, 191-217

Van Engel, W. A. (1958) The blue crab and its fishery in Chesapeake Bay: I. Reproduction, early development, growth, and migration. Comm. Fish. Rev. *20*, 6-17

Van Hook, R. I., Jr. (1971) Energy and nutrient dynamics of spider and orthopteran populations in a grassland ecosystem. Ecol. Monogr. *41*, 1-26

Van Pelt, A. F., Jr. (1956) The ecology of ants of the Welaka Reserve, Florida (Hymenoptera, Formicidae). Amer. Midl. Nat. *56*, 358-387

Van Raalte, C. D., Valiela, I., Teal, J. M. (1976) The effect of fertilization on the species composition of salt marsh diatoms. Water Res. *10*, 1-4

Van Raalte, C. D., Valiela, I., Teal, J. M. (1976a) Productin of epibenthic salt marsh algae: light and nutrient limitation. Limnol. Oceanogr. *21*, 862-872

Varley, G. C. (1967) The effects of grazing by animals on plant productivity. In: Secondary Productivity of Terrestrial Ecosystems. Petrusewicz, K. (ed.). Krakow: Polish Acad. Sci., 773-778

Vince, S., Valiela, I., Backus, N. (1976) Predation by the salt marsh killifish, *Fundulus heteroclitus* (L.) in relation to prey size and habitat structure: consequences for prey distribution and abundance. J. Exp. Mar. Biol. Ecol. *23*, 225-266

Virnstein, R. W. (1977) The importance of predation by crabs and fishes on benthic infauna in Chesapeake Bay. Ecology *58*, 1199-1217

Waldbauer, G. P. (1968) The consumption and utilization of food by insects. In: Advances in Insect Physiology. Beament, J., Treherne, J., Wigglesworth, V. (eds.). New York: Academic Press, Vol. 5, 229-288

Way, M. J., Cammell, M. (1970) Aggregation behavior in relation to food utilization by aphids. In: Animal Populations in Relation to their Food Sources. Watson, A. (ed.). Oxford: Blackwell, 229-247

Webb, J. L. (1914) Notes on the rice water weevil (*Lissorhoptrus simplex* Say). J. Econ. Entomol. *7*, 432-438

Webster, J. R., Waide, J. B., Patten, B. C. (1975) Nutrient recycling and the stability of ecosystems. In: Mineral Cycling in Southeastern Ecosystems. Howell, F. G., Gentry, J. B., Smith, M. H. (eds.). USERDA Symp. Ser. CONF-740513, 1-27.

Wells, H. W. (1957) Abundance of the hard clam, *Mercenaria mercenaria*, in relation to environmental factors. Ecology *38*, 123-128

Welsh, B. (1975) The role of grass shrimp, *Palaemonetes pugio*, in a tidal marsh ecosystem. Ecology *56*, 513-530

Westlake, D. F. (1963) Comparison of plant productivity. Biol. Rev. *38*, 385-425

Wetzel, R. L. (1975) An experimental study of detrital carbon utilization in a Georgia salt marsh. Ph.D. Thesis, Univ. Georgia, Athens

Wetzel, R. L. (1976) Carbon resources of a benthic salt marsh invertebrate, *Nassarius obsoletus* Say (Mollusca: Nassariidae). Estuarine Processes. New York: Academic Press, Vol. 2, 293-308

Wheeler, J. R. (1976) Fractionation by molecular weight of organic substances in Georgia coastal water. Limnol. Oceanogr. *21*, 846-852

Wheeler, W. M. (1910) Ants–Their Structure, Development, and Behavior. New York: Columbia Univ.

Whelan, T. (1974) Methane, carbon dioxide, and dissolved sulfate from interstitial water of coastal marsh sediments. Esturine Coastal Mar. Sci. *2*, 407-415

White, D. C., Davis, W. M., Nickels, J. S., King, J. D., Bobbie, R. J. (1979) Determination of the sedimentary microbial biomass by extractable lipid phosphate. Oecologia *40*, 51-62

Whitmore, T. C. (1975) Tropical Rain Forests of the Far East. Clarendon Press, Oxford

Whitney, D. E., Darley, W. M. (1979) A method for the determination of chlorophyll *a* in samples containing degradation products. Limnol. Oceanogr. *24*, 183-186

Whitney, D. E., Woodwell, G. M., Howarth, R. W. (1975) Nitrogen fixation in Flax Pond: a Long Island salt marsh. Limnol. Oceanogr. *20*, 640-643

Whittaker, J. B. (1971) Population changes in *Neophilaenus lineatus* (L.) (Homoptera: Cercopidae) in different parts of its range. J. Animal Ecol. *40*, 425-443

Widdel, F., Pfenning, N. (1977) A new anaerobic sporing, acetate oxidizing, sulfate-reducing bacterium, *Desulfomaculum* (emend) *acetoxidans*. Arch. Microbiol. *112*, 119-122

Wiebe, W. J. (1979) Anaerobic benthic microbial processes: changes from the estuary to the continental shelf. In: Ecological Processes in Coastal and Marine Systems. Livingston, R. J. (ed.). New York: Plenum, 469-486

Wiebe, W. J., Bancroft, K. (1975) The use of adenylate energy charge ratio to measure growth state of natural microbial communities. Proc. Nat'l Acad Sci. USA. *72*, 2112-2115

Wiebe, W. J., Pomeroy, L. R. (1972) Microorganisms and their association with aggregates and detritus in the sea: a microscopic study. Mem. Ist. Ital. Idrobiol. *29* (suppl.) 325-352

Wiedemann, H. U. (1972) Application of red-lead to the detection of dissolved sulfide in waterlogged soils. Zeitschr. Pflanzenernährung Bodenkunde *133*, 73-81

Wiegert, R. G. (1964) Population energetics of meadow spittlebugs (*Philaenus spumaris* L.) as affected by migration and habitat. Ecol. Monogr. *34*, 217-241

Wiegert, R. G. (1975) Simulation modeling of the algal-fly components of a thermal ecosystem: effects of spatial heterogeneity, time delays, and model condensation. In: Systems Analysis and Simulation in Ecology. Patten, B. C. (ed.). New York: Academic Press, Vol. 3, 157-181

Wiegert, R. G. (1975a) Simulation models of ecosystems Ann. Rev. Ecol. System *6*, 311-338

Wiegert, R. G. (1979) Ecological processes characteristic of coastal *Spartina* marshes of the south-eastern USA. In: Ecological Processes in Coastal Environments. Jeffries, R. L., Davy, A. J. (eds.). Oxford: Blackwell, 467-490

Wiegert, R. G. (1979a) Modeling coastal, estuarine, and marsh ecosystems: State-of-the-art. In: Statistical Ecology. Patil, G. P., Rosenzweig, M. (eds.). Vol. 12, 319-341

Wiegert, R. G. (1979b) Population models: experimental tools for the analysis of ecosystems. In: Ecosystem Analysis. Hain, D., Stairs, G., Mitchell, R. (eds.). Columbus: Ohio State Univ., 233-279.

Wiegert, R. G. (1980) Modeling salt marshes and estuaries: progress and problems. In: Estuarine and Wetlands Processes. Hamilton, P., Macdonald, K. B. (eds.). New York: Plenum, 527-540.

Wiegert, R. G., Evans, F. C. (1964) Primary production and the disappearance of dead vegetation on an old field in southeastern Michigan. Ecology *45*, 49-63

Wiegert, R. G., Evans, F. C. (1967) Investigations of secondary productivity in grasslands. In: Secondary Productivity of Terrestrial Ecosystems. Petrusewicz, K. (ed.). Krakow: Polish Academy of Science, 499-518

Wiegert, R. G., Wetzel, R. L. (1974) The effect of numerical integration technique on the simulation of carbon flow in a Georgia salt marsh. Proc. Summer Computer Simulation Conf., Houston, Vol. 2, 275-277

Wiegert, R. G., Wetzel, R. L. (1979) Simulation experiments with a fourteen-compartment model of a *Spartina* salt marsh. In: Marsh-Estuarine Systems Simulation. Dame, R. F. (ed.). Columbia: Univ. S. Carolina, 7-39

Wiegert, R. G., Christian, R. R., Gallagher, J. L., Hall, J. R., Jones, R. D. H., Wetzel, R. L.(1975) A preliminary ecosystem model of coastal Georgia *Spartina* marsh. In: Estuarine Research. Cronin, L. E. (ed.). New York: Academic Press, Vol. 2, 583-601

Wieser, W., Kanwisher, J. (1961) Ecological and physiological studies on marine nematodes from a small salt marsh near Woods Hole, Massachusetts. Limnol. Oceanogr. *6*, 262-270

Williams, A. B. (1965) Marine decapod crustaceans of the Carolinas. Fishery Bull. *65*, 1-298

Williams, P. M., Oeschger, H., Kinney, P. (1969) Natural radiocarbon activity of the dissolved organic carbon in the north-east Pacific Ocean. Nature *224*, 256-258

Williams, R. B. (1962) The ecology of diatom populations in a Georgia salt marsh. Ph.D. Thesis, Harvard Univ., Cambridge

Williams, R. B. (1963) Use of netting to collect motile benthic algae. Limnol. Oceanogr. *8*, 360-361

Williams, R. B. (1964) Division rates of salt marsh diatoms in relation to salinity and cell size. Ecology *45*, 877-880

Williams, R. B. (1966) Annual phytoplankton production in a system of shallow temperate estuaries. In: Some Contemporary Studies in Marine Sciences. Barnes, H. (ed.). London: George Allen and Unwin, 699-716

Williams, R. B. (1972) Nutrient levels and phytoplankton productivity in the estuary. Coastal and Estuary Management, Proc. Second Symp. Baton Rouge: Louisiana State Univ., 59-89

Williams, R. B., Murdoch, M. B. (1966) Phytoplankton production and chlorophyll concentration in Beaufort Channel, North Carolina. Limnol. Oceanogr. *11*, 73-82

Windom, H. L., Dunstan, W. M., Gardner, W. S. (1975) River input of inorganic phosphorus and nitrogen to the southeastern salt-marsh estuarine environment. In: Mineral Cycling in Southeastern Ecosystems. Howell, F. G., Gentry, J. B., Smith, M. H. (eds.). ERDA Symp. Ser. CONF-740513, 309-313

Winfrey, M. R., Zeikus, J. G. (1977) Effect of sulfate on carbon electron flow during methanogenesis in fresh water sediments. Appl. Environ. Microbiol. *33*, 281-295

Witkamp, M. (1973) Compatibility of microbial measurements. Bull. Ecol. Res. Comm. (Stockholm) *17*, 179-188

Wolf, P. L., Shanholtzer, S. F., Reimold, R. J. (1975) Population estimates for *Uca pugnax* (Smith, 1870) on the Duplin Estuary marsh, Georgia, USA. (Decapoda, Brachyura, Ocypodidae). Crustaceana *29*, 79-91

Wood, A. M. (1980) Biological and chemical factors affecting the response of natural phytoplankton communities to copper addition. Ph.D. Thesis, Univ. Georgia, Athens

Woodwell, G. M., Whitney, D. E. (1977) Flax Pond ecosystem study: Exchanges of phosphorus between a salt marsh and the coastal waters of Long Island Sound. Mar. Biol. *41*, 1-6

Woodwell, G. M., Whitney, D. E., Hall, C. A. S., Houghton, R. A. (1977) The Flax Pond ecosystem: Exchanges of carbon in water between a salt marsh and Long Island Sound. Limnol. Oceanogr. *22*, 833-838

Woodwell, G. M., Houghton, R. A., Hall, C. A. S., Whitney, D. E., Moll, R. A., Juers, D. W. (1979): The Flax Pond ecosystem study: The annual metabolism and nutrient budgets of a salt marsh. In: Ecological Processes in Coastal Environments. Jeffries, R. L., Davy, A. J. (eds.). Oxford: Blackwell, 491-511

Yoshinari, T., Knowles, R. (1976) Acetylene inhibition of nitrous oxide reduction by denitrifying bacteria. Biochem. Biophys. Res. Comm. *69*, 705-710

Zedler, J. B. (1977) Salt marsh community structure in the Tijuana Estuary, California. Estuarine Coastal Mar. Sci. *5*, 39-53

Zehnder, A. J. B., Brock, T. D. (1979) Methane formation and methane oxidation by methanogenic bacteria. J. Bacteriol. *137*, 420-432

Zehnder, A. J. B., Wuhrmann, K. (1979) Physiology of a *Methanobacterium* strain AZ. Arch. Microbiol. *111*, 199-205

Zeikus, J. G. (1977) The biology of methanogenic bacteria. Bact. Rev. *41*, 514-541

Zhilina, T. N., Zavarzin, G. A. (1973) Trophic relations between *Methanosarcina* and its accompanying cultures. Mikrobiologiia *42*, 266-273

Zieman, J. C., Odum, W. E. (1977) Modeling of ecological succession and production in estuarine marshes. Tech. Rep. d-77-35, U. S. Army Corps of Engineers

Zinder, S. H., Mah, R. A. (1979) Isolation and characterization of a thermophilic strain of *Methanosarcina* unable to use H_2-CO_2 for methanogenesis. Appl. Environ. Microbiol. *38*, 996-1008

Zingmark, R. G. (1978) An annotated checklist of the biota of the coastal zone of South Carolina. Columbia: Univ. S. Carolina

Index

Ecological Studies

Volume 30
Crassulacean Acid Metabolism
Analysis of an Ecological Adaptation
By M. Kluge and I. P. Ting
1978. xii, 210p. 112 illus. 21 tables. cloth
ISBN 3-540-**08979**-9

Volume 31
Physiological Ecology of the Alpine Timberline
Tree Existence at High Altitudes with Special Reference to the European Alps
By W. Tranquillini
1979. approx. 150p. 67 figures. cloth
ISBN 0-387-**09065**-7

Volume 32
Perspectives in Grassland Ecology
Results and Application of the US/IBP Grassland Biome Study
Edited by N. R. French
1979. approx. 215p. 60 figures. cloth
ISBN 3-540-**90384**-4

Volume 33
Methods of Studying Root Systems
By W. Böhm
1979. approx. 200p. 69 figures. cloth
ISBN 0-387-**09329**-X

Volume 34
Agriculture in Semi-Arid Environments
Edited by: A. E. Hall, G. H. Cannell and H. W. Lawton
1979. approx. 355p. 47 figures. cloth
ISBN 0-387-**09414**-8

Volume 35
Environmental Geochemistry
A Holistic Approach
By John A. C. Fortescue
1980. approx. 362p. 131 figures. cloth
ISBN 0-387-**90454**-9

Volume 36
Physiological Processes in Plant Ecology
Toward a Synthesis with Atriplex
By C. B. Osmond, O. Björkman and D. J. Anderson
1980. xi, 468p. 194 illus. 76 tables. cloth
ISBN 3-540-**10060**-1

Volume 37
The Soil Resource
Origin and Behavior
By Hans Jenny
1980. approx. 395p. 191 figures. cloth
ISBN 0-387-**90543**-X

Springer-Verlag
New York Heidelberg Berlin

VERMONT STATE COLLEGES
0 0003 0359510 2
DISCARD